A PRACTICAL GUIDE

TO THE

CLIMATES AND WEATHER

OF INDIA, CEYLON AND BURMAH

AND THE

STORMS OF INDIAN SEAS

A PRACTICAL GUIDE

TO THE

CLIMATES AND WEATHER

OF INDIA, CEYLON AND BURMAH

AND THE

STORMS OF INDIAN SEAS

BASED CHIEFLY ON THE PUBLICATIONS OF THE
INDIAN METEOROLOGICAL DEPARTMENT

HENRY F. BLANFORD, F.R.S., F.R.MET.S.

GYAN

Published by

Gyan Publishing House
5, Ansari Road
Daryaganj, New Delhi-110002
Phone: 011-47034999, 9811692060
E-mail: books@gyanbooks.com

Distribution Network
gyanbooks.com
India, USA, Canada, UK, Australia, France

ISBN: 978-81-212-9410-2 (PB)
First Published, 1889

2nd Impression 2023

Printed at: Gyan Press, Delhi.

A PRACTICAL GUIDE TO THE CLIMATES AND WEATHER OF INDIA, CEYLON AND BURMAH AND THE STORMS OF INDIAN SEAS

AUTHOR: HENRY F. BLANFORD

CLIMATES AND WEATHER OF INDIA

A PRACTICAL GUIDE

TO THE

CLIMATES AND WEATHER

OF INDIA, CEYLON AND BURMAH

AND THE

STORMS OF INDIAN SEAS

BASED CHIEFLY ON THE PUBLICATIONS OF THE INDIAN METEOROLOGICAL DEPARTMENT

BY

HENRY F. BLANFORD, F.R.S., F.R.MET.S.

LATE METEOROLOGICAL REPORTER TO THE GOVERNMENT OF INDIA
HONORARY MEMBER OF THE IMPERIAL AND ROYAL METEOROLOGICAL SOCIETY OF AUSTRIA
AND OF THE METEOROLOGICAL SOCIETY OF THE MAURITIUS
CORRESPONDING MEMBER OF THE GERMAN METEOROLOGICAL SOCIETY

London
MACMILLAN AND CO.
AND NEW YORK
1889

PREFACE

Owing mainly to the systematic work of the Meteorological Department, established by the Government of India in 1875, we now possess a far better knowledge of the weather and climate of India than of those of any other tropical country, and, in some respects, better than of those of many parts of Europe; thanks to the greater simplicity of the processes concerned, and to the prominence and regular recurrence of the more striking phases of the seasons. But to the great majority of the lay public, of those to whom information of this kind is a matter of daily interest and importance, this knowledge is as a sealed book. It may be found, indeed, embodied in the volumes periodically issued by the meteorological office, or scattered through the journals and transactions of various scientific societies: but all publications of this class are addressed to students and co-workers in the field; and besides being, in most cases, too technical in form to approve itself to the tastes and requirements of the general public, the literature is too voluminous, and the information it contains too diffuse, to be readily accessible to those who have neither the time nor the opportunity to search through so wide a field.

The want of some work that shall afford, in a compendious and apprehensible form, such information as is constantly in demand by those engaged in administration, in agriculture, sanitation, engineering works, and the like, and especially in the navigation of Indian seas, or by those who wish to follow, intelligently, the current reports of the weather issued daily at Simla and Calcutta, has long been obvious, and has been definitely expressed in the anniversary address of the President of the Asiatic Society of Bengal, delivered at the annual meeting of the Society in February 1887, in the following terms. After noticing the errors of the older storm literature, as represented by the works of Reid and Piddington, he remarks:—

"It is much to be desired that steps should be taken by some recognised authority, to prepare a simple and popular manual on the subject [of storms], based on the sounder knowledge which modern science has acquired. And indeed such a work is also needed for the meteorology of the land, to enable the lay public to understand and intelligently appreciate the mass of meteorological literature which the Government Gazettes and the entire press so lavishly place before us. We want something more popular than the very valuable 'Vade Mecum' of Mr. Blanford, not a scientific investigation of principles, but a brief, clear, and comprehensive explanation of the observations ordinarily made, their objects, uses, and mode of application."

It is to meet this want that the present work has been taken in hand. It is not addressed to meteorologists and physicists, although some of the data and facts adduced may be not without interest to those who are principally concerned with meteorology in its scientific aspect; but more particularly to agriculturists, medical officers, engineers, pilots, and other seafaring men, and to those others of the general public to whom the weather and the climates of India and of its seas are practical and not scientific objects of interest. I have therefore selected from the abundant materials now available, and from the general conclusions from these, worked out by Messrs. Eliot, Hill, and other Indian meteorologists, such portions as are of practical importance to inquirers of this class, and have endeavoured to set them forth in clear and concise language, avoiding, as a rule, all discussion of a technical character, and all technical forms of expression that have not become more or less familiarised to the public through the medium of the weather reports and other literature of a popular character. The explanation of these latter is one of the objects of the work.

The work consists of two parts. The first gives a concise general description of the kinds of observations made at Government observatories, and the information they directly afford. This constitutes what may be termed the elements of the climate and weather. The second deals with the practical applications of weather knowledge. In the first part, the intensity of the sun's heat in India, the air temperature, humidity, cloud, winds, and rainfall, together with the principal variations they present in different parts of the Empire, and the changes they undergo in the course of the year, the behaviour of the barometer and the methods of interpreting and utilising its teachings, and the constitution and nature of storms, are briefly described in separate sections. The second and more extensive part is devoted to a detailed notice of the local climates—first, those of the chief hill stations, and then those of the different provinces of the low country. Then follow the weather characteristics of the three Indian seasons, illustrated by the information furnished by the daily weather charts published at Simla; the tracks and seasons of storms on the seas around India, with some practical deductions for the guidance of seamen; and finally, the statistics of rainfall, evaporation, and wind pressure, which are chiefly required by engineers in dealing with questions of water supply, drainage, and the stability of structures. In this section an attempt has been made for the first time to estimate the average rainfall of some of the chief river basins, a datum of great importance to physical geographers, not less than to engineers.

In the Appendices are given, in a condensed form, some portions of the statistical data on which the above descriptions are based, much of which have been computed specially for this work. They include the more important climatic elements of 92 places in India, Burmah, and Ceylon, a list of well-recorded storms in the Bay of Bengal, with their approximate places of origin, tracks, and rates of movement, and tables of the average, maximum, and minimum monthly rainfall at 114 places.

Among the friends to whom I am indebted for information and assistance, I must especially mention Mr. James Forrest, Secretary of the Institution of Civil Engineers, to whose kindness I owe much of the information respecting engineering works and professional observations in India, Africa, and Australia, which will be found quoted in the final section of Part II. Also Captain Henry Toynbee, late of the London Meteorological Office, and his successor, Lieutenant C. W. Baillie, R.N., to whom I am indebted for valuable assistance in preparing the section on Storms in Indian Seas.

FOLKESTONE, *April* 1889.

CONTENTS

PART I.—ELEMENTS OF CLIMATE AND WEATHER

PART II.—CLIMATES AND WEATHER OF INDIA IN RELATION TO HEALTH AND INDUSTRY

APPENDICES

PART I

ELEMENTS OF CLIMATE AND WEATHER

The Heat of the Sun in India

Use of the Sun Thermometer.—The question is often asked of meteorologists, "What is the temperature in the sun?" The reply must be, "That depends entirely on the instrument with which you measure it, and on its surroundings;" in other words, the question, so put, does not admit of a definite answer. Still the fact remains that the sun is hotter in India than in England for example, hotter apparently in May than in January, and there must be some way of expressing this difference by a reference to the thermometer. And that, in point of fact, there is such a way stands in evidence in the reports of the chief observatories, published daily or weekly in the newspapers, which give, together with the temperature of the air in the shade, the reading of a thermometer exposed to the sun. This thermometer is an instrument specially constructed for the purpose. It has the bulb coated with lamp-black, and is enclosed in a large glass tube hermetically sealed after exhaustion, and it is exposed to the sun on a support four feet above the ground, as far as possible from buildings. Any other kind of thermometer would probably show a different temperature; and, indeed, thermometers of apparently similar construction are often found to differ many degrees in their readings in the sun, notwithstanding that, in the shade, they read alike. This latter discrepancy can be approximately ascertained and allowed for; but it is not possible to ascertain the exact effect due to variations in the surrounding objects and in the character of the ground beneath the instrument; and thus it is found that, under equally clear skies, the excess of the sun thermometer reading above that which shows the temperature of the air in the shade,—in other words, the heating effect of the sun, shown by the former instrument,—is greater at some stations than others.

Insolation Temperatures in India.—In India this difference is, as a rule, from 50° to 70° Fahr. It does not vary much in the course of the year so long as the sky is unclouded; but it is higher at hill stations than on the plains, in the cold season considerably higher, showing that, in opposition to the prevalent belief, the power of the sun is really greater on the hills, and that protection of the head and back is at least as necessary a precaution at Simla or Mussoorie (and still more so at greater elevations), as on the plains of India. At Leh, 11,500 feet above the sea, Dr. Cayley succeeded in making water boil by simply exposing it to the sun in a small bottle, blackened on the outside and placed inside an empty quinine phial to protect it from cooling by the wind.

As examples of the temperatures recorded with thermometers of the kind above described, on the hills and plains of India, the following table shows the highest readings registered in each month of 1885, by no means a hot year, at Simla, Murree, Lahore, Calcutta, and Madras; and also the greatest differences of these and the highest temperatures in the shade on the same day:—

[TABLE

	Murree.		Simla.		Lahore.		Calcutta.		Madras.	
	Temp. in Sun.	Excess.	Temp. in Sun.	Excess.	Temp. in Sun.	Excess.	Temp. in Sun.	Excess.	Temp. in Sun.	Excess
January .	122	79	134	80	128	62	137	55	146	61
February .	132	92[?]	136	84	152	76	145	64	146	62
March . .	144	74	140	75	168	79	155	63	161	69
April . .	138	82	146	75	172	79	159	62	150	59
May . .	145	83	144	72	171	73	165	66	150	55
June . .	151	72	152	70	175	68	162	71	151	53
July . .	155	71	146	70	173	69	160	71	153	54
August .	156	77	145	69	172	75	155	67	154	59
September	147	76	144	73	163	67	155	66	160	65
October .	139	68	143	72	157	66	155	67	149	60
November	133	69	131	70	147	60	146	62	148	62
December .	129	78	126	75	146	67	140	65	149	65

In England, the Rev. F. Stow, who has paid much attention to this subject, found that thermometers of the same kind as those used in India, and exposed in the same manner, seldom read above 140°, and states his belief that 154° is the highest temperature registered in five years. This would seem to show that the highest temperatures actually recorded in England do not differ very much from those of the hill station Murree, but their excess above the temperatures of the air would seem to be much less, and rarely, if ever, to amount to 70°.

This latter remark may apply equally to the case of Madras, where, indeed, the excess of the sun temperatures over those of the air is not greater than is occasionally recorded in the summer in England; but then the air temperature itself is 20° or 30° higher; and it would be a grave mistake to suppose that exposure to the sun in Madras can be incurred with the same impunity as exposure to the summer sun in England.

On the other hand, experience shows that the more fervid intensity of the sun's rays at the hill stations, in January and February, may be borne not only without danger but even without serious discomfort in a cool atmosphere, provided the head and spinal column are adequately protected. And it may be concluded that the temperature readings of the sun thermometer, which show the temperature of the air, plus the direct heating effect of the sun, notwithstanding the artificial and purely conventional character of the arrangements for determining them, probably afford a better criterion of the stress imposed on the animal system, than do the figures expressive of the sun's heating effect alone. The glare and intense light reflected from all surrounding objects, and especially from the ground bathed in Indian sunshine, are perhaps not less trying to the unacclimatised visitor to India, than the heat which they accompany.

Duration of Sunshine.—The greater intensity of the sun is, however, only one element of its distinctive action in India. Another of not less importance is the greater frequency and duration of sunshine. At St. Aubin's, in Jersey, where, according to a table of sunshine records published by Mr. R. H. Scott, the number of sunshiny hours exceeds that of any station in the British Isles, the average total of sunshine, on the mean of five years, was 1853 hours in the year, or rather over 5 hours per day. At Calcutta, on the mean of three years, it was 2732 hours, or an average of 7½ hours per day; and at Allahabad, also on the mean of three years, 3053 hours or an average of 8⅓ hours per day. Notwithstanding the greater length of the summer days in the British Isles, in no single month did the number of sunshiny hours in Jersey equal the May average of either Calcutta or Allahabad, and only from June to September, the rainy season of Northern India, did the total duration of sunshine in Jersey surpass that at the Indian stations. The following table, extracted from the Report on the Meteorology of India for 1885, gives the data on which these statements are based, and also, for comparison, the corresponding data on St. Aubin's for the average of five years :—

		Jan.	Feb.	March.	April.	May.	June.	July.	Aug.	Sept.	Oct.	Nov.	Dec.	Year.
Calcutta.	Hours of bright sunshine . . .	275	255	294	297	293	154	128	113	158	258	254	253	2732
	Average per day . .	8·9	9·0	9·5	9·9	9·4	5·1	4·1	3·6	5·3	8·3	8·5	8·1	7·5
	Percentage of possible sunshine . .	82	80	79	78	72	38	31	28	43	72	77	76	62
Allahabad.	Hours of bright sunshine . . .	263	269	309	307	317	213	162	148	233	286	296	250	3053
	Average per day . .	8·5	9·5	10·3	9·9	10·2	7·1	5·2	4·8	7·8	9·2	9·9	8·1	8·3
	Percentage of possible sunshine . .	80	86	84	81	77	52	39	37	64	81	91	77	70
St. Aubin's.	Hours of bright sunshine . . .	61	93	160	187	267	225	231	227	164	125	74	39	1853
	Average per day . .	2·0	3·3	5·0	6·2	8·6	7·5	7·4	7·3	5·5	4·0	2·5	1·3	5·1
	Percentage of possible sunshine . .	23	33	43	45	56	47	48	52	44	38	27	15	39

We have no records of sunshine from Southern India to compare with the above, but it appears from the estimates of the quantity of cloud which have been made for many years past in all parts of India, that, on the general average of the year, the skies are appreciably more cloudy in Southern than in most parts of Northern India, and especially Central and North-western India, and it may be inferred, therefore, that the quantity of sunshine is less. If we take 1000 as representing the total expanse of the sky, the average proportion of cloud in the Punjab, the North-west Provinces, Lower Bengal, and the plains of Madras are respectively as follow:—

Province.	No. of Stations.	Cloud Proportion.
Punjab . . .	9	283
N.W. Provinces .	9	303
Lower Bengal . .	7	453
Madras . . .	10	507

These figures, being based upon mere eye-estimates, cannot lay claim to the same precision as those of the duration of sunshine, which are derived from self-registering instruments, but the data of different stations are very accordant, and as the figures are the averages of, in most cases, from

14 to 17 years' observations, I believe they may be accepted with confidence.

The Temperature of the Air

Temperature Observations. Mean Temperature.—In what I have to say on this subject it must be understood that I speak of the temperature of the air outside the house, but in a place well shaded from the sun, where the air is not stagnant, but has free movement from all sides. In practice, at all Indian observatories, the thermometer is suspended under a roof, generally of thick thatch, which is supported either on posts or pillars, so that, while the ground beneath it is well shaded, the wind has free access to the instrument. The temperature shown by a thermometer in such a position rarely agrees with that of a house or even of a verandah, in either of which it is generally lower during the day and higher at night.

The observations, usually recorded under these circumstances, are the highest and lowest temperatures reached in the 24 hours, at whatever time they may occur, and, in addition, the readings of the actual temperature at certain fixed hours; most commonly 10 A.M. and 4 P.M. From these four observations it is usual to compute what is called the mean temperature of the day. This is assumed to represent the average of all the fluctuations during the 24 hours, not that of the four recorded readings only. And this result is arrived at by various devices founded on a more complete system of observations. As the consideration of these belongs to the technicalities of meteorological observation, they need not be discussed here. Suffice it to say that, although for any single day the result may differ by a degree or two from the real value, on the average of several days these errors should neutralise each other; so that, if the method adopted be valid, the mean temperature of a whole month should be well within a degree of the truth.

Diurnal Variation and Range.—In fine, cloudless weather, the changes of temperature during the day are remarkably regular in India. The air is coolest a few minutes before sunrise. No sooner does the sun appear above the horizon than the temperature mounts rapidly, the rate of rise gradually increasing up to about half-past 8 o'clock (somewhat earlier or later, according to the hour of sunrise). After this it continues rising, but at a decreasing rate, until it reaches its highest point about 2 P.M., or somewhat later. It then begins to fall, at first slowly, and afterwards more rapidly, till about sunset, when the fall is as rapid as was the rise in the forenoon. Later on the fall slackens, and after midnight continues slowly and steadily till a little before sunrise. The course of this variation is illustrated by Fig. 1, which represents the average diurnal curves for Calcutta in March and July.

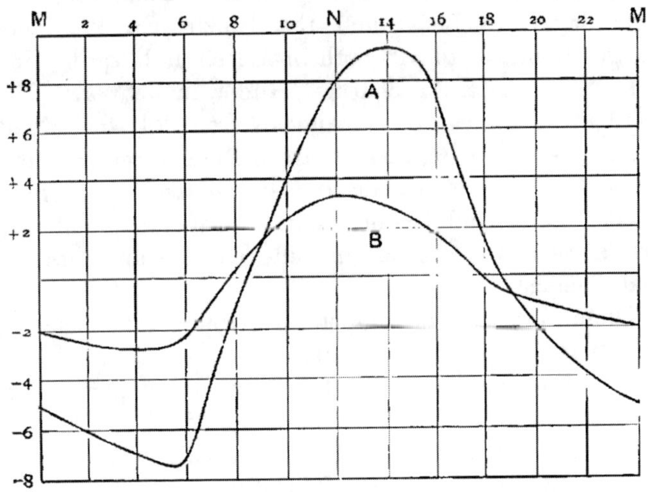

Fig. 1.—Diurnal Variation of Temperature at Calcutta.
A. March. B. July.

The range of temperature during the day depends chiefly on the dryness of the air and the clearness of the skies. It is much greater in the interior of the country, and especially in North-western India, than on the coasts and in the neighbourhood of the sea; and, as a general rule, is greatest in the driest spring months and least in the rainy season. In Sind and Baluchistan, and throughout the dry tract to the west of the Jumna and the Aravalis, the daily range of the thermometer is greatest in October or November, when the difference between sunrise and the afternoon averages not much less than 35° and sometimes exceeds 40°. In the northern districts of Bombay, in Kathiawar, Guzerat, Khandesh, and the Northern Mahratta country, the driest time of year is in the earlier months, when land winds from between west and north-west blow most steadily; and accordingly, in this part of India, the greatest range of the thermometer is in February or March, when it averages nearly 35° at such places as Poona and Malegaon, and in some years, on the mean of a whole month, is as much as 38°. In Lower Bengal, March is, on the whole, the driest month; but at Calcutta and Dacca the average diurnal range does not exceed 21° or 22°, and very rarely reaches 30° on any single day. In Behar and the North-west Provinces it averages from 28° to 32° both in March and April.

In the interior of Southern India, in February, March, and April, the mean daily range is from 24° to 30°, being slightly greater in February than in the succeeding months; but on the coast of Madras it is much less, not more than 19° at Madras, and 13° or 14° at Negapatam; and at this latter station it is greatest in June, when it amounts to 19° on an average.

These variations of temperature, which take place daily with much regularity, much exceed those experienced in most parts of Europe, and constitute an important element of the climate. In different parts of England the range varies from 10° to 20° in the summer, when it is greatest; and in Europe, only in the climate of the Spanish tableland does it amount to 27°, and has been observed in a single month as high as 31°.[1]

At hill stations, the range varies according to the position of the place. If, like Simla, Mussoorie, and most of Hima-

[1] Buchan, *Introductory Text-book of Meteorology*.

layan sanitaria, it is situated on a ridge, the difference of the early morning and afternoon temperatures is less than on the plains; but the same rule by no means holds good if it is on an elevated tableland like Quetta, still less when it is in a deep valley like Leh and Kailang. In some of these cases, in certain months, elevation brings about an exaggeration of the diurnal range. As these changes of temperature during the day are a matter of much importance to invalids, and indeed in their influence on health and comfort generally, it may be useful to give the mean diurnal range of each month at all those elevated stations at which observatories exist or have existed, and which afford, therefore, trustworthy data. I give for comparison also, in each case, the corresponding ranges at some neighbouring station on the plains.

MEAN DAILY RANGE OF THE THERMOMETER

	SIND FRONTIER.		N.W. HIMALAYA AND PUNJAB.						N.W. HIMALAYA AND N.W. PROVINCES.		
	Quetta (tableland)	Jacobabad (plain)	Leh (high valley)	Murree (hill ridge)	Chamba (valley)	Simla (hill ridge)	Kailang (high valley)	Lahore (plain)	Chakrata (hill ridge)	Ranikhet (hill summit)	Bareilly (plain)
Jan.	24	31	25	14	14	10	17	25	16	13	25
Feb.	21	29	26	14	17	17	22	25	16	15	26
Mar.	24	30	26	15	19	18	19	26	18	16	29
April	26	31	26	17	19	20	17	29	19	17	30
May	30	32	27	17	18	20	22	29	18	17	27
June	31	28	30	17	23	18	25	26	15	14	21
July	27	24	29	15	16	12	24	18	10	11	14
Aug.	28	22	30	14	13	11	23	18	10	11	13
Sept.	34	25	30	15	21	14	26	22	12	12	15
Oct.	35	33	29	16	25	17	28	31	15	15	23
Nov.	34	34	27	15	25	18	23	32	16	15	29
Dec.	29	32	26	14	22	17	21	28	16	14	27

	Sikkim and Assam.			Cl. Provinces.		Berar.		Guzerat.		Nilgiris.		
	Darjiling (hill ridge)	Shillong (tableland)	Dhubri (plain)	Pachmarhi (tableland)	Hoshungabad (plain)	Chikalda (hill ridge)	Amraoti (plain)	Abu (hill summit)	Deesa (plain)	Ootacamund (tableland)	Wellington (tableland)	Coimbatore (plain)
January .	9	20	20	25	28	17	28	16	32	31	21	23
February .	12	18	22	24	30	18	28	16	30	29	24	27
March .	13	17	22	25	32	19	31	17	31	27	21	27
April . .	15	16	18	23	31	20	31	18	30	21	20	24
May . .	13	13	13	20	29	21	29	18	27	18	17	21
June . .	10	11	9	14	19	15	21	15	20	12	14	18
July . .	10	10	9	9	11	8	14	9	13	10	13	18
August .	10	11	9	8	12	8	14	8	13	12	14	17
September	10	10	8	11	14	9	16	11	17	13	15	19
October .	12	14	11	17	22	13	23	15	27	14	13	17
November	13	17	16	23	28	15	26	17	32	16	13	17
December	12	19	18	25	27	17	26	17	32	22	17	19

Even at one and the same station, houses on the crest of a ridge have a smaller diurnal range of temperature than those on the lateral slopes. Thus, at Darjiling, the present observatory, which is on the crest of the ridge, shows a difference of night and day temperatures about 2° less than the old observatory, which for many years existed lower down on the side of the same hill (except from June to September, when the difference vanishes); and this, notwithstanding that the new observatory is more exposed than the abandoned site. The figures given above are those of the present observatory.

At most of the Himalayan sanitaria, and such places as Shillong, Chikalda, and Abu, the range of temperature approximates to that of midland and southern England in summer; but, as the table shows, it is greatest in the winter, whereas in Europe it is least at that season.

Irregular Changes with Weather.—Casual and irregular changes of temperature, such as accompany changes of wind and weather, are less frequent and for the most part less marked in India than in Europe and most extra-tropical countries. From northerly winds bringing the extreme heat or cold of Central Asia, India is shielded by the Himalaya; and only in rare cases, and such as are temporary and of short duration, are changes of wind accompanied with such striking changes in the warmth of the air as are commonly experienced in Europe, and still more conspicuously in Siberia and North America. The chief exceptions occur in the hot season, when a more or less prolonged spell of dry weather is followed by a burst of rain; as, for instance, at the onset of one of those afternoon squalls characteristic of this season, which, in Bengal, are known as *Nor'-Westers.* The cold wind which blows out in advance of one of these little storms sometimes causes the temperature to drop 15° or even 20° in less than as many minutes, but the full effect of the change lasts little longer than the storm, and even the cooling effect of the rain, which usually follows the wind, is in a great measure dissipated by another day's sunshine.

On the east coast of the peninsula, the setting in of the sea-wind, after a hot land-wind has been blowing for some hours, is also accompanied with a fall of 10° or 15°, but the effect is transitory, and is a relief rather than a trial to the bodily system.

More important than these are the rapid changes between consecutive days which accompany alternations of rain and fine weather in Upper India; changes especially frequent in the summer months, but also occurring at other times of year. This subject, which is one of great importance in relation to health, has hitherto received but little attention in India, although it is well known that chills are a fruitful source of fever and bowel complaints, maladies peculiarly prevalent in this as in most tropical countries. Perhaps the best mode of representing the important facts is that adopted by Dr. Hann, viz. tabulating the number of times in each month that a rise or fall of definite amount (*e.g.* 5° or 10°) takes place in the mean temperatures of successive days at different places. The daily and weekly weather reports furnish data which any one interested in the question may readily turn to account, and he may thus compare the vicissitudes of temperature in different seasons or at different stations with the corresponding health statistics. The following table, which shows the number of times that the change between any two consecutive days at Calcutta and Lahore amounted to 5°, 10°, and 15° in the years 1884 and 1885, may be taken as an example of this way of representing an important climatic feature, and it also illustrates the much greater variability of the Punjab climate as compared with that of Bengal. The cases of rise and fall of temperature are distinguished by the letters R and F.

			Jan.		Feb.		March.		April.		May.		June.		July.		Aug.		Sept.		Oct.		Nov.		Dec.	
			R.	F.	R.	F.	R.	F.	R.	F.	R.	F.	R.	F.	R.	F.	R.	F.	R.	F.	R.	F.	R.	F.	R.	F.
Lahore.	1884	5° to 10°	..	3	1	1	..	2	2	2	3	3	2	2	3	3	..	2	..	3	1	..	..	..	..	1
		10° to 15°	..	..	..	..	..	..	..	..	..	1	..	..	..	1	..	..	..	..	..	..	..	..	..	..
		15°+	..	..	..	..	..	..	..	..	..	..	..	..	..	1	..	..	..	..	..	..	..	..	..	..
	1885	5° to 10°	..	..	..	..	..	1	1	..	..	2	4	6	1	3	2	4	..	2	..	2	..	1	..	1
		10° to 15°	..	1	..	..	..	..	..	..	..	..	..	1	..	2	..	..	..	..	..	..	..	..	..	..
		15°+	..	..	..	..	..	..	..	..	..	..	..	..	..	..	..	..	..	..	..	..	..	..	..	..
Calcutta.	1884	5° to 10°	..	..	2	..	..	..	..	..	..	1	..	1	..	..	..	..	..	..	..	..	..	..	..	..
	1885	5° to 10°	..	..	1	..	..	..	..	..	..	1	..	1	..	..	..	..	..	..	..	..	..	..	..	..

In Calcutta there was no change amounting to 7° between any two consecutive days during the two years, and only seven amounting to 5°. At Lahore, during the same period, there was one fall of 15°, six between 10° and 15°, and twenty changes of rise and forty-four of fall, amounting to between 5° and 10°. Rapid falls of temperature are, therefore, between two and three times as frequent as rises, and, on the whole, greater in amount. In estimating the influence of such changes as these on health it must be borne in mind that, in most cases, they are superadded to a large part of the regular daily fluctuation, which may render the change to which the body is exposed more than double as great; and while it is unquestionable that prolonged residence in India renders the system very sensitive to changes of temperature, the thin clothing habitually worn affords but little protection against their injurious effects.

Annual Range.—The change of temperature in the

course of the year is very different in different parts of India, not only in magnitude but also in the duration and rapidity of the rise and fall. As regards magnitude, it is greater in dry parts of the country than in those which are damp and rainy; and the greater also, the higher the latitude, or, in other words, the greater the difference of the longest and shortest day. The greatest range is, therefore, at Leh, which is both the most northerly station and situated in one of the driest climates. The difference of the highest and lowest readings, recorded in the year at the Leh observatory, averages 94°, and has been as much as 103° in a single year. On the plains of the Punjab it varies, on a general average, between 80° (at Mooltan) and 86° (at Peshawar) at different stations, and in some years exceeds 92°. At the hill stations it is much less, viz. 69° at Murree, and 63° at Simla; and has not much exceeded 75° at the former and 69° at the latter within the last ten years. At Darjiling it is only 47°. At Quetta the average range in the course of the year is 84°, and the greatest in any one of the last eight years 92°; while at Jacobabad, on the border of the Cutchee desert, the average is 86° and the greatest 89°.

From the Punjab, which is the most northerly and, excepting Sind and Western Rajputana, the driest province of India, the annual range of temperature decreases eastwards to Bengal and Assam, and southwards to Travancore and Ceylon, where, at Point de Galle, it reaches its minimum. At Lucknow and Allahabad the average range is 76°, at Patna 68°, and at Calcutta it has diminished to 54°. Still farther east, in the damp province of Assam, it is 54° at Silchar and only 57° at the more northerly station Sibsagar.

In the interior of the peninsula, and particularly in the dry tract of the Deccan, it is greater than on either coast, and greater on the east than on the west coast. Thus it amounts to 62° at Poona, 61° at Secunderabad, 54° at Bellary, and 47° at Bangalore, while on the west coast it is only 34° at Bombay, 31° at Mangalore, 28° at Cochin; and on the east coast, 52° at False Point, 50° at Masulipatam, 48° at Madras, and 40° at Negapatam. Finally, in Ceylon the average annual range is only 25° at Colombo, 31° at Trincomallee, and less than 19° at Galle, where in some years it has scarcely exceeded 17°. Such is the uniformity of climate on the south-west coast of Ceylon, that the extreme range of temperature in the course of the year is only about as great as that experienced within 24 hours at Simla, and little more than half that of a single day at Jacobabad.

These great variations in the annual range of the thermometer depend more on the degree of the winter cold than on the varying intensity of the summer heat. Thus, the lowest temperatures hitherto recorded at Leh have fluctuated between 3° above and 17° below zero (29° and 49° below the freezing point), while the highest in different years have varied between 84° and 93°. On the Punjab plains the lowest winter temperatures have been between 24·7° (Peshawar, 1879) and 40° (Mooltan, 1884); in general, a little below the freezing point at the more northerly, and a degree or two above it at the more southerly stations; while the highest summer readings have varied between 112° and 122°.

At Allahabad the extreme summer temperatures are as high as in the Punjab, but the lowest winter reading recorded is 36°, and in one year did not fall below 42°. At Nagpur also, the summer readings are nearly as high, 117° in the shade having been reached in four of the last ten years; but in the coolest years the thermometer has not fallen below 43°, and sometimes not to 50°. And, even at Madras, 113° has been registered in the hottest months, although in the coolest 57·6° is the lowest temperature reached in the last ten years, and in many years it does not fall to 60°.

Madras, however, although situated on the coast, has a comparatively dry climate at the hottest time of year. In the damper provinces of India, in part of Bengal, in Assam, and on the west coast of the peninsula, the highest summer temperatures rarely or never reach 100°. At Calcutta, indeed, the thermometer sometimes rises to 105° or 106° in

April and May, while it falls to between 45° and 52° at the Alipore observatory in January and February. But in Cachar a reading exceeding 100° has been but once recorded in eleven years, and at Sibsagar this has occurred but twice in the same period. At Bombay in the same eleven years it has never reached 97°, and the highest recorded reading in thirty-seven years is 100·2°; at Cochin, notwithstanding its more southerly latitude, it has never marked 100°; and at Point de Galle, since the thermometers have had a proper exposure, it is doubtful whether it has ever risen above 90°, while the minimum has never fallen to 68°.

In comparing extreme temperatures, whether high or low, it must be borne in mind that such comparisons are valid only when the same method of exposing the instruments has been rigorously observed, and when also the errors inherent to almost every thermometer have been ascertained and corrected. It is easy to obtain a reading a few degrees higher in the daytime or lower at night by slightly diminishing the protection of the instrument. The object sought, since such is understood to be the meaning of a temperature reading, is to ascertain as nearly as possible the temperature of the ambient air, and to avoid the disturbing effect of radiation. In a site distant from buildings and not too much shaded by trees, the method of exposure adopted at Indian observatories very well fulfils the required conditions. And to ensure the accuracy of the figures quoted, I have restricted the discussion to such observations as have been recorded with all due precautions, and have rejected such as were registered prior to 1876 in the Punjab and some other parts of India, which, on account of improper exposure, I do not consider entitled to confidence.

Hottest and Coolest Months. Annual Variation.— The popular division of the year into a cold season, a hot season, and "the rains," holds good in the greater part of India, and fairly represents the more obvious phases of the climate. Such is the case in the North-west Provinces and

Oudh, Central India and eastern Rajputana, the Central Provinces, Chutia Nagpur, and Bengal, and, substituting the word "cool" for "cold," it may be held to apply to the northern and much of the eastern part of the peninsula, although the duration of each season varies not a little in different parts of this region. Wherever this is the case, the temperature continues rising from January or, in some years, February, only up to the setting in of the rains. The first drenching fall of rain sends the temperature down many degrees; and although it rises again whenever the rains are interrupted, it never attains the same intensity as in the long interval preceding the rainy season, when the whole land surface lay parched and bare of verdure under a torrid sun.

In North-western India, where the rains come to an end early in September, they are followed by a few weeks of warmer weather; but in the latter part of October the temperature falls rapidly, and in Bengal, if, as is often the case, the rain lasts well into October, a fall of temperature sets in soon after their conclusion and gradually brings on the cool season.

Thus, throughout Central India, and indeed everywhere east of the Aravalis and south of the Ganges, also in the north of the Peninsula and its eastern half as far south as the Godavery, and on the coasts of Bengal and Arakan, May is the warmest month in the year. In Sind, the Punjab, the region west of the Aravalis, and also on the plain north of the Ganges, where the rains seldom set in much before the latter part or quite at the end of June, or where they are too light and scanty to mitigate the heat, June is warmer than May; and, at the more remote stations of Quetta and Leh, the temperature goes on rising till July, as is generally the case in the temperate zone. For a different reason July is also the warmest month in Sikkim and Assam. In the Sikkim Himalaya, and in the valleys of Assam and Cachar, frequent showers keep down the temperature in April and May and neutralise the hot season. On the other hand, in the Western Deccan (at Poona and Malegaon), and in the south of the peninsula, except on the east coast, April is the warmest month in the year; and this is also the case in Ceylon and the Burman peninsula, save in the coast province of Arakan. In Burmah, as in Assam, this is due to the greater abundance of the May showers and to the earlier setting in of the monsoon rains; but in the southern provinces the changes of temperature are smaller, and in the Carnatic it is not until the October rainfall has cooled the plains and filled the irrigation tanks that there is any considerable mitigation of the heat.

From what has been said above, it will be apparent that from January to May the rise of temperature is greatest in Upper Sind, the Punjab, and the North-west Provinces, and diminishes steadily to the south and east. But the fall from May to July is greatest in the heart of the country, in the Central Provinces, where the dryness and heat of May are as great as in the Punjab, while the rains set in earlier and are more abundant. These variations are shown in the following table for all parts of India, together with the annual mean temperatures, those of the hottest and coolest months, the mean highest and lowest temperatures reached in the year, and the mean annual range. Some further particulars are given in the Tables of Appendix I.

[TABLE

MEAN AND EXTREME TEMPERATURES AND RANGES AT FIFTY-ONE STATIONS, CHIEFLY HILL STATIONS AND MILITARY CANTONMENTS. (Deg. Fahr.)

Stations.	Elevation in feet.	Mean Temperature of			Mean Highest Reading.	Mean Lowest Reading.	Annual Range.	Rise January to May.	Change May to July.
		Year.	Hottest Month.	Coolest Month.					
Kurrachee . .	49	77	87	65	107	45	62	20	− 1
Quetta . . .	5,501	58	77	40	99	15	84	26	+11
Jacobabad .	186	78	96	57	118	32	86	34	+ 3
Mooltan . .	420	76	94	54	114	34	80	35	+ 3
D. I. Khan .	573	74	93	52	117	31	86	35	+ 4
Leh	11,503	40	62	18	90	− 4	94	29	+15
Peshawar . .	1,110	70	89	50	115	29	86	30	+ 9
Rawalpindi .	1,652	69	89	49	114	29	85	32	+ 6
Murree . . .	6,344	56	71	39	93	24	69	26	+ 3
Sialkot . . .	829	73	91	52	117	34	83	33	+ 2
Lahore . . .	732	75	93	54	117	34	83	34	+ 1
Simla . . .	7,012	55	67	41	88	25	63	23	0
Delhi . . .	718	77	93	59	116	40	76	30	− 2
Meerut . . .	737	76	92	57	112	37	75	32	− 3
Agra . . .	555	79	95	60	116	40	76	34	− 7
Deesa . . .	465	80	92	67	112	40	72	25	− 9
Abu . . .	3,945	68	79	58	96	39	57	21	− 9
Neemuch . .	1,639	75	88	62	111	39	72	26	− 9
Ajmere . . .	1,611	74	89	58	112	34	78	31	− 7
Lucknow . .	369	78	92	61	114	38	76	31	− 6
Allahabad . .	307	78	92	61	116	40	76	31	− 7
Patna . . .	183	78	89	61	110	42	68	28	− 4
Darjiling . .	7,421	51	61	40	72	25	47	14	+ 7
Hazaribagh .	2,007	74	85	61	106	43	63	24	− 7
Berhampore .	66	78	85	65	109	46	63	20	− 2
Calcutta . .	21	78	85	65	102	48	54	20	− 2
Dacca . . .	22	78	83	66	100	48	52	17	0
Chittagong .	87	77	82	67	94	48	46	15	− 1
Sibsagar . .	333	73	84	59	99	42	57	19	+ 6
Cuttack . .	80	81	89	70	110	51	59	17	− 6
Saugor . . .	1,769	76	89	63	110	42	68	26	−11
Jubbulpore .	1,341	75	90	61	111	36	75	28	−11
Pachmarhi .	3,511	69	83	56	100	35	65	25	−12
Nagpur . .	1,025	79	93	67	115	46	69	24	−14
Poona . . .	1,849	78	86	72	106	44	62	13	−10
Bombay . .	37	80	85	74	95	61	34	11	− 4
Belgaum . .	2,550	74	81	71	102	51	51	8	− 9
Sholapur . .	1,590	79	89	70	110	47	63	17	−10
Secunderabad .	1,787	78	89	69	109	48	61	19	−12
Bellary . .	1,455	80	89	73	108	54	54	15	− 7
Bangalore . .	2,981	73	80	67	98	51	47	12	− 7
Wellington .	6,200	61	66	55	80	37	43	11	− 3
Madras . .	22	82	88	76	108	60	48	11	− 1
Trichinopoly .	275	82	88	76	106	60	46	12	− 3
Cochin . . .	11	80	84	79	95	67	28	3	− 5
Madura . .	448	82	86	77	105	62	43	9	− 2
Galle . . .	48	80	82	73	89	70	19	4	− 2
Rangoon . .	41	79	84	75	104	58	46	8	− 5
Moulmein .	94	78	83	75	99	58	41	7	− 5
Thyet Myo .	134	79	87	68	108	45	63	17	− 5
Toungboo . .	181	78	85	70	104	51	53	13	− 5

The Teachings of the Barometer

Effect of Elevation.—The weather indications formerly inscribed on barometers, such as "set fair," "change," "much rain," and "stormy," have been so generally discarded from modern instruments that it is scarcely necessary to warn the reader that they are quite inapplicable in India. Nowhere in the world would they be otherwise than misleading in any situation very much above the sea-level. The average reading of a barometer, at any place, depends on its elevation above the sea; it is the lower the higher the elevation of the place, the fall, roughly estimated, being about one inch for every thousand feet (or one-thousandth of an inch for every foot), provided the elevation does not exceed a few thousands of feet.[1] Until the average value for the place is approximately known, no inference concerning the weather can be drawn from a simple barometer reading.

Annual Fluctuation.—In India, a further cause of variation must be taken into account, viz. the annual fluctuation of the barometer, which is quite independent of weather vicissitudes. At most places on the plains of India the barometer stands highest in December or January, and from this time falls continuously till June or July; in August it begins to rise again, and continues to do so, amid all its fluctuations, till the end of the year. The whole rise and fall during the year, apart from such changes as accompany changes of weather, is less than a tenth of an inch in Ceylon, about three-tenths in Madras, about half an inch at Calcutta, and not quite six-tenths of an inch in the Punjab. At the hill stations it is less than on the plains. The average monthly readings at a number of stations are given in Appendix I.

[1] Thus, at the Alipore observatory, Calcutta, which is only 21 feet above sea-level, the average reading of the barometer is 29·79 inches. At Allahabad, 307 feet above the sea, 29·48 inches; at Lahore, 732 feet above it, 29·08 inches; and at Simla 7048 feet above it, only 23·24 inches, on the mean of the year.

Diurnal Fluctuation.—There is yet one further oscillation of the barometer which, in India, is remarkably regular, and must be duly taken into account before a change in the barometric reading can be appealed to as indicating a change of weather. Every day, with the utmost regularity, it rises from about 4 in the morning till half-past 9; then falls to about 4 or 5 P.M.; then rises again till 10 P.M.; and again falls till 3 or 4 A.M. These changes, which amount on an average to about a tenth of an inch, have nothing to do with the weather; but they must not be neglected, since they bear a very appreciable proportion to those that are produced by weather vicissitudes. Thus a fall of a tenth of an inch between 10 in the morning and 5 in the afternoon does not betoken any change, merely as regards weather, and so of the other diurnal changes. The course of the diurnal fluctuations at Calcutta in March and July are represented in Fig. 2.

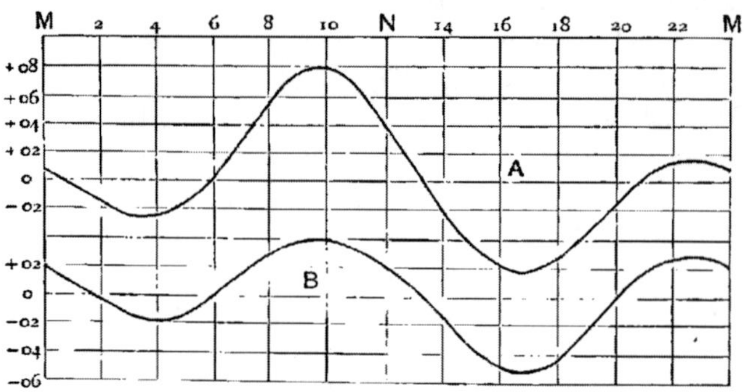

FIG. 2.—DIURNAL VARIATION OF BAROMETER AT CALCUTTA.
A. March. B. July.

Magnitude of Variations with Weather.—Lastly, it must be borne in mind by those who have used to consult the barometer as a weather-glass in England or other country in higher latitudes, that the oscillations which, in India, do betoken changes of weather, are much smaller than those to which such experience may have accustomed them; and it is for this reason that the several regular changes above noticed cannot be overlooked or ignored. In this country, a fall of three-tenths of an inch below the average of the time and place occurs only as a prelude to very disturbed weather, and it is only in violent cyclones and in the vicinity of the storm's centre that this amount is very greatly exceeded. In the centre of such a storm it has been known to fall more than two inches below the average of the place and time of year.

Temperature Correction.—One further remark on the action of the barometer is necessary before any comparison can be made between its readings and those published in the daily weather reports, or with such average values as are also periodically published, and some of which are given in the Appendix to this work. A mercurial barometer is affected by temperature as well as by the pressure of the atmosphere, and acts to some extent like a thermometer, rising with heat and falling with cold. The published readings of the barometer are, therefore, not such as are actually recorded, but are reduced to the height which the instrument would have marked if it had been at the temperature of melting ice, a condition which, it need hardly be remarked, it rarely indeed has anywhere on the plains of India. This correction is effected by tables computed for the purpose and given in most meteorological handbooks. These remarks apply to mercurial barometers only, not to aneroids, which, if properly compensated by the makers, should not require such correction. But every instrument, whether aneroid or mercurial barometer, generally has some error, which can be ascertained only by comparing its readings with those of a standard instrument, and this should never be neglected. Barometric registers that have been kept for years at the cost of much trouble and self-denial, are of little value to science unless their correction is known, and also unless the temperature of the instrument has been recorded whenever a reading is taken. The barometer many years in use at the Calcutta observatory is used as the standard for all Indian stations.

Barometric Changes with Weather.—With these preliminary remarks we may pass on to consider the manner in which the readings of the barometer may be interpreted as prognosticating the character of the weather. In addition to the fluctuations just described, which are regularly recurrent and can, therefore, be predicted with confidence, the pressure of the air, which immediately affects the barometer and the variations of which are all that it really indicates, oscillates more or less irregularly at all times of year, increasing for two or three days, and then falling in like manner. The rise and fall of the barometer in these movements ordinarily amount to between one and two-tenths of an inch, and are greater in Northern than Southern India. They do not necessarily betoken any important change of weather, though generally there is more cloud about the sky when the pressure is low than when it is higher; and it is at its low phase, in the hot season, that thunder-storms and dust-storms most frequently occur. In the cold season, in Northern India, a fall is frequently accompanied with an easterly or southerly wind, the wind that precedes and brings the winter rains of Northern India. In the rainy season a fall of greater amount precedes one of those bursts of heavy rain, that throughout this season alternate with intervals of fine weather.

When a cyclonic storm is generated far down in the Bay of Bengal, which occurs chiefly in the months of May, October, and November, the barometer on the north of the bay and the surrounding land is often but little affected, and sometimes even rises while the storm is in course of formation; and it is not until it begins to move northward that any fall sets in. Even then the fall is slow, until perhaps a few hours before the storm reaches the neighbourhood of the place, when it becomes more rapid. Examples of this will be given in the second part of this work. In judging of the probability of a cyclone, other indications must be taken into account, more particularly the direction of the wind and the aspect of the sky. These also will be noticed in another part of this work.

Distribution of Pressure. Barometric Depressions.—So far I have noticed only such indications as are afforded by the behaviour of the barometer at a single station, and where no information can be obtained of what is taking place in other parts of the country. Before the days of telegraphic weather reports, this, together with the local signs of the weather, was all the practical meteorologist had to rely on for framing his conclusions, and under these circumstances the barometer was little more than an auxiliary guide, far from having the importance which it possesses in these days of telegraphic reports and weather charts.

For, while very little of value can be inferred from the simple reading of a barometer, even when compared with the average of all its readings at the same place and time of year and day, a knowledge of the distribution of atmospheric pressure over an extensive country, such as India, affords a firm basis for meteorological reasoning. Primarily, it is some difference in the pressures of different parts of the country and the surrounding seas that determines the courses of the winds, and it is the character of the winds as damp or dry that determines that of the weather. The place where the pressure is least, in other words, where the barometer stands lowest, is usually termed a *barometric depression* or *minimum*, and the immediate neighbourhood of this depression is the place where rain is most likely, and when it falls most abundant. This last statement is, indeed, not to be taken without some qualification, but the exceptions are readily recognisable, and will presently be more particularly specified: as a general rule, the maxim holds good.

Reduction to Sea-level.—It has been stated on a previous page that the reading of the barometer depends on the elevation, falling about one-thousandth of an inch for every foot above sea-level. Differences of readings, due to those of elevation must therefore be allowed for and corrected before the distribution of atmospheric pressure over the surface of a country can be known for the purpose of meteorological reasoning. In the published weather reports this allowance has been made; all barometric readings of stations on the plains of India are given, not as recorded, but reduced to the values they would have at the common level of the sea surface. These pressures are, of course, purely imaginary, and in the case of stations on the hills of India would be probably false and misleading; but when restricted to such places as are not more than a couple of thousand feet above the sea, the readings, thus reduced, afford a fair idea of the differences of atmospheric pressure that are effective in determining the winds, to which they bear a well-ascertained relation.

Barometric Charts. Isobars.—To render the facts of pressure distribution evident at a glance, the method now universally practised is to lay them down on a chart. The preparation of such a chart is very simple. The several barometric readings, duly corrected for temperature and reduced to sea-level, are pencilled against the stations to which they severally refer. A series of lines is then drawn between or over these, passing through all those points at which the pressure is the same, the points selected being those where it is the exact tenth, or, in Indian charts, the twentieth of an inch. The interspaces between any two consecutive lines show where the pressure is of intermediate value, and the distance between them is the horizontal distance within which the pressure increases or decreases by one-tenth (or one-twentieth) of a barometric inch. The lines therefore represent differences of pressure in the same way that the contour lines of a map show differences of elevation and the slope of the ground, enabling one to gather correct notions of the form of its surface. They are technically termed *isobaric lines*, or simply *isobars*. A *barometric depression* is the space enclosed within the isobar of lowest value, and a barometric elevation, or as it is usually termed, a *barometric maximum*, is that surrounded by the isobar of highest value. Other terms, which will be noticed when we have to speak of the winds, are also used to denote these two important regions.

Barometric Gradients.—Another term often employed in weather reports and requiring explanation is the *barometric gradient*. This term, which has been adopted from engineers, means the rate at which the atmospheric pressure decreases between any two places. It may be expressed either by the number of miles in which the pressure falls one-tenth of an inch, or by the amount of decrease per hundred miles. On any barometric chart its value, expressed in the former manner, may be found at once by simply measuring on the scale of the chart the distance between two

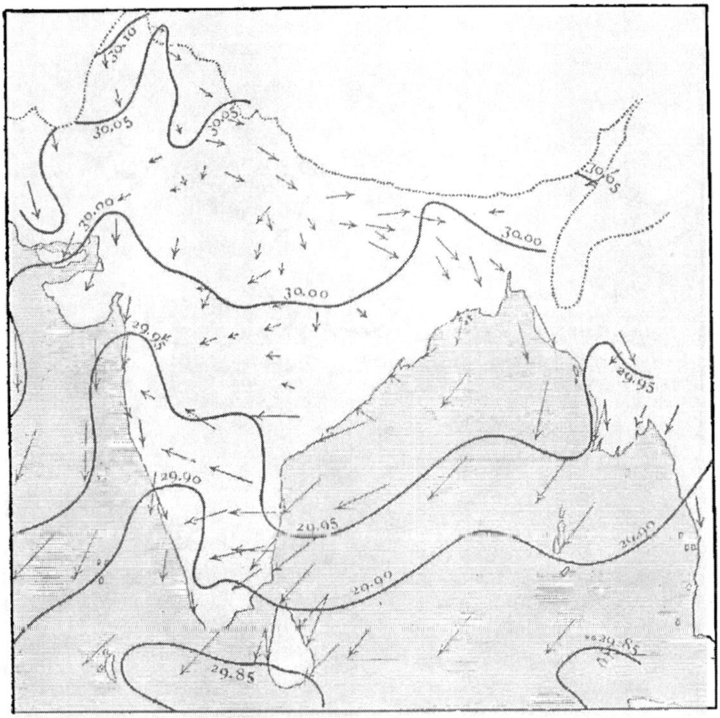

Fig. 3.—Average Barometric and Wind Chart for January.

isobars differing one-tenth of an inch in value. It is an important datum, because, just as the velocity of a stream of water is the greater, the greater the slope down which it flows, so is the strength of the wind the greater the higher the barometric gradient where it blows. Wherever, therefore, the isobars of different values approach each other most nearly, there do the strongest winds prevail. Other relations of the winds to the isobars and the distribution of pressure generally will be described in the next section.

Persistent and Temporary Barometric Differences.—It would be beside the purpose of this work to discuss at any length the circumstances that give rise to these inequalities of pressure, which the barometer proves to exist, and

FIG. 4.—AVERAGE BAROMETRIC AND WIND CHART FOR JULY.

which are represented on the weather charts issued from day to day. And indeed, in some of the most important cases, we are not yet able fully to explain them. But it is necessary to point out the distinction between such as last for months together and are intimately related to the monsoons, and such as are temporary only and regulate the variable phases of the weather, and more especially those that accompany storms.

As illustrations of the former class I give two Figures (3 and 4) showing the average distribution of atmospheric pressure in January and July.[1] In the first of these, the seat of the highest pressure is in the north-western corner of India, and it decreases steadily southwards to two regions, one to the south of Ceylon, the other near Sumatra, where are two incomplete isobars, indicating the places of lowest pressure on the chart. In the July chart, the seats of highest and lowest barometer have exchanged places. The former now occupies the equatorial sea, the latter Upper Sind and portions of the adjacent tract, and throughout India and its seas the barometric gradient is reversed. In the interval between January and July this change is brought about entirely by the changes of temperature; the seat of lowest pressure in both cases being also the seat of the greatest warmth, and that of highest pressure the region where the atmosphere is coolest. Amid all the changes that take place from day to day in the course and position of the isobars, with very rare exceptions, these leading features remain undisturbed; and, as will be seen in the next section, it is these that determine the general course of the monsoons.

Now barometric depressions of this class sometimes form an exception to the rule given above, viz. that rain is most likely, and when it falls, heaviest in the neighbourhood of the depression. Those represented in the January chart conform to the rule, since they are situated over the sea where the supply of vapour is abundant, and the air therefore damp and capable of yielding rain. But in July, and

[1] It is necessary to point out that in these charts the barometric values of the isobars are all somewhat lower than would be assigned to them by the actual readings, corrected only for temperature and elevation as described in the text. A further correction has been applied for gravity, but as this is important only to the physicist, it need not be specially noticed here. The charts show the relative distribution of pressure accurately, and, except in the values of the isobars, the chart differs but little from that which would result from the uncorrected data.

indeed all through the summer monsoon, when the seat of the lowest pressure is the driest land tract in India, the most copious rainfall takes place at a considerable distance from it, viz. in Bengal and Assam, along the face of the Himalaya, and over a broad zone extending across India from the Gulf of Cambay to the Bay of Bengal (see Fig. 6); and also, owing to local circumstances, on the west coasts of India and Burmah.

Of the irregularities of pressure which are only temporary, it is necessary to notice only the depressions, and of these only such as accompany storms. Slight local falls of the barometer occur at all seasons, and have no important significance. Owing to incessant changes of temperature, the atmosphere is never at rest through its whole depth, and every movement is accompanied by some small variation of pressure. Of most of these it is unnecessary to speak. But when, for some days together, the pressure falls continuously over some tract of land or sea, a movement of the air is set up around, which may become strong enough to generate a storm, the constitution and nature of which will be described further on. Such a storm may either exhaust itself in the place where it originated, or may move forward and travel thousands of miles, its centre being marked by the barometric depression which originated it, and which indeed forms one of its leading features. Illustrations of such storms are given in Figs. 16 to 23 in the latter part of this work.

When such a storm approaches any place, as a matter of course the barometer falls continuously until the storm either reaches or passes it; but in order to know whether any observed fall is due to the approach of such a storm or to some other cause of less importance and significance, it is necessary to be provided with charts of the pressure on successive days, such as are published by the meteorological offices of Simla and Calcutta; unless, indeed, other weather signs, of which more anon, be sufficiently distinct and characteristic to indicate the real character of the disturbance.

Summary.—These remarks may suffice, for the present,

to enable the reader to intelligently interpret the teachings of the barometer. But, before leaving the subject, I will briefly recapitulate the different points that demand attention. First, the readings of the barometer vary with the elevation of the place, and to such an extent that very little of value can be inferred from them until that elevation is known to the nearest foot or two. Secondly, in India, the barometer falls from January to July, and rises during the remainder of the year. This fall and rise are related to the changes of the monsoons. Thirdly, it undergoes a double oscillation daily of about one-tenth of an inch. This has no reference to changes of weather, but is too considerable to be left out of account in using the barometer as a weather-glass. Fourthly, the mercurial barometer is affected by changes of temperature like a thermometer, and in India, where these daily changes sometimes amount to 40°, the rise and fall of the barometer owing to this cause alone is very appreciable. Tables are published for correcting it, and in order to render the readings made at different temperatures comparable with each other, it is usual to reduce all to their corresponding values at the standard temperature of the freezing point. Fifthly, every barometer has some error, sometimes a large error, inherent to the instrument. This is to be ascertained by comparing the instrument with some acknowledged standard. And further to render its readings comparable with those of other barometers at different elevations, it is necessary to reduce all to the same level, generally the level of the sea.

The oscillations which depend on changes of weather are comparatively small in India, not exceeding a few tenths of an inch, except near the centre of violent cyclones. But oscillations of one or two-tenths of an inch, extending over a few days, follow each other incessantly at all times of year, and do not necessarily betoken any change of importance. As a rule, the fall takes place with a wind from the sea, which is damp, and rises with one from the land, which is dry.

The full meaning of any barometric change can only be

known and understood when it is taken in connection with those in other parts of the country. These are shown by barometric charts, the method of preparing and interpreting which has been described above, and will be further noticed in a subsequent part of the work.

The interest of barometric readings centres entirely in their relations to the weather and the physics of the atmosphere. As an element of climate, the variations of atmospheric pressure are far too small to be sensible to the human system, except indeed at great elevations, where, owing to the decreased pressure, the tenuity of the air is such as seriously to tax the heart to effect the necessary aeration of the blood; but to persons of good constitution, and with no special feebleness of that organ, this is not felt at the elevation of 8000 feet, which is above that of the hill stations of India, most of them being situated between 6000 and 7500 feet, and some lower.

The Winds

Lightness of the Winds in India.—Next to its sunny skies and its notorious and somewhat oppressive heat, perhaps no feature of the Indian climate is more characteristic than the prevailing lightness of the wind. Many a one who has watched the laborious processes by which the Indian ryot raises water to irrigate his patch of sugar-cane or paddy must have asked himself the question, "How is it that the same mechanical skill that has enabled the village carpenter to construct a Persian wheel, or the simple but effective turbine and its gearing, wherewith the hillman drives his primitive mill, has never been turned to economise the work of irrigation, by substituting for the labour of the ryot or his cattle some simple motor driven by the wind?" The explanation is probably to be found less in his want of mechanical skill than in the insufficiency and uncertainty of the motive power at his command. It is true that on the coasts of India, and especially the west coast, the wind has, as a rule, sufficient strength to be so utilised, and such is also the case in the Deccan during the summer monsoon; but in Northern India it seldom has sufficient force, except perhaps in the spring months for some hours of the afternoon, when the hot winds blow, and these are by no means regular, and sometimes fail for many days together. At best, there are but few parts of India where windmills could be usefully employed, and hence probably the fact that they are practically unknown. As illustrating the average rates of the wind's movement in different parts of the country, the stations in the following table have been selected from those for which anemometric data are given in the Meteorological Report for 1885. The height of the anemometer above the ground is given in each case, and those stations only are quoted at which the instrument is comparatively unobstructed. The table shows the number of miles per day registered by the anemometer, the average being in all cases derived from many years registers.

AVERAGE DAILY MOVEMENT OF THE WIND

Stations.	Anemometer feet above ground.	Miles per Day.											
		Jan.	Feb.	Mar.	April	May	June	July	Aug.	Sept.	Oct.	Nov.	Dec.
Peshawar	21	57	78	85	90	109	103	102	94	75	63	51	60
Lahore	28	44	55	72	80	77	83	77	69	48	38	34	35
Jacobabad	44	47	74	91	97	117	113	131	96	75	57	40	36
Kurrachee	22	232	247	283	337	424	451	428	412	351	231	191	223
Deesa	62	207	215	214	229	315	420	347	292	206	160	184	194
Ajmere	43	48	63	85	131	174	204	169	129	114	66	37	36
Bareilly	48	68	98	105	111	125	120	100	86	73	47	43	51
Benares	76	68	90	111	120	121	122	113	98	92	60	51	57
Sutna	26	94	113	132	167	190	229	221	184	158	102	77	85
Patna	35	49	68	82	106	111	100	85	77	71	45	36	40
Calcutta	54	71	93	132	180	172	147	124	126	110	78	70	73
Silchar	25	49	67	78	84	83	74	74	72	65	52	50	45
Nagpur	36	69	87	101	122	156	176	177	138	112	83	65	63
Buldana	24	128	148	178	226	297	278	241	217	206	122	121	124
Bombay	53	243	257	276	273	246	397	459	383	285	232	238	239
Poona	73	132	157	194	238	338	401	404	342	283	137	137	138
Belgaum	47	272	228	269	320	436	601	686	512	418	240	278	288
Secunderabad	31	90	87	116	120	164	291	318	239	168	82	78	72
Bangalore	32	91	82	84	84	128	203	208	183	162	92	89	92
Madras	50	150	122	152	191	224	218	202	179	159	123	166	184
Negapatam	27	132	83	82	114	164	180	175	142	126	88	126	163
Thyet Myo	52	57	100	124	174	132	191	158	144	113	72	88	113
Rangoon	49	104	98	121	137	125	131	136	121	98	77	103	123
Port Blair	31	174	123	110	126	183	258	286	248	239	160	175	173

Wind Measurement.—At land observatories the wind is measured by means of the anemometer—an instrument which consists of four hemispherical cups, mounted on light horizontal cross arms, which, being set in rotation by the wind, communicate the movement through a train of wheel work to an index, which shows on a graduated dial, either the number of revolutions or the distance travelled by the wind. The instrument is set up usually on the highest point of a building, if possible above and free from all obstructing objects, such as trees, higher buildings, etc., around. This condition cannot, however, be always secured, and at some observatories the register of the wind from certain quarters is interfered with to an extent which can only be vaguely estimated.[1] In some places it is mounted on the top of a pole, but as these are of various heights, and every additional foot of elevation reduces the retarding effect of ground surface friction on the wind, under the most favourable circumstances, the anemometric registers of different observatories are only roughly comparable with each other. The direction of the wind, as shown by the wind-vane, is also very much influenced by purely local conditions.

In the above table no register is included that has been obtained at a less elevation than 20 feet above the ground, or where obstructions are known to exist in the immediate neighbourhood of the recording instrument. At Calcutta, Nagpur, Bombay, Deesa, Kurrachee, Poona, Belgaum, Madras, and Rangoon, and at some other stations not included in the above list, both the movement and direction of the wind are recorded continuously on a sheet or strip of paper set in movement by clockwork, and at these stations all changes of rate and direction can be determined from hour to hour or any shorter interval, but at the majority of the observatories the anemometer dial is read off once or twice only in the 24 hours, and the direction at the time of observation noted by the wind-vane.

[1] A list of the elevations of all anemometers at the Indian observatories, with notes of their position and that of obstructions, if any, was given in the *Report on the Meteorology of India in* 1885, p. 113.

Beaufort's Numbers.—At sea, it has long been the practice to estimate the *force* of the wind instead of its rate of movement, and to express it by the series of numbers 1 to 12, according to a scale proposed by Admiral Beaufort and given below. These are known as *Beaufort's numbers*. By comparing the forces thus estimated at Holyhead pier lighthouse with the registers of the autographic anemometer on the dome of the lighthouse, an approximate evaluation of Beaufort's numbers, in terms of the rates of movement, has been obtained by Mr. R. H. Scott. These are given in the right hand column of the following table:—

Force. Beaufort Scale.			Approximate Velocity. Miles per hour.
0. Calm			0 to 5
1. Light air	or just sufficient to give steerage way		6 ,, 10
2. Light breeze	or that in which a well-conditioned man-of-war, with all sail set, and clean full, would go in smooth water from	1 to 2 knots	11 ,, 15
3. Gentle breeze		3 ,, 4 ,,	16 ,, 20
4. Moderate breeze		5 ,, 6 ,,	21 ,, 25
5. Fresh breeze	or that to which she could just carry in chase, full and by	Royals, etc.	26 ,, 30
6. Strong breeze		Single-reefed topsails and top-gallant sails	31 ,, 36
7. Moderate gale		Double-reefed topsails, jib, etc.	37 ,, 44
8. Fresh gale		Triple-reefed topsails, etc.	45 ,, 52
9. Strong gale		Close-reefed topsails and courses	53 ,, 60
10. Whole gale	or that with which she could scarcely bear close-reefed main topsail and reefed foresail		61 ,, 69
11. Storm	or that which would reduce her to storm staysails		70 ,, 80
12. Hurricane	or that which no canvas could withstand		80 and upwards.

These numbers will be sometimes referred to in the section on "Storms in Indian Seas" in the second part of this work.

Wind Pressure.—A third mode of expressing wind force is by the pressure it exerts per square foot of surface; and this is the form in which it is of most use to architects and engineers. The pressure is sometimes computed from the rate of movement, but the result so obtained is of little practical value, since it gives only the mean and not the maximum pressure, and stormy winds always blow in gusts which may vary from 5 or 10 to 50 lbs. in the course of a few seconds. It is better shown by a pressure gauge, of which several forms have been devised. That which registers the pressure autographically at the Calcutta observatory, known as Osler's anemometer, exposes the surface of an iron plate one foot square, which is maintained facing the wind and transmits its pressure to springs of known resistance. The pressure thus indicated is, however, not the same that would be exerted by the same wind on each square foot of an extended surface, such as the face of a building, and differs with the form of the plate as well as with its dimensions. In the opinion of Professor Stokes and others who have studied the subject, such a plate indicates a higher pressure than would be exerted by the same wind on each square foot of a masonry structure. But it may represent more nearly that on a lattice girder, which, like the pressure plate, allows the air to pass and form eddies on the rear face.

Diurnal Variation of Wind.—Everywhere in India, as far as observation enables us to verify the fact, except perhaps in certain positions on the mountains, the movement of the wind is more rapid in the daytime than at night, and most rapid at the hottest time of day. Indeed it varies very nearly with the temperature of the air. The same thing has been observed in other countries, but, like most of the regular periodical changes of the atmosphere, it is so marked in India, that it must be recognised by any one whose attention is once directed to it. The air is most still just before sunrise, indeed at certain times of year absolutely calm; and in and about large cities, where, in this stagnant condition, it becomes charged with mephitic gases and vapours, it is extremely questionable whether the Indian habit of taking exercise before sunrise is not rather prejudicial than otherwise to health. But the rising sun at once stimulates its movement, which increases rapidly in the earlier hours of the forenoon, and about 2 P.M., as a general rule, the wind blows with its greatest force, afterwards declining at almost exactly the same rate as the temperature falls. The hot winds of Northern India in the spring months are essentially diurnal winds, springing up in the forenoon, blowing with some strength during the heat of the day, and sometimes not falling to a lull till long after sunset. They are land winds, and derive their heat from the ground, which is much hotter than the air, while their dryness is such that paper becomes brittle, and furniture and woodwork of all kinds shrinks and tends to fall to pieces by the loss of the moisture which, in more ordinary conditions of the air, all such organic substances hold absorbed in an insensible form.

On the coasts, as for instance at Bombay, the sea winds follow a very similar course of variation during the drier months of the year, but, at the height of the monsoon, any diurnal variation in the force of the wind is less appreciable and hardly to be recognised by the casual observer.

In some parts of the country, at certain seasons, over and above this variation in the rate of its movement, there is a well-marked change of the wind's direction in the course of the day. On the coasts, as is well known, in the long lull between the monsoons, and generally in calm states of the atmosphere, a land wind at night and in the early morning alternates with the stronger sea breeze of the warmer hours. At Calcutta, sixty-eight miles from the sea, the variation is of a different character. In the spring months, during the hottest hours of the afternoon, the wind frequently blows from the west, being in fact the land wind of the Upper Provinces, and is characterised by the heat and dryness of that wind in a somewhat mitigated degree; but, in the evening, it yields place to a cool southerly wind, in fact, a late sea breeze, blowing up from the wide estuary of the Hooghly. This latter does not reach the more inland stations, such as Kishnaghur and Berhampore, and such stations, therefore, do not enjoy the refreshing change which ushers in the Calcutta dinner hour, and affords some hours of comparative coolness and rest. Easterly winds are most frequent in the early hours of the morning, chiefly in October and November, and are too often tainted with the unsavoury exhalations from the Calcutta sewage and refuse, which are discharged in that direction on the borders of the Salt Lake.

In the valleys of the Himalaya, as in all mountainous countries, there is a tendency to an up-valley wind in the daytime and a down-valley wind at night; but, as a rule, this tendency is shown, not so much by a distinct reversal of the directions in the course of the day, as by the exclusive prevalence of one or the other wind at its appropriate hours, alternating with a comparative calm. The up-valley wind is felt most strongly on the passes, where, as General Strachey has described it, it sometimes blows in the afternoon with such force as to present a serious obstacle to travellers. On the other hand, the down-valley wind is strongest where the great valleys debouch on the plains. At Hurdwar, at the mouth of the Ganges valley, where it is known as the *Dadu*, it sometimes blows with such force, in the early morning hours of the cold weather, as to blow down the tents of persons camping in the vicinity.

In point of fact, there is perhaps no station in any part of India which has not a characteristic diurnal variation in the direction of the wind. Such has been shown to exist by Mr. Hill at Allahabad, by Mr. F. Chambers at Kurrachee, and by Mr. C. Chambers at Bombay; and it is equally evident in the records of the anemographs at Lucknow and Nagpur, published monthly by the Meteorological Office. But, in most cases, it is not manifested by a distinct change of the wind, but only in a more or less slight modification of the direction prevalent at the season, and has a scientific rather than a general interest.

Annual Variations. The Monsoons.—The reversal of the prevailing winds, which takes place twice regularly in the course of the year, has long been familiar under the name of *the monsoons* to all who have any knowledge of the east. It holds good in all parts of the country, but the directions are by no means everywhere those by which the monsoons are commonly known. The terms *south-west* and *north-east* are, in fact, truly descriptive, only in the Bay of Bengal and the western half of the Arabian Sea. And for this reason, in dealing with the land, it is better to substitute for them the equally distinctive but less misleading designations, *summer* and *winter monsoons*, as will be done throughout this work.

Figs. 3 and 4 (pp. 25 and 26) exhibit the average wind directions, both on India and the seas around, in the months of January and July, when the two monsoons are respectively at their height. The relative steadiness of the wind is indicated by the lengths of the arrows, some of the longest representing 90 per cent of the observations and upwards.

The winds are lightest in the cold season of Northern India, that of the winter monsoon. In the Punjab, even in the daytime, on nearly half the days at this season, there is not wind enough to stir the cups of the anemometer; and all down the Gangetic plain, while a calm state of the atmosphere is very common, light zephyrs are the chief movement of the air, until the sun of March and April has sufficiently heated the ground to set up active convection and give rise to the hot west winds. On the higher ground south of the Ganges and Jumna, the movement is somewhat greater and more northerly, and the same is the case in Lower Bengal; but, as the table on a preceding page shows, up to the end of February, at nearly all stations it is less than 100 miles a day. In the peninsula at this season the wind is easterly, between north-east and east in November and December, and more easterly and southerly in January and February, while, on the west coast, it is light from north and north-west.

At times indeed, and in some seasons more than in others, the light northerly and westerly winds of Northern and Central India are suspended for many days together, and replaced by light winds from the opposite quarters. These are warmer, damp and oppressive, and the vapour brought up by them is condensed, first clouding the Himalaya and then the plains, and finally turning to rain, the cold weather rainfall of Northern India.

In the north of the Bay of Bengal the north-east monsoon is very light, and often an absolute calm prevails for days together. In the heart of the bay, and from the mouths of the Godavery southwards, it blows more steadily, and is often strong to the south of Madras, producing a rough sea and, up to January, stormy weather off Ceylon and in the south of the bay. On the Arabian Sea it is very light near the Indian coast, but becomes stronger and steadier farther west, blowing from between north and north-east.

In the north-west of the bay, on the coasts of Bengal and Orissa, the winter monsoon may be said to have ceased by the end of January, but light northerly winds continue some weeks longer down the Arakan coast, and in the middle and south of the bay it lasts well into March, becoming less steady and by degrees shifting towards the east. In the Arabian Sea, on the other hand, it backs through north-west to west, as the barometer falls over the land in consequence of the increasing heat; but the weather remains fine, and the Arabian Sea is seldom disturbed by stormy weather, such as, after March, sometimes occurs in the Bay of Bengal in the interval between the winter and summer monsoons.

On the coast of Bengal southerly winds begin to set in as early as the end of January, at first merely as afternoon sea breezes. Little by little they penetrate farther inland, blowing night and day, and also extend their range down the coast and farther out to sea. In the east and south of the peninsula the east and north-east winds of the cool season also veer to south-east; but in Northern India, on the plains of the Ganges and the higher plains and hilly country to the south of that river, the westerly winds strengthen and become hot and drier. In the meeting of these and the sea winds are generated the thunder-storms and dust-storms characteristic of the spring season.

Up the Gangetic plain along the foot of the Himalaya, and over the crests of the lower hills, easterly winds advance earlier than on the plains to the south, feeding the diurnal storms, often accompanied with hail, which are common in the spring, and adding copiously to the snows of the higher ranges; and, recurving from the hills, they are often felt as north-east winds on the plains of the Ganges, as at Purneah, Patna, Benares, Allahabad, and Lucknow. Meanwhile, in Western India, north-west winds, sometimes changing to west, blow very steadily in Rajputana, Guzerat, Khandesh, Berar, and the north-west of the peninsula; and these being very dry, rain seldom falls in this part of India during the spring; nor indeed, except in the rare case of a winter disturbance in some part of this region, or of a cyclone crossing from the Bay of Bengal, during the whole interval between the cessation of one summer monsoon and the beginning of the next. But in the south and east of the peninsula, where the winds of the earlier months are east and south-east, little thunder-storms occur, as in Bengal. Fig. 15 shows the average wind directions in the month of May.

In May, while light west and north-west winds still sweep the Arabian Sea, in the Bay of Bengal south-west winds have sprung up, but somewhat unsteadily and often interrupted by calms; although this condition is less common than in March and April. The summer monsoon, accompanied with heavy rainfall, now sets in on the west coast of Ceylon and Travancore, and, in the eastern peninsula, on the coast of Tenasserim, and in the course of two or three weeks has taken possession of the bay and the Arabian Sea up to their northern coasts. It invades India from these seas in two main currents. That from the Arabian Sea blows on the west coast from directions between west and south-west, sweeps over the western Ghats, on which it discharges a very heavy rainfall, and continuing its course steadily across the Deccan, holds undisputed possession of the peninsula up to the end of August. In the interior its direction is most commonly west, especially in the Deccan, where it is strongest and steadiest; but in the south it often changes to south-west, and, at slack intervals in all parts of the peninsula, sometimes to north-west.

The other branch sweeps from south-west up the greater part of the bay; a part passing directly over Burmah and Assam towards the eastern Himalaya. The remainder curves to south in Bengal, and farther recurving along the face of the Himalaya, becomes an easterly wind up the Gangetic plain. To some extent it penetrates the north-eastern part of the peninsula, but this is exceptional until near the end of the season, and it sweeps past the Carnatic, withholding from that province more than a scanty share of the rainfall which it carries abundantly to Bengal, Assam, and, in a minor degree, other parts of Northern India.

In the Punjab, during the summer monsoon, the winds are somewhat variable. Easterly winds predominate, and these are partly drawn from the current blowing up the Gangetic plain. But the westerly branch of the monsoon from the coasts of Sind, Cutch, and Guzerat blows across Rajputana as a south-west wind, and in part coalesces with the former in the northern Punjab. And, at intervals, the easterly wind yields place to a dry wind from the opposite quarter. This takes place during breaks in the rains; and, in seasons of drought, west and north-west winds hold possession of North-western India for weeks together, penetrating eastwards across Central India and down the Gangetic plain, even to Bengal, and southwards to the Deccan and Hyderabad. Such was the case in the disastrous years of 1876 and 1877, and during the prolonged suspension of the rainfall in 1880 and 1883.

In Lower Sind the summer winds are chiefly south-west; in Upper Sind south-east or south; but the atmosphere is often calm, and rarely indeed does rain visit this arid tract. On the Baluchistan tableland, which rises to 5000 feet and upwards immediately to the west of Sind, the prevailing wind throughout the season is from north-west. The same wind would doubtless prevail in Upper Sind, were it not for the mountain barrier and the fact that winds do not readily descend from a higher to a lower level. But there is little doubt that, descending in a very flat air cataract and brought down by convective intermingling with the lower atmospheric layer, it is this current that feeds the dry west winds of North-western India.

We have seen that while a west wind holds almost undisputed sway in the peninsula up to the Satpura range, an easterly wind is only less dominant along the northern border of the Gangetic plain and on the Himalaya. The tract between these, which comprises the greater part of the loop of the isobar 29·50, on the July chart on Fig. 4, p. 26, is debatable ground in which sometimes one, sometimes the other, predominates; more often the stronger westerly current. The distribution of the rainfall in this region will be described later on; but it may be here mentioned that, throughout the monsoon, the cyclonic storms which are generated in frequent succession in the north of the Bay of Bengal (sometimes also over the Gangetic delta, or in Chutia Nagpur), as a rule, travel along this debatable zone. Such, for instance, was the case with the storms illustrated on Figs. 16 to 23. Further details on this subject may be reserved until we come to describe more fully the natural history of these storms.

On reference to the table on page 31, comparing the rates of movement in the summer with those of the winter months, it will be apparent that the summer monsoon is a much stronger current than its winter correlative; and in India this fact is recognised in popular language, since it is often spoken of distinctively as "the monsoon," the claim of the winter monsoon to the same designation being, for the moment, tacitly ignored. On land, in fact, the winter monsoon blows so short a time, and is so unsteady and frequently interrupted, that it is far from impressing the landsman in the same vivid manner as the seaman, who has perhaps to beat up the bay against a steady head wind; and his attention is rather turned to that feature of the season which is far more striking and important than the direction of its gentle wind. In his consciousness of awakened energy, the expatriated European feels that it is also the cold season, refreshing and invigorating, and affording him a climate, than which, as experienced in Northern India, Italy itself can offer nothing more delightful. The thinly clad native, inured to heat, and living in a draughty hut, with perhaps a single meal a day of less stimulating food, is less enraptured with the delights of the cold weather, but he is not less conscious than his European brother of this its most characteristic feature, which he feels in the early morning in benumbed limbs and torpid faculties, and which he endeavours to meet by swathing his head and mouth in a fold of his body cloth, and cowering over the embers of his little fire, till the warmth of the ascending sun restores him for some hours to his state of normal activity.

But in spite of this tacit recognition of its superior force and steadiness, it would be a mistake to suppose that the summer, any more than the winter monsoon is a persistent and unvarying movement of the atmosphere. In Northern India, at least, the wind changes frequently both in direction and strength; and even a calm state of the atmosphere, though far less frequent than in the winter months, is by no means so rare as to be regarded as phenomenal, even in Bengal and the eastern part of the Gangetic plain, while farther to the north-west, and more particularly on the hills and in their neighbourhood, the record of calms constitutes a very appreciable and, at some stations, a large proportion of the wind registers, throughout the summer monsoon. It has already been mentioned that, in some seasons, it is excluded, for weeks together, from a great part of its ordinary domain, which is then invaded by dry winds from the opposite quarter; and these prolonged periods of its suspension are but a magnified form of those interruptions, intervals of sunny weather known as "breaks," which alternate with bursts of rain as an ordinary feature of the season. In fact, the goal of the monsoon, the place of low barometer to which its course is directed, is constantly changing. Generally it lies beyond the limit of the tropic on the western side of India, or not far within it on the eastern side. The seat of lowest pressure in Sind is pretty stable, but very little of the sea wind ever reaches it; it is rather to the zone indicated by the loop of the isobar 29·50, on the July chart Fig. 4, or still farther to the south-east, that the main volume of both branches of the monsoon is directed, and within this tract the seat of lowest pressure is constantly shifting.

At times, indeed, both branches of the monsoon relax their strength, and fair weather with a moderate wind prevails for days or even weeks over a great part of the adjacent seas. On the Arabian Sea this occurs during those invasions of the continent by north-west winds above spoken of; on such occasions the mail steamers cross from Bombay to Aden with fine weather and light winds, similar to those usually experienced before the monsoon sets in.

Towards the close of the season, in September, the rain-bearing wind ceases to penetrate to North-western India, and is directed rather towards the north-east of the peninsula and the adjacent portion of the Bay of Bengal; the winds of Bengal become more easterly, and those of the northern provinces of the peninsula more northerly. By degrees, the seat of low barometer moves farther south, till, in the latter part of October and November, it lies over the bay, opposite the coast of Madras. This is the season of the Carnatic rains. In November, the winds are very variable in the south of the bay, between Sumatra and Ceylon, shifting to all quarters from south-west round by north and east to south-east, and often calm. In December calms are very common in the south-east of the bay, off Sumatra, and at no time does the north-east monsoon blow so steadily in latitudes south of Ceylon as in the Bay of Bengal at this season.

Relations of Winds to Regions of High and Low Pressure.—For those whose only object, in consulting these pages, is to gain some general idea of the character of the winds at the different seasons of the year, the above description may suffice. But for those who desire to go farther, to understand the meaning of weather reports and charts, and something of the principles on which weather prediction is based, and for seamen, to whom it is a matter of paramount importance to gather as much knowledge as possible of the weather round about them, and especially of their position in a storm, from the indications afforded them by their barometer and the wind direction, it will be necessary to go somewhat farther into the subject, and to show at greater length than has yet been done, how the winds are related to regions of high and low barometer.

The preceding description of the two monsoons, taken in connection with the distribution of pressure and the prevalent wind directions exhibited on the *barometric* charts on Figs. 3, 4, and 15, has already afforded a striking illustration of the general fact that winds blow from a place where the barometer is high to that at which it is relatively low. But it has also shown that the course they follow is not direct, but, on the contrary, very circuitous. And since this course is no local peculiarity, but that which winds always

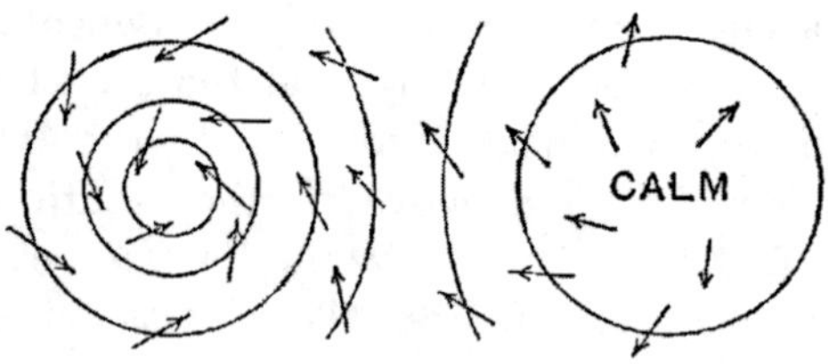

FIG. 5.—CYCLONE AND ANTICYCLONE [Diagram].

follow under the like circumstances in the northern hemisphere, it will be desirable to describe it more particularly in relation to the seats of high and low barometer generally. Except that in a cyclone the action is more concentrated and localised than in the case of the monsoons, the description will equally apply to both.

To make the facts as clear as possible I will, in the first place, exhibit them in the form of a diagram, which may afterwards be compared with the results of actual observation, represented on the storm charts, Figs. 16 to 23, or with the general description of the monsoons, in connection with the two charts of average pressure and wind distribution on Figs. 3 and 4. Let the right hand circle in Fig. 5 be supposed to represent a region where the barometer is highest, and the left hand circle that where it is lowest, the intermediate lines indicating a number of isobars of intermediate gradations of pressure. Then the winds that blow outwards from the region of high barometer, if in the northern hemisphere, instead of blowing directly outwards like the spokes of a wheel, all take an increasingly oblique course to the right of the radial direction, as shown by the arrows, and those that blow in towards the region of low barometer all blow obliquely to the right of that course which leads directly to the centre. The result is that the former describe a series of spirals curving round the seat of high barometer in the same direction as the movement of the clock hands; the latter, a series of spirals circulating round the seat of low barometer against the direction of the clock. The latter represents the movement of the winds in a cyclone. They do not blow, as was formerly supposed, in a circle, but as is represented on the diagram. For proof of this the reader has only to refer to the charts on Figs. 16 to 23, where the arrows represent the winds observed and the small circular isobar the eye of the storm.

When, as in the case of the monsoons, neither the seat of highest barometer nor that of lowest barometer is a circle or any other regular figure, the course of the winds is not in such regular spirals as those shown in the diagram. It is determined by the directions of the isobars, and, unless diverted by some local condition, is always obliquely to the right of a perpendicular to the isobar towards which it is directed. The amount of obliquity depends on a number of circumstances, the enumeration of which would be foreign to the purpose of this work. As a general rule, the stronger the wind, the more acute is the angle at which it meets the isobar, and the nearer do they approach parallelism. Near the centre of a cyclonic storm, in the North Atlantic, the spirals become so flat that they approximate to the circles which formerly they were supposed to be, but they are never truly circular. In the Bay of Bengal this is less marked, and I have not been able to detect more than a very slight decrease of the radial component in the published storm charts.

Around a region of high barometer the pressure decreases slowly, and, when represented on a chart, the isobars are far apart. In other words, the barometric gradient is low, and, as has been stated in the previous section, under these circumstances the winds are light, and radiate more directly outwards than where the pressure decreases rapidly. This is why, when the barometer stands high, indicating that the seat of maximum barometer is in the neighbourhood, the winds are generally light, as is the case in North-western India in the winter months. On the other hand, when it shows a very low atmospheric pressure, it frequently happens that a cyclonic storm is in the neighbourhood, and around such a storm the barometric gradients are high and the winds strong.

The system of winds around a region of low barometer, even when less strong than in the storms, originally termed cyclones, are frequently spoken of as a *cyclonic system* and those around a seat of high barometer as an *anti-cyclonic system.*

These remarks, which are of general application, will suffice for the present. The whole subject of storms, and the indications which are of most importance to the mariner, together with a further subject equally essential to his guidance, viz. the direction in which storms travel, will be treated of more conveniently and at length in the subsequent section on storms and in the second part of this work.

Dampness and Dryness

Relative and Absolute Humidity.—The dampness and dryness of the air, as these terms are generally understood in ordinary language, have reference, not so much to the quantity of moisture which it holds suspended, as to the readiness with which it parts with this moisture to objects capable of absorbing it, or, on the other hand, deprives them of that which they hold absorbed. Now this depends quite as much on the temperature as on the quantity of vapour present in the air. The atmosphere of Great

Britain in the winter, when the thermometer stands, let us say, at 40°, is usually very damp; but if this same air were at the May temperature of Calcutta it would be an exceedingly dry atmosphere. Dampness and dryness, in this sense, are the conditions most important to us, both in respect of our own bodies, and also as affecting vegetation of all kinds; and only, therefore, in this sense, will it be necessary here to discuss the subject. In technical language, it is spoken of as the *relative humidity* of the air, in contradistinction to the *absolute humidity,* which has reference only to the amount of water vapour present, independently of the temperature.

It is convenient to have some way of expressing numerically the degree of dampness (of which dryness is of course the correlative), and the plan universally adopted is to regard completely damp air, usually termed "saturated" air, as a standard, and to express the degree of dampness as a percentage of the saturating quantity. Thus, when it is stated that the relative humidity of the air is 50, the meaning is that it holds just half the quantity of water vapour that it is capable of taking up at that temperature.

Methods of Observation.—The humidity of the air is an object of observation at all observatories, and not the least important. As an element of climate it is probably one of the most important. The method adopted is an indirect one, and requires attention to some points of detail, which need not be insisted on here. But it is very simple in practice, and when properly attended to is found to give a very near approximation to the truth. It consists in taking the readings of two thermometers suspended at a little distance from each other, under the same conditions as other thermometers (see page 6), one of which shows the temperature of the air, the other, having its bulb covered with a single fold of wet muslin, the temperature of a freely evaporating wet surface. From these two readings the humidity is found by the use of a special table.

Variations of Humidity in India.—One characteristic

feature of the climate of most parts of India is its great variability with respect to humidity. In the rains, during a spell of very wet weather, the air is sometimes saturated, or has a relative humidity of 100. But when a hot wind is blowing, such as is common in April and May in the Upper Provinces, it is frequently less than 10, and has been known to fall to 3; as dry a condition as is probably to be met with anywhere in the world. At Agra, in May 1884, on the mean of the whole month, it was only 20, and at Roorkee but little higher. In Scotland, Mr. Buchan states that 73 is about the lowest humidity that occurs in the driest month, which, there also, is May.

Fortunately for us, when the heat is very great, the air is always dry; fortunately, I say, because it is the rapid evaporation from the skin, evaporation which can take place only in a dry atmosphere, that enables us to withstand the heat. As some of those who have made the passage down the Red Sea in August or September may have experienced, an atmosphere at 90° which, though not saturated, is, at all events, very damp, is insupportably oppressive, but the hot winds of the Upper Provinces, with a temperature between 112° and 118°, if not exactly agreeable, are borne without serious inconvenience.

Diurnal Variation.—The humidity of the air varies during the day almost exactly in the opposite manner to the temperature. The atmosphere is most damp in the early morning before sunrise; in fact, the fogs that are so common in the early morning in the cold weather testify that the air is then saturated, for fog cannot exist in air that is not in that condition; and the dissipation of the fog soon after sunrise affords evidence of the increasing dryness.[1] The driest time of day is the same as that of the highest temperature, and the humidity increases in the later hours of the afternoon and evening, as the temperature falls.

[1] A remarkable exception to this ordinary sequence occurs in Assam, which I once had the opportunity of verifying personally. Its probable explanation will be given farther on (p. 55).

Dryness and Dampness dependent on the Winds.—The dryness of the warmer hours of the day is not due solely or, to speak more correctly, not directly to the warming of the air. It is partly an effect of the movements set up in the lower atmosphere by the greater heating of that layer which rests immediately on the ground, in consequence of which the whole atmosphere is stirred up eventually to a height of some thousands of feet in dry weather; and thus, while the damper and more heated air ascends, drier and originally cooler air descends from a higher level to replace it. This action, which takes place in all countries more or less in dry warm weather, is developed in India in a very high degree, and most so in the dry atmosphere of the Upper Provinces. This, in fact, is the real explanation of the extreme dryness of the hot winds above referred to. In substance, they are part of an air current coming originally from the western mountains and blowing far above the earth's surface, but brought down to the lower level by the action just described. Originally it is indeed much cooler than the air which it displaces, but it is speedily heated, partly as a necessary consequence of its descent,[1] and partly by contact with the ground glowing in the sunshine. It is not improbable that the well-known healthiness of the hot season is in some measure due to this stirring up of the atmosphere, the removal of all mephitic exhalations, and the invasion of the lower atmosphere by the purer air drawn from the high level of the mountain zone.

The part played by land winds in India is only beginning to be fully recognised. As has been set forth in the preceding section, they occur at all seasons of the year, while in none are they exclusively prevalent. They are characterised partly by their direction, which, in Northern India, is always westerly and, to some extent, northerly, but chiefly by their dryness, of which some instances have already been given. When they penetrate to Southern India they are

[1] Dry air, descending through the free atmosphere, is heated 1° for every 183 feet of its descent, in virtue of the compression which it undergoes as it becomes subject to the increasing weight of the superincumbent atmosphere.

more northerly, but retain somewhat of their westing, except in the cold season, when they are between east and north-east.

Winds that come originally from the sea, however great the expanse of land they may have traversed, unless, indeed, like the westerly summer monsoon of Southern India, they have been driven to a great height over a mountain range, are relatively damp winds. As an almost invariable rule in India, except temporarily and locally, they are more or less southerly; but while in Western India, in the greater part of the peninsula, and in Assam, Arakan, and Burmah, they combine this direction with more or less westing, in the Gangetic plain and the neighbourhood of the eastern coast they are from south-east or even more easterly.

Geographical Distribution of Humidity.—Different parts of India are characteristically dry or damp, accordingly as one or the other of these winds is most prevalent. North-western India, including all the northern portion of the Bombay Presidency, is more particularly the domain of the former; North-eastern India, Assam, and the western part of the Burman peninsula, together with the western margin of the Indian peninsula, that of the latter. The following table shows the mean humidity of the several provinces of India. It must, however, be remarked that different parts of one and the same province sometimes afford a considerable range of variation:—

MEAN ANNUAL HUMIDITY OF THE SEVERAL PROVINCES OF INDIA

Provinces.	Rel. Hum.	Provinces.	Rel. Hum.
Ladak	49	Orissa	76
N.W. Himalaya (lower ranges)	62	Central Provinces (southern)	59
Nepal Valley	72	Berar and Khandesh	53
Sikkim Himalaya	84	Konkan and Surat	72
Baluchistan	50	Deccan	55
Sind[1] and Cutch	50	Hyderabad and Bellary	54
Punjab Plains	55	Northern Circars and Godavery	71
Rajputana	48	Carnatic	67
Guzerat[1]	47	Mysore	66
Central India Plateau	50	Malabar and Coorg	79
Nerbudda Valley	55	Ceylon	80
N.W. Provinces and Oudh	59	Arakan	80
Behar	65	Pegu	77
Bengal	80	Tenasserim	80
Assam	80	Bay Islands	81
Chutia Nagpur	57		

[1] Exclusive of the coast.

Hence the driest parts of India have an average humidity a little below 50, the dampest about 80. Mountains, but not high plains, have a damper atmosphere than the plains of the same part of the country.

Annual Variation.—In those parts of the country that are intermediate between the permanently arid and the permanently damp provinces, the variation in the course of the year is as great as between these geographical extremes, and indeed greater. The rainy season which, in most parts of India, lasts from June to September, is also the dampest in nearly all parts of the country. But this is not the case on the north-western frontier, nor in the Himalaya north of the snowy range. At Quetta, Peshawar, Leh, and even in Lahoul, the winter is the dampest time of year, as is also the case in Western Europe, and indeed as a rule in extra-tropical countries. On the plains of Northern India August is the dampest month, the mean humidity being a little below 90 on the plains of Bengal and Assam, and as much as 95 at Darjiling, which is nearly the dampest station at which an observatory exists. It is above 80 everywhere to the east of Allahabad, but not farther westward, except on the hills. On the plains of the Punjab, south of Lahore and Umballa, and throughout Western Rajputana and Sind, the mean humidity of August remains below 70. At Quetta it is only 43, and at Leh, 48.

In the north of the peninsula, July and August are about equally damp, their average humidity being about 85 in the provinces east of Nagpur, and 87 at Bombay and in the Konkan generally. But in the Deccan, to the east of the Ghats, it is everywhere below 80, and in the drier parts below 70. Farther south, in Malabar and Travancore, it is about the same as in the Konkan or a little higher, between 87 and 89; and at Mercara in Coorg, on the crest of the Ghats, as high as 97 in July and 95 in August. This station, in the rains, is damper therefore than even Darjiling. On the high plateau of the peninsula it is 77 at Bangalore, and only 62 or 63 at Bellary. Throughout the interior and eastern half of the peninsula south of Orissa, September is damper than either of the preceding months; and as we proceed farther south, the dampest season falls later and later, being in October and November at Masulipatam, in November at Madras, and in November and December from Negapatam southwards as far as Hambantotte in the south coast of Ceylon; thus following the rainfall. In Burmah it is generally in July and August, as in the north of the Indian peninsula.

The driest month in the year is November or December on the coasts of Bengal and Orissa, February or March in the interior of Bengal and Assam. Everywhere to the west of Bengal April is the driest month; except in the Punjab and Upper Sind, where it is deferred till May, and at Quetta, Peshawar, and Leh till June. With local exceptions, the average humidity of April is below 30, from Agra southwards, all over Rajputana and Central India, the Central Provinces as far south as Nagpur, in Berar, Khandesh, the Deccan, and Guzerat, except on the coast; and at some stations in the heart of this tract, such as Akola, Sutna, and Indore, below 25. In Upper Sind and Western Rajputana, and in the Punjab and North-west Provinces, except in the neighbourhood of the hills, also in Chutia Nagpur and the country south of the Lower Ganges, it is between 30 and 40, but eastwards it increases rapidly, and in the drier parts of Bengal averages little below 60, though on certain days it may fall to 30 or even lower. At the Alipore observatory, in a suburb of Calcutta, the average of February and March is 69; but in the more closely-built suburb of Chowringhee only 66, showing that small differences of position may make an appreciable difference in the dryness of the air. In the valleys of Assam and Cachar it is 70 or higher; at Sibsagar, in Upper Assam, as high as 79.

On the west coast of the peninsula the driest time of year is between December and February, when the humidity falls to 70 or a little lower; at Rutnagiri apparently to 60 on the average of the driest month, and at Mangalore to 62. At these two stations the observatory is situated on high ground, which is not the case at the other stations on this coast. They afford another instance of the variation of the dampness of air due to small changes of site. In the Deccan and Hyderabad March and April are the driest months, and this is also very generally the case in the south-east of the peninsula, except on and near the coast, where the driest period is as late as June and even July; but throughout the Carnatic the air is comparatively dry in all these months, the humidity varying between 61 and 64 on the coast to a little above 50 in the interior, in the driest months. At Trincomalee, on the east coast of Ceylon, July is the driest month in the year.

In Burmah it varies between January at Mergui, in the south of Tenasserim, and March or April in the north of Pegu, where, at Thyat Myo, the humidity averages only 56. Even at Rangoon it falls to 62 in February. Observatories have only lately been established in Upper Burmah, and the only exact information that we have of its climate is furnished by a register kept at Mandalay for the first nine months of 1879 and the last two months of 1886. In the former year, March was the driest month, when the mean humidity was as low as 43, and that at 4 P.M. only 22. September was the dampest, when it averaged 87. In this respect, therefore, the climate of Mandalay would seem to be comparable with that of the drier districts of northern Bengal. It is certainly drier than any part of Pegu.

From this brief review it results that different parts of India exhibit as great diversity in the seasons of dampness and dryness as they do in their degrees, and it is driest on the east coast of Ceylon and the Carnatic when it is dampest in the north of the peninsula and on the west coast.

Haze, Fog, and Cloud

Haziness of the Sky. Dust Haze.—However sunny the sky of India, it is hardly to be described as an azure sky. Of cloud indeed, save sometimes in the rainy season, and at such a station as Darjiling, the amount is not such as to cause depression of spirits; but he who wends his way to India in the hope of living under a deep blue heaven, like to that of Italy or Greece, will surely reap disappointment. The artist who seeks to depict a cloudless Indian sky will deal sparingly with his cobalt and ultramarine if he would work truthfully, and not in the spirit of the engraver in a well-known story who introduced palm trees in a sketch of Punjab scenery. Except now and then, in a fine interval of the rains, or at the hill stations of the Himalaya in September and October, the pallor of the celestial vault is its ordinary and persistent characteristic; whether it be due to thin diffused cloud matter at a high level, or, as seems equally probable, to the haze of impalpable dust, which is carried up by the daily commotion of the heated air, and darkens and sometimes even extinguishes the setting sun before it has dipped to the horizon. So dense is this haze in the hot season, and to such heights does it rise, that even at Simla, the opposite mountain ridge, only four or five miles distant, is ofttimes invisible through the murky atmosphere. That the pallor of the sky is, however, partly due to very attenuated cloud, seems probable from some observations of Mr. Hennessey at Mussoorie. In the course of a long series of actinometric measurements of the sun's heat, it was necessary to watch closely the appearance of the aërial vault, and he frequently descried a thin gauzy veil, which he designated as "steam-cloud," creeping over the otherwise cloudless expanse, reducing the depth of the sky tint, and appreciably also the action of the solar rays on the instrument.

On the plains the haze is almost constantly present, and it is the exception rather than the rule, in the dry season, that the slopes of the Himalaya are visible from a distance of twenty or thirty miles. As far as my own experience enables me to form an opinion, the atmosphere of Southern India is less hazy than that of Northern India, but even there a very clear atmosphere is exceptional.

Fog.—Fog is restricted to the damper provinces and the cold season, and in them as a rule to the early morning.

It is sometimes thick over the rivers, and also in the extreme north of the bay off the Sunderbans, impeding navigation; but in Bengal it is dissipated an hour or two after sunrise. In Assam, however, it often lasts till mid-day, and sometimes presents a peculiarity of behaviour to which my attention was first drawn by Sir Charles Elliott and Mr. C. J. Lyell. This is that the atmosphere remains perfectly clear till some time after sunrise; it then rapidly fills with dense fog, and so remains for several hours, when at length the fog breaks up and becomes dissipated under the increased warmth of the sun. During a short visit to Assam in December 1884, I once witnessed this at Gauhati on the bank of the Brahmaputra. The night was clear and starlit, but the heavy drip from the eaves of the bungalow told that the air was completely saturated, depositing dew copiously. The ground on which the station stands is high, but immediately in front flows the broad stream of the Brahmaputra, and there are swamps and tanks in other directions around. The air was perfectly still, unstirred by the slightest breath, and remained clear until the sun had risen about half an hour, when, in the course of a few minutes, everything around was obscured by a dense fog. . As I was leaving the station the same morning I had no opportunity of investigating the matter, but the probable explanation appeared to me to lie in the facts of the stillness of the atmosphere during the night, of the alternations of land and water surfaces, and of the saturation of the air probably to a considerable height. It is well known that a land surface cools more by radiation under a clear sky than one of water, and the air over each such surface would approximate to the temperature of that on which it rests. Owing to the entire absence of wind, these air masses might remain unmingled, though at different temperatures, until set in motion by the sun. But it is a well-known physical fact that two masses of air at different temperatures, and both completely charged with invisible vapour, when intermingled, can no longer hold the whole quantity in suspension. The excess must therefore be deposited as fog.

These suggestions require verification, which might be supplied by any one who would suspend two minimum thermometers half an hour before sunrise at the same level, one midway over a tank or other water surface, the other over the land at a sufficient distance from the shore, both being sheltered to prevent radiation from the instrument. It would thus be ascertained whether the difference of temperature is such as is assumed in the hypothesis. For this purpose the minimum temperatures would suffice.

Forms and Nomenclature of Clouds.—To an experienced eye the forms and movements of the clouds and the general aspect of the sky are eloquent of impending weather; but save seamen, and among intelligent and observant nations, farmers and herdsmen, there are but few who have learned their language and can rightly interpret it, while such knowledge as these possess is for the most part empirical and little capable of being harmonised with the facts and translated into the language of physical science. Among meteorological observers of the ordinary class it is certainly rare; it cannot be learned from books, and unlike the art of managing and reading a thermometer or a barometer, cannot be imparted in a few easy lessons. The first difficulty in the way of such teaching is the want of a sufficiently exact language to designate the protean forms in which cloud masses present themselves to the eye; forms which, for one and the same cloud, are different when seen overhead and when seen laterally at a distance; and second, the difficulty of identifying the particular kind of cloud intended from a mere verbal description. Owing mainly to this second difficulty, the art of cloud observation, as treated in text-books, and still more in practice, is involved in endless confusion, and confusion is inimical to all advance in knowledge.

At the beginning of this century, Luke Howard distinguished some of the principal and most characteristic simple forms of cloud, giving them names which have obtained universal acceptance. Two of these, applied respectively to the highest and to the most common and familiar low cloud form, are easily referred to their proper objects, and are not often misapplied. These are first, *cirrus*, feather cloud or mare's tails, the wisps of feathery or fibrous cloud often seen against the blue sky far above any other kind of cloud; and, second, *cumulus* or heap cloud, the fine weather cloud of the daytime, with a flat base and a rounded upper surface. But a third name of a simple form, viz. *stratus* or layer cloud, applied by its author simply to a fog bank, whether resting on the ground or floating at some moderate height above it, has been indiscriminately applied to clouds of very different character at great heights in the atmosphere. It is really characteristic of the night and of the cold weather, always on or near the earth, and the use of the term should properly be restricted to such.

So far, however, it is comparatively plain sailing. But when we come to deal with the innumerable forms which occur at all intermediate levels, and which are either modifications of these or transitional forms, or compounded of them and other simple forms, the difficulty of distinguishing and identifying the several varieties becomes very great; and very much yet remains to be ascertained respecting their mode of formation and their movements, all of which are essential to enable us to interpret their meaning and to draw any trustworthy inference as to what they portend.

That this state of ignorance, which is a reproach to meteorological science, will be allowed to endure much longer is, however, unlikely. For some years past, a few excellent observers have devoted themselves to the special study of clouds. Not to name others, Mr. Clement Ley in England, Professor Hildebrandsson at Upsala, and the Hon. Ralph Abercromby in the course of two journeys round the world, have not only studied and recorded their diverse forms with the aid of photography, but what is still more important, have established their connection with those other weather conditions in which they occur, and have shown how certain important inferences may be drawn from their appearance.

Of the results at which these observers have arrived,

only a few preliminary sketches have as yet been published; but the following outline of a scheme proposed by Messrs Hildebrandsson and Abercromby represents what will probably be the basis of the future classification of clouds, and recognises the very important distinction between such as are characteristic of fine and such as betoken or accompany rainy weather. It may be remarked in connection with this scheme, that Mr. Abercromby, as the result of his widely extended experience, finds that the forms of clouds are similar in all parts of the world, the chief distinction being that certain of them are more common in some latitudes and places than others.

Messrs. Hildebrandsson and Abercromby's Proposed Scheme of Cloud Classification

α. *Discrete tending to rounded forms (principally in dry weather).*	β. *Extended and sheet-like forms (rainy weather).*
A. Highest clouds, mean height 30,000 feet.	
1. Fibre cloud [*Cirrus* or mare's tails].	2. Thin cloud veil [*Cirro-stratus*].
B. Medium elevation 13,000 to 20,000 feet.	
3. Small globular cloudlets, shining white like silk, 20,000 feet [*Cirro-cumulus*, mackerel sky]. 4. Larger globular, like white wool, 13,000 feet [*Cumulo-cirrus*].	5. Thicker ash-coloured or bluish-gray sheet, 17,000 feet [*Strato-cirrus*].
C. Lower clouds, 5000 to 7000 feet.	
6. Great rounded masses or rolls of gray cloud [*Strato-cumulus*].	7. Ragged sheets of gray cloud from which rain commonly falls [*Nimbus*].

D. Clouds in ascending air currents.

8. Heap cloud [*Cumulus*]. Summits at 6000 feet; bases at 4500 feet.
9. Storm (thunder) clouds [*Cumulo-nimbus*]. Summits 10,000 to 16,000 feet; bases 4500 feet.

E. 10. Elevated fogs. Below 3500 feet [*Stratus*].

The elevations given[1] are furnished by measurements made

[1] The originals are given in meters, which I have converted into feet in round numbers, as great exactitude is not claimed for them, nor indeed possible.

in Northern Europe. It is probable, nay, almost certain, that they would have to be extended in India to greater heights. Not only does this seem to follow from the consideration that at the ground surface, and at equal heights above it, the temperature is higher in India than in Europe, and that temperature must be one principal condition in determining the character of the cloud; but it is also confirmed by direct observation. It is common at Simla to see clouds of the D class shrouding the summits of the snowy range, and rising some thousands of feet above its highest peaks, which range to 20,000 feet; and the plane of the cloud bases evidently slopes regularly upwards from over the plains towards the snowy range. But this does not affect the general validity of the scheme, which shows how cloud forms become modified as they sink lower in the atmosphere, and also how certain forms are characteristic of dry or wet weather respectively. Scud, or torn masses of rain cloud, is regarded as only a slight modification of Nimbus borne along by a gale. It often descends very low; in cyclones only a few hundred feet above the earth's surface. The tops of thunder clouds are often seen tailing off laterally in a fibrous tuft like *cirrus*. This, Professor Hildebrandsson calls *false cirrus*, and believes that like true *cirrus* it is composed of ice spicules, not of water corpuscules.

The Teachings of Clouds.—According to the above scheme the inferences to be drawn from the behaviour of the clouds are obvious. If they show a tendency to form sheets, at first high and thin but gradually becoming darker and more opaque, showing that they are increasing in thickness, sinking at the same time lower in the atmosphere, it is a sign of probable rain. But if, on the other hand, such a sheet is breaking up into separate masses, and sunlight begins to be visible through the thinner portions, it is a sign of the weather's clearing. These signs are in full accord with universal experience.

Besides these, the clouds teach other lessons. The movements of the higher clouds tell us what winds are blowing in those elevated regions of the atmosphere which are otherwise inaccessible to observation. To a casual observer this movement may seem to be very slow; but this is deceptive, and owing to the great height at which it takes place. As a fact, it appears from actual measurements made by Dr. Vettin at Berlin, the high feathery cirrus travels at an average rate of forty miles an hour, or about three times that of the surface winds; and some such result might perhaps have been expected when we bear in mind how small is the retarding friction at great heights in the atmosphere.[1] It is only of late years, and only at two stations in Northern India, viz. Calcutta and Allahabad, that the higher cloud movements have been systematically observed in India. Opportunities for observing such do not often present themselves, and the following tables enumerate the total results of the observations of six years at Calcutta and eight years at Allahabad:—

MOVEMENTS OF CIRRUS CLOUDS OBSERVED AT ALIPORE (CALCUTTA) DURING THE SIX YEARS 1881 TO 1886

Months.	Days of Movement from								Total.
	N.	N.E.	E.	S.E.	S.	S.W.	W.	N.W.	
January	0	0	0	0	0	13	27	2	42
February	1	0	0	0	0	14	16	3	34
March	2	0	0	0	0	5	18	8	33
April	0	0	0	0	0	4	8	3	15
May	1	0	2	1	2	4	3	1	14
June	0	0	2	1	6	5	4	0	18
July	0	0	3	0	7	6	0	0	16
August	0	0	3	7	3	7	0	0	20
September	0	3	3	9	2	7	3	3	30
October	2	1	1	0	3	5	4	8	24
November	3	1	2	0	0	18	27	1	52
December[1]	3	2	0	0	0	7	27	5	44

[1] The recently-published report on the eruption of Krakatoa shows that the ash cloud emitted by the volcano was carried westward round the world in an elevated stratum of the atmosphere at the rate of from 70 to 76 miles an hour.

[2] For December, five years only.

MOVEMENTS OF CIRRUS CLOUDS OBSERVED AT ALLAHABAD DURING THE EIGHT YEARS 1878 TO 1885

Months.	Days of Movement from								Total.
	N.	N.E.	E.	S.E.	S.	S.W.	W.	N.W.	
January . .	1	0	0	0	0	27	44	20	92
February . .	2	0	1	0	3	18	55	9	88
March . .	1	0	2	3	3	28	58	8	103
April . .	2	0	0	1	1	10	31	15	60
May . . .	0	0	0	0	0	8	22	7	37
June . . .	0	6	6	4	2	15	24	7	64
July . . .	0	4	10	3	3	7	8	3	38
August . .	2	6	10	2	1	7	9	4	41
September .	2	4	11	5	3	6	9	3	43
October . .	0	1	1	1	0	7	5	1	16
November .	1	0	0	0	1	4	6	2	14
December . .	0	0	0	0	1	21	17	3	42

Hence, as far as the occasional high cloud affords any basis for judgment, the winds in the upper part of the atmosphere appear to be chiefly west and south-west over both Calcutta and Allahabad, except in the summer monsoon, when they are variable, from all quarters at Allahabad, and at Calcutta chiefly southerly. When a cyclone is forming in the bay, cirrus cloud from the south-west is often seen passing over Bengal before there is any other distinct indication of its existence.

Cloud Proportion.—The average cloudiness of different parts of India, at different seasons of the year, varies very much with the dampness of the atmosphere, as already described in the previous section. The usual mode of recording it is to estimate the proportion of the sky that is covered by cloud, irrespective of the kind of cloud. The whole expanse of sky overhead, that is, so much as is included in an angle of 60° on either side of the zenith, is designated by the number 10, and the amount of the cloud is estimated as so many tenths of that expanse. This is the meaning of the figure expressing cloud proportion that stands against each station in the daily reports. A completely clear sky is thus indicated by 0, an entirely overcast sky by 10. The average cloudiness of a day, a month, or a place is found in the usual way, by taking the arithmetical mean of all the individual observations included in the period or relating to the place in question, care being taken that the observations are so timed as to represent equally or in due proportion all the phases of variation.

Distribution.—The comparative serenity of the skies of most parts of India is shown by the fact that, except in the Carnatic, Ceylon, the southern parts of Pegu and Tenasserim, Assam, and Sikkim, the mean cloud proportion is everywhere below 5. In Scotland, according to Mr. Buchan, it is between 6 and 7. In Sind and the southern Punjab it is below 2, and in the remainder of the latter province, except on the hills, scarcely exceeds 3, on the average of the whole year. In Rajputana it varies between 2 and 3·5, and in the remainder of India between 3 and 5. The cloudiest stations at which observatories exist would appear to be Hambantotte, on the south coast of Ceylon, and Sibsagar in Upper Assam, at both of which the mean proportion amounts to 7 or more. In the case of Sibsagar this is partly due to the fogs of the cold weather, which are properly reckoned as cloud. In the summer months alone, Mercara, on the crest of the Western Ghats, is the cloudiest station at which an observatory exists, the proportion, from June to August, being between 9·5 and 10.

The cloudiest month in the year in by far the greater part of India is July, the proportion at most stations being between 7·5 and 9; and August is nearly as cloudy. In the western Punjab and Upper Sind, which are scarcely ever reached by the summer rains, it is February or March, but here the proportion does not much exceed 3 on the average of any month. In Baluchistan, where March is also the cloudiest month, it amounts to nearly 5. In the extreme south of the peninsula, and on the east coast of Ceylon, the last two months of the year are more cloudy than any of the summer months, and also the dampest and rainiest. But in the western half of the island, and also in Malabar, June is the cloudiest.

The season of greatest serenity is more variable, ranging in different parts of India from October to March. In the Punjab and Sind October is the clearest month, the average proportion of cloud being below 0·5 at the drier stations, and little exceeding 1 in the north-east of the tract. In Guzerat, Rajputana, Central India, and the Gangetic plain, the skies of November are slightly clearer; and in Bengal and Assam both months are surpassed by December and January. In these provinces and months the average cloud is between 1 and 2. In the peninsula, again, the first three months of the year are the clearest, in general, February; and as far down as Bellary and Dharwar, the skies are then as clear as in Bengal; but farther south the proportion does not fall much below 2·5 in the clearest month, and in Ceylon it exceeds 4. In Burmah also the first three months of the year are the clearest, the proportion in the least cloudy month averaging about 0·5 at Toungnoo and as much as 3 at Mergui.

The Rainfall

The Contrasts of Indian Rainfall.—What has already been said of the variations of the climate, in respect of dampness and cloudiness, will have prepared the reader to expect a great diversity in the rainfall of different parts of India; and, in point of fact, no country in the world furnishes such contrasts as Northern India. In the east we have the station of Cherrapunji in the Khasi hills, with an annual fall of from 500 to 600 inches; and, in the west, Jacobabad, with an annual average of $4\frac{1}{2}$ inches, and in some years less than 1 inch.[1] Between these extremes we

[1] The rain-gauge is so well known an instrument that any detailed description is scarcely needed. The essential parts are—a funnel into which the rain falls, and the area of the aperture of which regulates the quantity received; a receiver, in which the water is collected; and a graduated glass to measure the quantity. The glass must be adapted to the area of the funnel. It is of such capacity that when filled to the topmost mark it holds half an inch (or one inch) of rainfall, and the graduation shows the tenths

have every possible variation, and the contrasts afforded by large provinces are only less striking than that of these prominent examples. In Southern India the diversity is also very great, but here the relative positions of the dry and wet tracts are reversed. It is the west coast of the peninsula that receives the heavy rainfall of the summer monsoon, while it is withheld from the plains of the south-east coast, and still more from certain portions of the high plains of the interior; and the transition from the zone of heaviest rainfall to that where it is least and lightest is extraordinarily rapid. Thus, on the crest of the Western Ghats are the two stations, Baura Fort and Mahableshwar, with an average fall of 251 and 261 inches respectively. Only sixty-five miles distant from the former is Gokak, the annual average of which is no more than 22 inches, or less than one-tenth of the former; and this is representative of a broad zone extending, with but slight interruption, all down the peninsular to the east of the Ghats, in which the rain-fall averages from 18 to 25 inches only in the year. See Fig. 6.

Scarcely less remarkable is the occasional heaviness of the falls, even at places where the average rainfall is not by any means excessive. That upwards of 40 inches in the 24 hours should have been once recorded at Cherrapunji will perhaps hardly seem surprising, but falls nearly as great (from 30 to 35 inches), in the same interval, have occurred on more than one occasion on the plains of the Ganges valley, at places where the average of the whole year is not more than from 40 to 65 inches; and even in the extremely arid province of Sind as much as 20 inches fell on one day in 1866 at Doorbaji, where the annual average is probably less than 6 inches.

Another noteworthy peculiarity in the rainfall of India is the variability of different parts of the country in respect of the regularity of the rainfall. As will presently be shown more in detail, as a general rule, those provinces

and hundredths of an inch. An inch of rainfall means that if collected on an impervious level surface it would form a sheet of water one inch deep.

which have the lowest rainfall are also those in which it is most precarious, and the most rainy parts of India are also those in which the rainfall is most regular. But there are some striking partial exceptions to the rule; or rather, among the provinces with a moderate average rainfall, some are subject to great vicissitudes, while in others it is remarkably regular. Thus, Nagpur and other parts of the Central Provinces south of the Satpura range have a very regular rainfall, and never suffer from droughts of any magnitude, although the annual average fall does not exceed 50 inches, whereas the North-west Provinces, with a general average of 36 inches, have repeatedly been subject to disastrous droughts; not less than seven times in the course of the present century.

It is not my purpose, in this work, to discuss the physical circumstances that determine these and other peculiarities in the rainfall of the country. To the physicist and geographer they present problems of the greatest interest, and afford a most instructive field of study. But they have lately been discussed at length in a somewhat bulky volume,[1] issued by the Meteorological Department of the Government of India, and the reader, whose interest in the meteorology of the country attaches rather to its physical than its statistical aspects, may be referred thereto for such information as we possess on the subject. The bearing of these rainfall variations on the agriculture and general welfare of the population are more germane to the objects of the present treatise, and only in this connection need they be here discussed. I shall first describe the general distribution of the rainfall over India, then its relations to the seasons, and lastly, the more important vicissitudes to which it is subject.

Geographical Distribution.—The chart of the average rainfall of India (Fig. 6) shows that the west coasts of both peninsulas are fringed with a belt of very heavy rainfall, where, as may be seen in Fig. 4, the westerly summer monsoon blows full upon the land. In both, at a distance

[1] *Indian Meteorological Memoirs*, Vol. iii. "The Rainfall of India."

of 50 miles or less from the sea, a hill range, such as in many countries would be called mountains, runs parallel with the coast, and over the whole intervening space the rainfall is very high, varying from 100 to 250 inches on the coast of India, and probably of larger amount on the coast of Arakan. In the western peninsula this zone of high rainfall contracts rapidly to the north of Bombay,

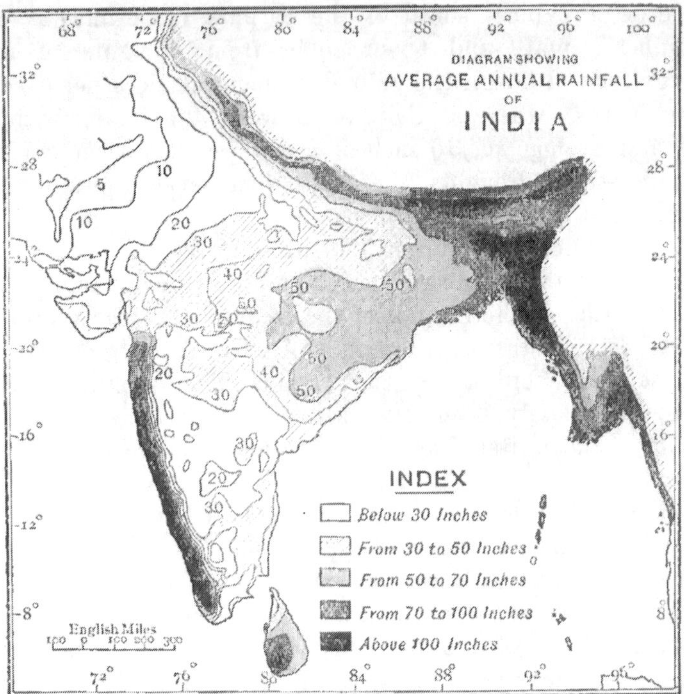

FIG. 6.

and dies out at the end of the Ghats, south of the Tapti. But in the eastern peninsula it is continued northward beyond the shores of the Bay of Bengal, and beyond the tropic, through eastern Bengal and Cachar, and along the face of the Khasi hills to the gates of Assam, where it links on to the similar zone, which runs east and west along the face of the Himalaya. In a westerly direction, this latter zone, as defined by a rainfall of over 100 inches, has not been actually traced beyond Sikkim, owing to the absence of rainfall registers from this part of the range; but it probably extends uninterruptedly to the farther extremity of the Himalayan frontier of Nepal, and is taken up again in the far north-west by the southern slopes of the Dhaoladhar range, overlooking the valley of Kangra. Eastward, there is little doubt that it extends to the upper end of the valley of Assam, where the rainfall in the valley itself, at Dibrugarh, amounts to more than 100 inches.

The zone of heavy rainfall on the face of the Himalaya does not extend far into the mountains. It is broadest in Sikkim, where it includes a strip of the plains of northern Bengal; and, in a northerly direction, up to the valley of the Great Rungit, and perhaps somewhat beyond, the annual fall exceeds 100 inches. At Katmandu, in the Nepal valley, the rainfall is only 57 inches. In the north-west Himalaya, beyond the Nepal frontier, it is barely 100 inches on the outermost slopes of the range (except, as already mentioned, on the Dhaoladhar), and it decreases rapidly as we penetrate farther into the hills, until, beyond the first snowy range, we enter on the dry region of Tibet, represented by Leh, with an annual fall of only 2½ inches.

From the Gulf of Cambay and the Sabarmati river, a band of moderately high rainfall (between 30 and 70 inches) stretches eastwards across India, rapidly broadening out and increasing in average amount in the direction of Bengal. It occupies all North-eastern India, extends to the valley of the Ganges on the north, and on the south covers the drainage basin of the Godavery and its tributaries, down to the delta of that river and the Kistna. From Allahabad an offshoot runs up to the north-west, including the Gangetic plain north of the Ganges, and continues through the northern Punjab and along the face of the Himalaya as far as the Indus valley. In the south of the peninsula are two outlying tracts with a rainfall above 30 inches; one on the Eastern Ghats, east of Kurnool; the other, including all the central districts of the Carnatic and linked on to the zone of high rainfall of the Western Ghats, along the face of the Eastern Ghats; and again, farther south, across the hill groups between Trichinopoly and Madura.

The remainder of India has an annual rainfall below 30 inches. It consists of (1) the dry north-western region, which includes Rajputana, Sind, Cutch, part of Guzerat, the Punjab (with the exception of the strip of the plains bordering the Himalaya), and nearly the whole of the Gangetic *doab* (the plain between the Ganges and Jumna); and (2) the peninsular dry region, which comprises Khandesh and the valley of Berar, the Western Deccan, the southern districts of Hyderabad, the ceded districts of the Madras Presidency, the northern Carnatic, Mysore, Coimbatore, and the plain in the south-east of the peninsular, between Tanjore and Cape Comorin. In the former the rainfall decreases steadily in a westerly direction, till it reaches its minimum in the Cutchee desert, between Jacobabad and the hills of Baluchistan. In the latter are several isolated tracts, in which the rainfall is somewhat below 20 inches, the principal of which are a large area around Bellary, and a smaller one in Coimbatore, between the Nilgiris and the Shevaroy hills. Also, on the narrow strip of coast plain bordering the Gulf of Manaar, in the extreme south-east of the peninsula, facing Ceylon, the rainfall is equally low.

In Burmah, from the crest of the Arakan range, and from the mouths of the Irawadi, the rainfall rapidly decreases towards the interior of Pegu, and falls still lower as one ascends the river towards Mandalay. But it is comparatively high on the Pegu Yoma, east of the Irawadi delta, on the hills around the source of the Sitang, and on the range west of the Salwin. From Upper Burmah our present information is insufficient to show the features of its distribution. But rain-gauges were established at a number of stations of the newly-acquired province, even before it was fairly pacified, and in the course of a few years it will be possible to extend the rainfall chart of the Burman peninsula at least to Manipur and Bhamo.

The average rainfall of the whole of India up to the foot

of the mountain barrier, including Assam and Cachar, but not the Burman peninsula, has been computed at 42 inches in the year; that is to say, such would be the depth of the sheet of water, if the whole quantity that falls on this area were equally distributed over its whole surface, instead of being concentrated in certain provinces to the great deprivation of others. Of the general character of its distribution some idea may be gathered from the table given below. In this table the country is partitioned out in a number of rainfall provinces, partly conforming to administrative divisions, but with such modifications as to avoid bringing together parts of the country which differ very greatly in their average amounts.

AVERAGE ANNUAL RAINFALL OF THE PROVINCES OF INDIA AND BURMAH

Rainfall Provinces.	Area in Square Miles.	Number of Stations.	Average Rainfall. Inches.	Local Variation. Inches.
Punjab Plains	120,000	29	22	6 to 36
N.W. Provinces and Oudh	83,500	45	36	25 ,, 50
Rajputana (Eastern only)	67,000	19	28	14 ,, 63[2]
Central India States[1]	91,000	21	42	32 ,, 55
Behar	30,000	14	43	39 ,, 18
Western Bengal, Chutia Nagpur, etc.	38,000	10	49	43 ,, 61
Lower Bengal	54,000	29	66	54 ,, 112
Assam and Cachar	61,000	17	94	69 ,, 475[2]
Orissa and Northern Circars	27,000	16	47	31 ,, 70
Central Provinces, South	61,000	19	51	43 ,, 79[2]
Berar and Khandesh	43,000	11	35	21 ,, 69[2]
Guzerat	54,500	14	33	18 ,, 47
Sind and Cutch	68,000	10	9	4 ,, 19
North Deccan	48,000	14	29	18 ,, 49
Konkan and Ghats	16,000	13	141	74 ,, 261[2]
Malabar and Ghats	18,000	8	114	74 ,, 132
Hyderabad	74,000	19	32	23 ,, 43
Mysore and Bellary	58,000	17	29	18 ,, 36
Carnatic	72,000	40	36	20 ,, 62[2]
Arakan	11,000	6	156	105 ,, 214
Pegu	32,500	7	73	46 ,, 123
Tenasserim	10,500	4	171	142 ,, 196

[1] Including Jhansi, Saugor, and Damoh, and the Nerbudda valley.

[2] In these cases one or more hill stations have a much higher rainfall than any at the lower levels.

Seasonal Distribution.—By far the greater part of the rainfall is brought by the summer monsoon; and over a great part of Western India, including nearly the whole of the Bombay Presidency, it is only during this season, viz. from June to October, that there is any rain of importance. This is the chief season of rainfall in Burmah, Assam, Bengal, the North-west Provinces, the greater part of the Punjab, Central India, Rajputana, the Central Provinces, Hyderabad, the Deccan, Berar, Guzerat, and the whole of the west coast of the peninsula. But the south-east of the peninsula, including the Carnatic and the Eastern Ghats, and the eastern half of Ceylon, receive only an occasional shower so long as the main body of the monsoon is directed towards Northern and Central India, and their season of heaviest rainfall comes later, viz. from October to December, when north-east winds are beginning to blow in the north-west of the bay, and both the incipient north-east monsoon and the residue of the southerly current are drawn towards the Carnatic and the southern half of the bay. In these months the seat of the chief rainfall moves southwards, till at length, by the end of December, the north-east monsoon has established itself in the bay, and, in the south, rain is restricted to a zone a few degrees north of the equator, in which it falls, more or less, all through the year.

In the Punjab and Rajputana the summer rainfall is much interrupted. It seldom penetrates to the Indus or to the Himalaya beyond the first snowy range. In the interior of this range, and also on the mountains and high plains beyond the north-western frontier, in Afghanistan and Baluchistan, the latter part of the winter and the earlier spring months are the season in which the chief additions are made to the snows; and on the outer hills and plains of the Punjab, rain, sometimes heavy, occurs at intervals throughout this season. It extends also, though less frequently and heavily, to the Gangetic plain, and sometimes, but still more rarely, to the whole of Northern India, including Bengal. Occasionally it occurs even in the Northern Provinces of the peninsula, but this is exceptional.

In the spring, when sea winds blow on the south and east of the peninsula, and very steadily and sometimes strongly in Lower Bengal and Cachar, thunder-storms are not infrequent, and sometimes the showers are heavy and prolonged. Chiefly is this the case in the damp province of Assam and Eastern Bengal; and it is in consequence of these frequent falls of rain that Assam and Cachar are so favourably distinguished as tea-growing provinces. The showers become more frequent as the season advances, and the May rainfall of Assam is not very far short of that of June, the first month of the monsoon rains.

In Southern India, too, where the spring rains are known as the "mango showers," they amount to a total of three or four inches, falling chiefly in the months of April and May. But from the Bombay Presidency, north of Dharwar, the spring rainfall is almost entirely withheld. This is the season of north-westerly winds, as has been explained on a previous page, and these, being dry land winds, bring no rain.

Sometimes the rain of the spring thunder-storms takes the form of hail, and, though not very common on the plains, it is very destructive, the hailstones being generally large, from the size of marbles up to ice concretions two or three inches across. On the Himalaya hailstones occur more often than on the plains, but as far as I can speak from personal experience the hailstones are generally of moderate size. There are, however, recorded instances of very large hailstones on the hills.

Thus, there is no time of year in which rain does not fall in some part or other of India. No sooner have the Carnatic rains come to an end in December, than the winter rains are setting in in the most northern province; and while these are prolonged into the spring months, the spring rainfall of the south and east of the Indian peninsula, of Bengal, Assam, and the Southern Provinces of Burmah, is already preparing the fields for the spring sowings, and bringing out flushes of young leaves on the tea bushes of Assam, Chittagong, and Sikkim. In the interior of Southern India these showers become less frequent in June; but on the south-west coast the monsoon rains set in at the end of May, and reach Bombay in the first or second week of June; while, in Assam, the spring rainfall graduates into the steadier downpour of the summer monsoon. In September this monsoon ceases to blow up to North-western India, and after concentrating a diminished rainfall in Bengal and the north-east of the peninsula, gradually transfers it southwards to Madras, and finally to Ceylon and the equatorial sea, when the southerly current has completely yielded place to the winter monsoon of the bay.

Statistics of Monthly Rainfall, Quantity and Frequency.—The following two tables, one giving the average monthly rainfall of a number of important stations in all parts of India and Burmah (to the nearest tenth of an inch), the other the average number of rainy days in each month, will serve to illustrate the above description, and show in somewhat greater detail the relative proportions in which the rainfall is distributed over India, and also through the year. The second table is based on the registers of the last 11 years; the first, in most cases, on those of much longer periods. A more comprehensive table of the rainfall is given in Appendix III.

AVERAGE MONTHLY RAINFALL OF EIGHTY STATIONS IN INDIA, CEYLON, AND BURMAH

Stations.	Elevation in Feet.	Rainfall in Inches.												
		Jan.	Feb.	March.	April.	May.	June.	July.	Aug.	Sept.	Oct.	Nov.	Dec.	Year.
Kurrachee	49	0·6	0·3	0·2	0·2	0·1	0·2	3·1	1·8	0·9	0·1	0·1	0·2	7·8
Hyderabad	66	0·2	0·1	0·1	0·2	0·1	0·4	2·8	3·2	0·8	..	0·1	..	8·0
Quetta	5,501	1·6	1·9	2·4	1·3	0·5	0·1	0·8	0·6	0·2	0·1	..	0·4	9·9
Jacobabad	186	0·2	0·2	0·3	0·2	0·1	0·1	1·4	1·4	0·3	..	0·1	0·1	4·4
Mooltan	420	0·4	0·3	0·5	0·3	0·5	0·4	2·2	1·3	0·8	0·1	0·1	0·3	7·2
D. I. Khan	573	0·4	0·7	0·9	0·8	0·4	0·6	1·8	1·6	0·6	0·1	0·1	0·3	8·3
Leh	11,503	0·2	0·2	0·2	0·1	0·1	0·2	0·5	0·4	0·2	0·5	..	0·1	2·7
Peshawar	1,110	1·6	1·2	1·8	2·0	0·7	0·3	1·7	2·0	0·8	0·2	0·6	0·6	13·5
Rawalpindi	1,652	2·4	2·0	1·9	2·3	1·6	1·7	7·4	7·3	3·2	0·6	0·9	1·1	32·4
Murree	6,344	2·8	3·4	3·7	4·3	3·8	2·4	11·0	14·0	6·1	2·2	1·7	1·2	56·6
Sialkot	829	1·4	1·8	1·9	1·6	1·2	3·2	11·6	9·1	3·2	0·6	0·4	0·8	36·8
Lahore	732	0·7	1·1	1·1	0·6	0·9	1·8	7·4	4·6	2·4	0·6	0·2	0·5	21·9
Umballa	902	1·4	1·5	1·1	0·6	1·0	4·0	11·6	8·6	4·1	0·6	0·2	0·7	35·4
Simla	6,953	2·8	2·7	3·0	2·8	4·7	7·9	19·3	18·1	6·0	1·4	0·3	1·1	70·1
Delhi	718	1·0	0·5	0·7	0·4	0·7	3·4	8·5	6·9	4·5	0·5	0·1	0·4	27·6
Roorkee	887	2·0	1·4	1·0	0·4	1·2	5·1	12·5	12·3	5·1	0·6	0·2	0·4	42·2

Average Monthly Rainfall of Eighty Stations in India, Ceylon, and Burmah—Continued

Stations.	Elevation in Feet.	Rainfall in Inches.												
		Jan.	Feb.	March.	April.	May.	June.	July.	Aug.	Sept.	Oct.	Nov.	Dec.	Year.
Meerut	737	1·0	0·7	0·7	0·4	0·8	3·6	9·2	7·2	4·0	0·5	0·1	0·3	28·5
Agra	555	0·5	0·3	0·2	0·2	0·7	2·9	9·8	6·7	4·3	0·4	..	0·2	26·2
Deesa	465	0·1	0·2	0·1	0·1	0·2	2·2	9·8	8·5	3·3	0·8	0·1	..	25·5
Abu	3,945	0·2	0·4	0·1	..	1·0	5·1	22·2	22·5	9·1	2·1	0·2	0·2	68·1
Jodhpur	1,274	0·3	0·2	..	0·1	0·6	1·3	3·6	5·2	2·3	0·2	0·1	0·1	14·0
Neemuch	1,639	0·1	0·2	0·1	0·1	0·5	3·9	11·2	10·4	5·5	1·0	..	0·2	33·2
Indore	1,822	0·4	0·3	..	0·1	0·6	6·8	10·4	7·8	8·1	1·2	0·2	0·2	36·1
Ajmere	1,611	0·2	0·3	0·4	0·1	0·7	2·5	6·9	7·3	3·4	0·3	0·1	0·3	22·5
Jhansi	855	0·5	0·2	0·4	0·1	0·3	4·0	13·6	10·5	5·2	0·8	..	0·2	35·8
Lucknow	369	0·8	0·3	0·3	0·1	0·9	5·0	10·8	10·4	7·1	1·4	..	0·5	37·6
Allahabad	307	0·8	0·4	0·4	0·2	0·3	4·6	11·9	9·6	6·7	2·3	0·2	0·2	37·6
Benares	267	0·7	0·5	0·4	0·2	0·5	5·0	12·8	10·7	6·5	2·1	0·1	0·1	39·6
Gorakhpur	256	0·7	0·5	0·4	0·3	1·5	7·7	13·3	11·8	8·8	3·0	0·2	0·1	48·3
Patna	179	0·7	0·5	0·3	0·3	1·6	7·1	11·0	10·1	7·9	2·9	0·2	0·2	42·8
Darjiling	6,912	0·7	1·2	2·4	3·7	7·1	24·1	30·5	26·0	17·8	6·4	0·2	0·2	120·3
Jalpaiguri	270	0·4	0·4	1·7	3·7	11·2	28·6	26·3	25·9	24·3	5·5	0·1	0·1	128·2
Hazaribagh	2,010	0·4	0·8	0·7	0·4	1·6	8·3	12·6	12·7	8·0	3·4	0·3	0·2	49·4
Bhagulpore	159	0·5	0·7	0·4	0·8	2·5	8·3	11·2	10·7	7·8	4·1	0·2	0·1	47·3
Berhampore	66	0·4	1·0	1·0	1·9	4·8	9·7	10·3	10·8	9·8	5·3	0·3	0·1	55·4
Calcutta	18	0·4	1·0	1·3	2·3	5·6	11·8	13·0	13·9	10·0	5·4	0·6	0·3	65·5
Dacca	15	0·3	1·1	2·5	5·8	9·2	13·3	12·8	12·4	10·2	5·2	0·7	0·2	73·7
Chittagong	87	0·4	1·2	1·9	4·6	9·2	23·8	22·2	20·5	14·1	5·7	1·6	0·6	105·8
Silchar	87	0·6	2·6	7·9	13·0	15·7	19·1	20·6	18·2	14·2	6·4	1·0	0·7	120·0
Cherrapunji	4,455	0·6	2·6	9·0	29·6	50·0	110·0	120·5	78·9	57·1	13·6	1·8	0·3	474·0
Gauhati	370	0·6	0·9	2·5	5·8	10·1	12·9	12·7	11.2	8·1	3·1	0·6	0·3	68·8
Sibsagar	333	1·1	2·2	4·4	9·9	11·1	14·1	15·6	16·0	11·7	5·2	1·3	0·6	93·1
Cuttack	80	0·4	0·6	1·1	1·5	3·2	10·7	12·6	11·2	9·8	5·8	1·0	0·5	58·4
Samblapur	451	0·6	0·6	0·7	0·5	1·6	13·0	17·7	15·2	8·7	2·4	0·3	0·4	61·7
Saugor	1,769	0·6	0·5	0·2	0·2	0·6	6·3	16·8	11·2	7·3	1·3	0·4	0·7	46·1
Jubbulpore	1,351	0·6	0·5	0·5	0·2	0·5	8·5	18·6	13·8	8·2	1·5	0·4	0·3	53·6
Pachmarhi	3,504	0·5	0·3	0·4	0·3	0·6	10·8	20·3	18·2	15·1	1·9	0·4	0·7	78·0
Nagpur	1,025	0·6	0·4	0·6	0·5	0·8	8·8	13·3	8·9	7·8	2·3	0·4	0·5	44·9
Amraoti	1,213	0·5	0·2	0·3	0·2	0·6	6·9	8·8	7·0	5·3	1·6	0·2	0·6	32·2
Dhulia	1,000	0·3	0·1	..	..	0·4	4·8	4·8	4·0	4·6	2·0	0·5	0·4	21·9
Poona	1,819	0·2	..	0·2	0·6	1·6	5·6	6·6	4·1	4·3	4·1	0·8	0·2	28·3
Bombay	37	0·1	..	..	..	0·5	20·8	24·7	15·1	10·8	1·8	0·5	0·1	74·4
Matheran	2,200	0·1	..	..	..	0·8	35·7	84·4	50·7	31·0	5·3	0·6	0·1	208·7
Mahableshwar	4,540	0·4	0·1	0·4	0·9	1·4	47·3	102·1	68·6	32·9	5·8	1·1	0·4	261·4
Karwar	44	0·2	..	0·1	0·3	2·9	34·5	37·7	21·7	12·0	5·2	1·7	0·2	116·5
Belgaum	2,550	0·1	..	0·5	2·0	2·8	9·3	15·2	9·0	3·7	4·7	1·2	0·3	48·8
Jalna	..	0·1	..	0·1	0·1	0·8	7·2	6·8	5·0	7·5	3·4	0·8	1·4	33·5
Sholapur	1,819	..	0·1	0·3	0·7	1·2	4·6	4·3	6·0	7·5	3·7	0·7	0·4	29·5
Secunderabad	1,787	0·3	0·2	0·7	0·7	1·4	3·7	6·0	5·7	5·2	3·3	0·8	0·3	28·3
Bellary	1,455	0·1	..	0·6	0·8	1·8	1·8	1·3	2·3	3·7	3·9	1·0	0·3	17·6
Bangalore	2,981	0·2	0·1	0·6	1·3	5·0	3·2	4·0	5·9	6·3	6·4	1·9	0·7	35·6
Wellington	6,200	0·8	0·3	2·0	2·9	4·1	3·6	3·2	4·0	4·7	9·8	8·5	4·1	48·0
Bimlipatam	30	0·3	0·5	0·2	0·2	2·0	3·2	3·6	4·4	6·3	8·1	2·5	1·3	32·6
Masulipatam	10	0·3	0·1	0·3	0·1	1·7	4·4	5·6	6·0	6·5	8·8	4·0	0·7	38·5
Rajamundry	68	0·2	0·3	0·3	0·9	3·3	4·5	7·2	6·6	7·1	6·5	1·7	0·2	38·8
Cuddapah	477	0·1	..	0·2	0·3	1·6	2·6	3·5	5·1	5·8	5·3	3·1	0·7	28·3
Madras	22	1·0	0·3	0·4	0·6	2·2	2·1	3·8	4·4	4·7	10·8	13·7	5·1	49·1
Trichinopoly	275	1·0	0·5	0·7	1·8	3·8	1·3	2·2	4·4	5·3	7·8	5·2	3·1	37·1
Cuddalore	20	1·0	0·3	0·4	0·9	1·5	1·4	2·3	5·1	4·7	8·1	14·1	5·7	45·5
Madura	448	0·7	0·4	0·6	2·0	2·8	1·6	1·7	4·7	4·5	8·7	5·1	2·2	35·0
Mangalore	52	0·2	0·1	0·1	2·0	8·1	37·8	37·9	23·1	11·3	8·0	1·9	0·5	131·0
Cochin	11	0·9	0·7	2·1	4·4	12·7	30·7	22·7	12·4	9·4	12·1	5·1	1·9	115·1
Trincomalee	175	6·2	2·4	1·3	1·6	2·2	1·9	2·2	4·2	4·6	8·9	13·1	13·2	61·8
Colombo	40	3·0	1·7	5·5	8·8	13·2	8·2	5·5	4·5	4·9	12·9	12·7	6·4	87·3
Galle	48	4·4	3·3	4·7	8·7	11·6	8·2	5·7	5·3	7·6	13·0	11·5	6·7	90·7
Rangoon	41	0·2	0·1	0·1	1·8	10·9	18·4	21·3	18·6	16·0	8·1	3·4	0·1	99·0
Moulmein	94	..	0·1	0·1	3·0	19·7	38·4	43·9	43·0	30·3	8·4	1·5	0·1	188·5
Thyet Myo	134	..	..	0·1	0·7	5·3	7·9	8·0	8·5	7·8	4·9	2·3	..	45·5
Tounghoo	169	..	0·2	..	1·5	6·6	13·4	17·5	18·1	11·6	7·4	1·4	0·2	78·1
Akyab	15	0·1	0·2	0·5	1·6	12·2	51·6	51·0	38·6	23·0	12·4	3·9	0·6	195·7

MONTHLY AVERAGE OF RAINY DAYS AT EIGHTY STATIONS IN INDIA, CEYLON, AND BURMAH

Stations.	Rainy Days.												
	Jan.	Feb.	March.	April.	May.	June.	July.	Aug.	Sept.	Oct.	Nov.	Dec.	Year.
Kurrachee .	1	2	1	1	...	1	6	4	2	...	1	...	19
Hyderabad .	1	1	...	...	1	1	5	5	2	...	...	...	16
Quetta . .	7	7	8	8	3	1	2	2	1	...	...	1	40
Jacobabad .	1	3	1	1	1	...	3	4	...	...	...	...	14
Mooltan .	1	1	1	1	1	2	3	2	1	...	...	1	14
D. I. Khan .	2	3	3	3	2	2	4	3	2	1	1	1	27
Leh . .	4	5	2	1	2	1	3	3	1	...	...	2	24
Peshawar .	3	4	4	5	3	1	3	3	2	1	2	2	33
Rawalpindi .	5	5	5	6	5	4	9	11	6	3	2	3	64
Murree . .	4	5	8	10	9	7	14	18	9	4	3	1	92
Sialkot . .	2	4	3	3	3	4	9	9	4	1	1	1	44
Lahore . .	2	3	2	3	3	3	7	6	4	1	1	2	37
Umballa .	2	2	1	1	2	3	9	7	4	1	...	1	33
Simla . .	3	5	6	6	9	10	21	22	12	2	1	2	99
Delhi . .	2	2	2	1	3	4	10	9	5	1	...	1	40
Roorkee .	3	4	3	2	3	6	15	14	7	1	1	2	61
Meerut . .	2	2	2	1	3	5	11	11	6	1	...	1	45
Agra . .	1	1	2	1	3	4	13	10	7	1	...	1	44
Deesa . .	...	1	...	1	1	4	13	12	8	1	...	...	42
Abu . .	1	1	1	1	3	8	22	22	15	3	1	1	79
Jodhpur .	1	1	...	1	3	3	7	8	4	...	...	...	28
Neemuch .	1	1	1	...	3	8	15	14	11	2	...	1	57
Indore . .	1	1	...	...	3	11	20	19	15	4	1	1	76
Ajmere . .	1	2	1	1	4	4	13	11	17	1	1	1	47
Jhansi . .	1	1	1	1	2	7	15	14	9	1	...	1	53
Lucknow .	2	2	2	1	2	6	16	14	9	2	...	1	57
Allahabad .	2	2	1	1	2	7	16	17	11	3	...	1	63
Benares .	2	2	1	1	2	7	18	18	12	4	1	1	69
Gorakhpur .	1	2	1	1	3	7	13	14	9	3	...	1	55
Patna . .	2	2	1	1	4	10	18	18	12	4	...	1	73
Darjiling .	3	4	5	9	19	22	27	26	21	6	...	1	143
Jalpaiguri .	1	1	3	6	15	21	23	21	17	6	...	1	115
Hazaribagh .	3	3	3	1	8	15	24	23	17	7	2	1	107
Bhagulpore .	2	2	1	2	5	12	19	18	12	4	...	1	78
Berhampore .	1	2	2	3	9	13	19	19	15	6	...	1	90
Calcutta .	1	3	3	5	12	18	25	26	19	8	1	2	123
Dacca . .	1	3	5	9	15	19	22	22	16	6	2	1	121
Chittagong .	1	2	5	6	13	21	25	23	17	9	4	1	127
Silchar . .	2	6	13	17	20	22	24	24	18	8	2	1	157
Cherrapunji .	2	4	10	20	25	26	29	28	24	12	3	1	184
Gauhati .	2	4	9	14	19	20	20	18	13	6	1	1	127
Sibsagar .	7	9	14	19	20	21	23	21	18	11	3	3	169
Cuttack .	1	2	2	3	9	16	23	23	19	9	3	1	111
Sambalpur .	1	2	2	2	5	14	24	24	17	4	1	1	97
Saugor . .	2	1	...	1	2	11	19	17	13	3	...	1	70
Jubbulpore .	2	2	1	2	3	12	19	18	11	3	1	1	75

MONTHLY AVERAGE OF RAINY DAYS AT EIGHTY STATIONS IN INDIA, CEYLON, AND BURMAH—Continued

Stations.	Rainy Days.												
	Jan.	Feb.	March.	April.	May.	June.	July.	Aug.	Sept.	Oct.	Nov.	Dec.	Year.
Pachmarhi	2	2	1	2	4	14	25	23	17	5	1	2	98
Nagpur	2	1	2	2	4	15	21	18	15	4	1	1	86
Amraoti	1	1	1	1	3	14	18	15	13	5	1	2	74
Dhulia	...	...	...	...	1	8	14	11	13	5	1	1	54
Poona	...	...	1	2	2	13	21	20	15	8	2	1	85
Bombay	1	1	1	...	2	20	29	26	21	7	1	...	109
Matheran	...	...	...	...	2	25	31	30	27	10	2	...	128
Mahableshwar	1	...	1	2	4	27	31	30	28	12	4	1	141
Karwar	...	...	...	1	5	26	30	26	20	13	4	...	125
Belgaum	...	1	3	6	7	21	28	25	22	13	5	3	134
Jalna	...	...	1	1	2	13	16	13	14	5	3	2	70
Sholapur	1	1	1	3	4	14	16	17	14	8	3	2	84
Secunderabad	1	...	3	4	5	12	17	18	15	8	4	2	88
Bellary	...	...	1	3	5	6	8	8	9	9	5	1	55
Bangalore	1	...	2	3	10	13	14	17	13	13	7	3	96
Wellington	3	2	6	7	14	15	14	16	13	18	16	10	134
Bimlipatam	1	1	1	1	4	5	6	9	11	7	4	1	51
Masulipatam	1	1	1	1	4	8	15	16	16	12	6	2	83
Rajamundry	...	1	1	1	5	7	11	11	9	7	3	...	56
Cuddapah	...	...	1	1	3	6	4	10	8	7	5	1	46
Madras	3	...	1	1	2	10	14	15	12	14	14	9	85
Trichinopoly	1	...	1	1	5	3	3	7	8	11	9	6	55
Cuddalore	3	...	1	1	2	6	0	11	9	11	14	10	77
Madura	1	...	1	2	5	3	3	7	8	14	9	5	58
Mangalore	...	...	...	3	8	27	30	28	22	15	7	2	142
Cochin	1	2	5	8	17	27	27	22	19	19	12	5	164
Trincomalee	10	3	4	4	6	4	2	9	7	16	19	20	104
Colombo	6	4	10	13	20	17	12	13	14	21	18	13	161
Galle	12	8	11	14	21	22	20	21	20	22	18	16	205
Rangoon	...	...	...	3	19	26	29	25	27	12	10	...	151
Moulmein	...	...	...	7	18	28	29	23	27	15	5	1	158
Thyet Myo	...	...	...	2	11	22	27	23	20	11	3	1	120
Tounghoo	1	...	...	4	12	24	26	27	22	14	4	1	135
Akyab	1	...	1	2	14	28	30	29	22	12	4	2	145

Average Heaviness of the Rainfall.—If the quantities of rainfall at the several stations in the first of the foregoing tables be divided by the number of rainy days in the second, it will be found that the average fall of each rainy day is between six and seven tenths of an inch on the plains of Bengal and the North-west Provinces, and in parts of Central India and the Central Provinces. In the wetter parts of Assam, on the coasts of Arakan and Tenasserim, at Pachmarhi, and on the west coast and Western Ghats of India, it is higher; at Cherrapunji as much as 2·6 inches, and at Akyab and Moulmein, Mahableshwar and Matheran, from 1 to 2 inches on each rainy day. But in the Deccan and Mysore, and also in the Indus valley, it is only between three and four tenths, and at Leh but one-tenth of an inch. This last is about the same as the general average of the British Isles. It results, therefore, that, day for day, rain is from three to seven times as heavy on the plains of India as in Western Europe, and that the common proverbial expression, "it never rains but it pours," is a literal description of the ordinary rainfall of India. This is as true of such places as Jacobabad and Mooltan, where there are no more than about fourteen rainy days in the year, as of Point de Galle, where it rains on 205 days on an average.

In consequence of this character of Indian, in common with tropical, rainfall generally, it is less penetrating in proportion to its quantity than in countries where much of it falls in a state of fine division, allowing time for its absorption by the ground. Instead of feeding perennial springs, and nourishing an absorbent cushion of green herbage, the greater part flows off the surface and fills the dry beds of drains and watercourses with temporary torrents. In uncultivated tracts, where jungle fires have destroyed the withered grass and bushy undergrowth, and have laid bare the soil and hardened its surface, this action is greatly enhanced; and while all perennial water supplies which depend on the absorbed rain are either greatly reduced or altogether suppressed, a rainfall which, if husbanded by nature and art, would suffice for the agriculture and domestic requirements of the population, is thrown into the nullahs and rivers, and not only is wasted and lost for any useful purpose, but by producing floods, becomes an agent of destruction. Under any circumstances, the character of the rainfall is hardly compatible with its economical storage and expenditure in any high degree; and much more, therefore, than in temperate regions is it incumbent on us to safeguard such provident arrangements as nature has furnished for the purpose.

Excessive Rainfalls.—Seeing, then, that on the plains of Northern and Central India the average fall of each rainy day is not less than six or seven tenths of an inch, such enormous falls as those referred to on page 64 become less surprising as an exceptional occurrence. The following are a few instances of remarkable quantities recorded within 24 hours, in the years 1875-1885, but by no means all of those exceeding 12 inches, which would be much more numerous. A more comprehensive list will be given in the second part of this work.

Cherrapunji, Khasi Hills	40·8 inches.	14th June 1876.
Purneah, Northern Bengal	35·0 „	13th September 1879.
Nagina, Rohilkhand	32·4 „	18th September 1880.
Danipur, „	30·4 „	„ „ „
Najibabad, „	28·5 „	„ „ „
Hardwar, Dehra Doon	19·5 „	„ „ „
Rewah, Central India	30·4 „	16th June 1882.
Sutna „ „	22·1 „	„ „ „
Chandbally, Orissa	20·3 „	16th September 1879.
Delhi, Punjab	19·5 „	9th September 1875.
Gurgaon „	10·5 „	„ „ „
Allahabad, N.W. Provinces	15.5 „	29th July 1875.
Pilibhit, Rohilkhand	13·6 „	2d August 1879.
Mozufferpore, Bengal	12·5 „	17th June 1883.

Although, in their extreme intensity, such falls are generally local, they are only part of a less copious but still heavy fall, often extending over thousands of square miles, and lasting over two or three days. Such was the case on the 18th September 1880 when, in addition to the torrential rainfall in the Bijnore district (in which are three of the four stations enumerated above), a quantity, exceeding 10 inches in two days, was registered over a large part of the adjoining districts of Saharunpore, Mozuffernagar, Meerut, Moradabad, Pilibhit, and Kumaon; and also on the 13th September 1879, when 5, 6, 9·5, and 12 inches, respectively, were registered in adjacent districts around Purneah. There would seem also to be a tendency to a repetition of heavy falls in about the same part of the country in the same season. Thus, only nine days before the above enormous fall of 35 inches took place at Purneah, one of 11·5 inches had been registered at the same station, and only five days previously one of 5 inches. In 1875 the excessive fall of September in the eastern districts of the Punjab had been preceded in August by very heavy falls at Sialkot, Jullunder, and Lahore, of 11 inches in two days, 18·8 inches in three days, and 11·7 inches in three days respectively. Some other similar cases have been recorded in the Annual Reports on the Meteorology of India.

These excessive falls are always the result of cyclonic storms, not indeed such as are of destructive violence as regards the wind, but the long-lived cyclonic storms which occur chiefly in the rains, storms in which the barometer is not greatly depressed, and only recognisable in their true character when the barometric readings and the winds are laid down on charts in the manner already described.

Another noteworthy point is that they have frequently occurred in years of partial drought, as if, in such seasons, the whole energy of rain formation were concentrated in the storms, to the deprivation of parts of the country not reached by them. Thus in 1875, although in the Punjab it was one of the wettest years on record, the rainfall was very deficient in Bengal and all over the southern half of the peninsula. 1876 was the year that produced the last great peninsular famine, and in both 1880 and 1883 there was a prolonged suspension of the rains over all North-Western and Central India and a part of the peninsula.

Annual Variability.—It has already been stated that the variability of the rainfall, that is to say, the degree in which that of any given year is liable to deviate from the local average, either in excess or defect, is much greater in some provinces than in others. This liability is proportionally greatest in the driest part of the country, viz. in Sind, and is least in Assam, which is one of the wettest. In the former, the mean annual deviation is as much as 37 per cent of the general average, in the latter only 5 per cent, although in certain years these amounts are greatly exceeded. Next to Sind, the North-west Provinces and Oudh are subject to the greatest vicissitudes, the mean annual deviation being 23 per cent; and within the last 22 years it has been as much as 47 per cent above and 39 per cent below the average. Berar and Khandesh, Hyderabad, Behar, and in a somewhat less degree, Guzerat, the Deccan, Mysore, and the Carnatic, also have a comparatively precarious rainfall. But all the older provinces of Burmah, viz. Arakan, Pegu, and Tenasserim, also Lower Bengal, Chutia Nagpur, and the Central Provinces, rank next to Assam in point of regularity. Although certain portions may sometimes suffer from drought, there has not been any extensive and disastrous famine in any of them during the present century.

In no single year has the rainfall been either deficient or excessive over the whole of India; and a recent critical examination of its vicissitudes during the last 22 years has brought to light some curious relations in the way in which certain provinces are prone to vary alike, while others vary in the opposite direction. Thus the Northern Provinces of the peninsula, Khandesh, Berar, the Central Provinces, and Orissa, usually all share either an abundant or a deficient rainfall; and, what was hardly to be expected, Assam and Eastern Bengal vary in the like manner with Orissa, while Lower Bengal, which lies between them, varies in the opposite direction. Again, when the rainfall is excessive in Bengal, it is usually light and deficient on the Bombay side and in Hyderabad. Many of these peculiarities admit of, at all events, a probable explanation; they seem to depend partly on the unseasonable prevalence of dry land winds in Western India, partly on the course taken by the cyclonic storms of the rainy season. But others are at present quite inexplicable, such, for instance, as the fact that, in five years out of six, the rainfall of Arakan is deficient or excessive when that of Sind varies in the opposite direction. On a general average, about two-thirds of India varies in one way, and one-third in its opposite. But in 1864, 1868, and 1873 the rainfall was deficient, and in 1878 excessive, over a very much larger proportional area. The average rainfall of India being taken, as stated on page 69, at 42 inches, the greatest deficiency during the last 22 years was about 6·6 inches, viz. in 1868, and the greatest 6·3 inches in 1878.

The Question of Cyclical Variation.—Much has been written of late years of the supposed variation of rainfall in a cycle of 11 years, coincidentally with the now well-known variation of sun-spots. Without venturing to pronounce any opinion on the question whether any such variation exists in the rainfall of the globe generally, it is certain that that of India, as a whole, has not displayed it during the last 22 years. But while this is true of the country *as a whole*, the rainfall of the Carnatic, which chiefly occurs at a later season than the heavy rains of other parts of India, certainly showed a somewhat striking fluctuation in the 11 years, 1864-1874, and again, though less regular, in the next 11 years, 1875-1885. It is still a matter of opinion whether this was fortuitous or otherwise, a question which can be decided only by further experience. According to such evidence as is before us, there would appear to be a mean difference of not less than 14 inches between the years of lightest and heaviest rainfall in this cycle, the general average of the province being 36 inches.

Some evidence has been adduced by Mr Hill of a similar fluctuation in the winter (but not the heavier summer) rainfall of Northern India.

Droughts.—The interest of this last observation lies in the fact that a heavy winter and spring rainfall takes the form of heavy snow on the Himalaya, and, as far as our limited experience goes, seasons in which the accumulation of snow has been excessive, have always been followed by a more or less prolonged suspension of the summer rainfall, as a rule in Northern India, but sometimes in some portion of the peninsula. It must, however, be added that there have also been serious droughts which, as far as our information goes, have not been preceded by any unusual snowfall on the mountains in the earlier months of the year, so that, even if this be a cause of drought, which many considerations render probable, it is not the sole determining agency. The immediate cause of drought is the unseasonable persistence of dry land winds to the exclusion of the rain-bearing wind of the summer monsoon.

Finally, it appears from the list of droughts furnished in the Report of Indian Famine Commissioners, that on no less than five occasions in the course of the last century, a serious drought in the peninsula has been followed by one in Northern India in the next year. The following is a list of the cases:—

{ 1782. Drought in Bombay and Madras.
{ 1783. Drought in Upper India.
{ 1802. Drought in South Hyderabad and Deccan.
{ 1803. Drought in North-western Provinces and Central India.
{ 1823. Drought in Madras.
{ 1824. Drought in Bombay, chiefly Guzerat and North Deccan.
{ 1832. Drought in northern districts of Madras, etc.
{ 1833. Drought in north part of Bombay, Rajputana, Punjab, and North-west Provinces.
{ 1876. Drought in all parts of Madras, Deccan, etc.
{ 1877. Drought in North-west Provinces, Central Provinces, and Punjab.

Storms

Thunder-Storms and Dust-Storms.—Any violent wind, accompanied with rain, hail, or snow, may be termed a storm, and in India the use of the term is somewhat further extended, since it is often applied to dust-storms with little or no rain. In these storms, however, which are common in the hot weather in the Punjab and North-western India, the coolness of the wind, and that of the atmosphere after the storm is over, is hardly to be accounted for otherwise than by supposing that rain is always formed by the cloud overhead, but being re-evaporated before it reaches the earth, the storm is often apparently rainless. A complete transition may be traced between dust-storms of this character and the nor'-westers of Bengal, which are accompanied by heavy rain. Even in the dry climate of the Eastern Punjab and North-west Provinces, a slight sprinkling of rain often follows the dust-storm, and in the damper atmosphere of Bengal the dust-storm is, as a general rule, only the first stage of a nor'-wester, and is followed by heavy rain and sometimes hail. Atmospheric disturbances of this class are individually local, though they generally occur simultaneously at different places over a large area of country. On their approach, the barometer is much disturbed, but the action is spasmodic, consisting of a rapid rise, followed by much irregular oscillation, and is very different from that which portends the approach of a cyclone, which is a steady continuous fall. For a day or so before a dust-storm or nor'-wester the barometer is, as a rule, somewhat lower than the average of the time of year; and indeed they are most common in the north and east of a region of barometric depression. But the depression of the barometer is not very great, and although the wind squalls are sometimes violent at the beginning of the storm, the whole character of the weather is so unlike that which betokens the approach of a cyclone, that a very little experience suffices to enable any one to distinguish between them.

Both dust-storms and nor'-westers almost always occur in the afternoon or evening, towards the end of a close warm day, either with little wind or with a damp wind from the sea. But, in the latter case, this lulls some time before the storm approaches, and for an hour or more the air is calm, until violent gusts of cool air blow out in advance of the storm, raising clouds of dust. Sometimes these gusts blow with great force, blowing down trees or breaking their branches; and on many occasions pressures of from 30 to 50 lbs. to the square foot have been registered by the Calcutta pressure-gauge.

These storms are of precisely the same nature as the thunder-storms of an European summer, except that they are more violent. The accompanying diagram (Fig. 7) of a summer squall, taken from a treatise by Dr. Möller,[1] exactly represents the leading features of an Indian nor'-

[1] *Meteorologische Zeitschrift*, 1884, p. 238.

wester. The strong gusts that blow out in front of the storm, *a*, *a*, and which constitute the dust-storm, are air dragged down by the falling rain, and their coolness is due to the partial evaporation of the water drops; therefore to the same cause as that of the air blown through a *tattie*.

The conditions for the production of a storm of this character are that the lower layer of the atmosphere is warm and damp; that immediately above it, perhaps at a height of 4000 or 5000 feet, dry and cooler; and the

FIG. 7.—SECTIONAL DIAGRAM OF A SQUALL. AFTER MÖLLER.

cloud is formed by the ascent of the former, penetrating the latter and becoming cooled by its ascent, the reverse of the process noticed on page 49. Sometimes, when the storm is seen from a distance, this process may be watched, the motion of the air that streams in under the cloud bank in front of the storm (indicated by the short arrows, *b*, in the figure) being rendered visible by the little cloud tufts that appear in it before it reaches the main body of the storm cloud, and their horizontal motion being gradually arrested, are then carried up and absorbed in the cloud bank above.

The storm almost always advances from the west or north-west; whence its name. This movement is due to the upper air current, which is the same as the west land wind of the Upper Provinces, blowing above the damp air that comes originally from the sea. Of the existence of this upper current ocular demonstration is afforded by the fact that the upper part of the storm cloud (and, in fine weather, of the *cumulus* cloud, of which the storm cloud is only a more developed form) is almost always inclined eastward, and also by the *false cirrus* (see page 59 and *c*, *c*, Fig. 7) which advances at a great height as a sheet of *cirro-stratus*, with a hard, well-defined margin in advance of the body of the storm cloud. Looking out westward across the open expanse of the Calcutta maidan, I have often watched the advance of these storms, and long before the publication of Dr. Möller's description and figure, had arrived at the conclusion that the process of their formation is exactly such as is represented in his diagram.

Storms of this class are short-lived, and are not an indication of any general or important disturbance of the atmosphere, however great their local violence. They are local eddies between opposite air currents, the one above, the other below, and do not affect the general movements of the atmosphere to any great distance from the storm region.

Cyclones and Cyclonic Storms.—Very different is the other more lasting and sometimes more destructive class of storms, now known as cyclones and cyclonic storms. I use these two terms alternatively, as recognising the fact that, while the disturbances respectively denoted by them are essentially identical in constitution and mode of origin, there is a marked difference in the degree of their violence and destructiveness. The term *cyclone* is so familiarly associated in men's minds with winds of destructive violence, and frequently with the still more disastrous accompaniment of the storm-wave, that it will be convenient in a work addressed to practical men, to whom such differences are all important, to restrict its application to storms of this kind, and to call the milder though not less extensive disturbances which are experienced in the summer monsoon, and those also which bring about the winter rains of Northern India, by its modified derivative, *cyclonic storms.*

The essential features common to both have already been described in the section on the winds (page 44). Over a limited region, more or less circular or oval in form, the atmospheric pressure, as indicated by the low barometer, is lower than anywhere for a great distance around. Into this barometric depression the air pours from all sides, following courses which, as we have seen, are neither directly towards the centre, nor, as was formerly believed, in concentric circles around it, but intermediate between the two, in converging spirals. In the northern hemisphere these turn around the centre in a left-handed direction as in Fig. 8,[1] against the hands of the clock, or in the direction in which one turns an ordinary screw in order to extract it. In the southern hemisphere they are right-handed, or move with the clock hands, as in Fig. 9. The strength of the winds, as has been already explained, varies with the steepness of the barometric gradient, or the rapidity with which

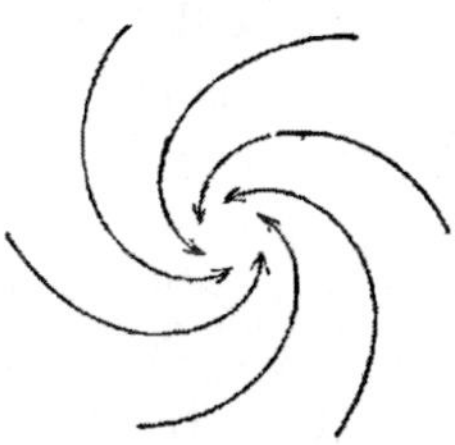

Fig. 8.—Tropical Cyclone. Northern Hemisphere.

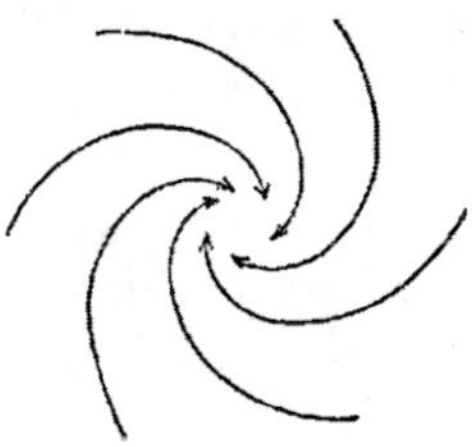

Fig. 9.—Tropical Cyclone. Southern Hemisphere.

the pressure falls from the outside towards the eye and centre of the storm. Hence, when such a storm is approaching, the winds become fiercer, the more rapidly the barometer is falling. In the centre of the storm the air is calm or the winds are light and variable, notwithstanding that the barometer is then at its lowest. But immediately around they blow with their greatest violence, and become by degrees more moderate as the distance from the centre increases. In the centre of a cyclone the barometer often marks an inch, and sometimes as much as two inches, below its readings at a distance of a few hundreds of miles. In a cyclonic storm the depression does not exceed a few

[1] The spirals in Fig. 8 represent the course of the winds in a cyclone in the north of the Bay of Bengal. Those in Fig. 9 the winds of a cyclone in the same latitude in the southern hemisphere.

tenths of an inch, and although at sea the winds are locally strong and squally, sometimes indeed attaining to hurricane force to east and south of the centre, yet on land they seldom blow with much violence: the central barometric depression is larger, and the calm area less defined; but the rain accompanying them is often very heavy, and it is in such storms that those torrential falls take place, some examples of which have been quoted on a previous page.

Another distinction of much practical importance between cyclones and cyclonic storms is that cyclones, notwithstanding their fierceness, are, as a rule, less lasting than the milder form of storm, when they have reached the land; and are often broken up and dissipated when they encounter even a low range of hills interposed in their path. Such was the case, for instance, with the cyclone which, on the 1st November 1876, passed over Backergunge in Eastern Bengal, submerging the large islands at the mouth of the Megna, and sweeping some 100,000 human beings to destruction; but was dispersed and annihilated on meeting the Tipperah hills, which are not more than about 2000 feet high. And cyclones which cross the Coromandel coast generally break up and disappear on reaching the Eastern Ghats, which are also little, if at all, over 2000 feet in height. Even if not extinguished, they are much reduced in force when crossing the plateau of the interior, and become reinvigorated only on reaching the Arabian Sea, which they may then traverse for a distance of many hundreds of miles beyond.

On the other hand, cyclonic storms which are formed at the head of the bay, or over Bengal or Chutia Nagpur during the rains, traverse the hilly country of Chutia Nagpur and the Central Provinces without any great loss of energy, and sometimes pass right across Central India, Guzerat and Cutch, and out to the Arabian Sea, off the coast of Baluchistan. Others again travel to the northwest, generally taking four or five days in the transit, and sometimes lasting much longer. In September 1884, a storm of this kind, formed in the west of the bay about the

16th of the month, crossed the Ganjam coast on the 20th, then travelling slowly on a north-west course, by the 24th had passed the Satpuras, and the centre lay between Jubbulpore and Saugor; in two more days it had reached a position between Jhansi and Agra, after which its farther progress was stayed, and it remained virtually stationary till the end of the month, and indeed till the 1st or 2d October, giving very heavy rain to Rohilkund and the Meerut division of the North-west Provinces. The life of this storm therefore exceeded a fortnight.

One further point of distinction between cyclones and cyclonic storms is that the former are restricted to the intervals between the monsoons or to their beginning and ending. The latter occur in frequent succession all through the summer monsoon, and, as above mentioned, also in the winter months in Northern India, bringing the winter rainfall. Their movements are, however, quite different at the two seasons. The storms of the late spring and summer almost invariably move on a course between north and west. Some few autumnal storms have taken a north-north-east course up the Bay of Bengal, and those which reach the shores of Bengal often curve to north-east in the final part of their course before breaking up. But the cyclonic storms of the winter and early spring months, that make their first appearance in North-western India, if not stationary, always move eastward, and sometimes, though rarely, a little to the south of east. Cyclones of great violence are comparatively rare. In the Bay of Bengal they are certainly not more numerous than about two in the year on an average, perhaps less; and in the Arabian Sea are even less frequent, as far as our records go. Whereas, during the summer monsoon, cyclonic storms sometimes succeed each other at the rate of four or five in the month.

Cyclones, as experienced on the coasts of Madras, Bengal, and Burmah, always come from the sea; probably because it is only at sea, where there is a wide expanse of open water, that circumstances admit of their developing a high degree of severity. Otherwise there is no reason, why a cyclonic storm originally formed over the land and passing out to sea should not develop into a cyclone. And indeed, more than one instance is on record, in which a cyclone formed in the bay, has reached the Coromandel coast, and in its passage across the peninsula has been so far mitigated that it could only be described as a cyclonic storm; but after crossing to the Arabian Sea it has again developed its former intensity, and as an unquestionable cyclone has been encountered by vessels making the passage between Aden and Bombay. Such was the case with the cyclone that passed over Madras at the end of November 1885.

There is, then, no fundamental distinction between cyclones and cyclonic storms. Mr. Eliot, who has paid much attention to this subject, chiefly in connection with the requirements of the Bengal storm-warning system, has distinguished them simply as *large* and *small* storms, the former being characteristic of the periods of transition between the monsoons, the latter of the monsoon season. And from his point of view this terminology is unobjectionable. But inasmuch as the latter class of storms, when laid down on the Indian weather charts, are found to cover quite as large an area as the more severe cyclones, these terms might be open to misconception, and those here proposed seem therefore preferable.

These general remarks on cyclones and cyclonic storms may suffice for this part of the work. In the second part they will be described more in detail, with respect to the regions affected by them, to their relative frequency in different months, the circumstances and place of their origin, and the courses they generally follow at each season, and finally the antecedent conditions which are premonitory of their occurrence.

Tornadoes.—It remains to notice a class of storms which resemble cyclones in the vortical movement of their winds, and which, within the narrow limits of the areas affected by them, surpass these and all other storms in violence and consequent destructiveness; but in the circumstances of their origin, and in their brief duration, are more nearly akin to ordinary thunder-storms. Tornadoes or whirlwinds, as these are termed, are comparatively rare in India, there being only four or five recorded instances known to me, and all these were in Bengal. Possibly some of the dust-storms of North-western India, described by Baddeley,[1] may be of the same character; they have apparently a similar vortical movement, but not the same intensity, and at present it is at least uncertain how far the conditions of their formation are identical with those of tornadoes.

The few recorded tornadoes have all occurred in the spring months, between March and May. They have varied from about 200 yards to half a mile in diameter, and while the track of their passage has been marked by complete devastation, by trees torn up by the roots, or stripped of their largest branches, by houses unroofed and blown into ruins, and by human beings, cattle, and even more ponderous objects being carried into the air and transported to considerable distances, immediately beyond its limits everything has remained undisturbed and uninjured. And the track of destruction has not been continuous. It appears that whirlwinds, at least such as are of small diameter, only reach the ground intermittently, and in the intervals pass overhead without much affecting objects on and near the ground. There is some evidence that such was the case in a tornado that originated over the Jamùna river in Eastern Bengal on the 26th March 1875, and also in the more recent tornado that passed over Dacca on the 7th April 1888, of which an excellent description has been published by Dr. Crombie in the *Journal of the Asiatic Society of Bengal.*[2]

It is hardly possible to draw further general conclusions from these few instances, and it may therefore be useful to consider such as are afforded by storms of the same kind in the United States of America, where they are of frequent occurrence, and have been made the subject of careful study.

[1] *Phil. Magazine*, 1850, and *Journal As. Soc. Bengal*, 1852.

[2] Vol. lvii., 1888, Part ii. No. 2.

In the single year 1884, 171 tornadoes were recorded and reported in the States of the Union, and have been discussed in a Memoir by Lieut. J. P. Finlay. The most important conclusions of a general character, apart from questions of geographical and seasonal distribution, etc., which have chiefly a local interest, may be summarised as follows: Tornadoes occur when a barometric depression is traversing the States, usually one of more or less oval or elongated form, having its major axis north and south, and always in a definite part of the depression. From the charts which accompany the Memoir, this appears to be the south or south-east of the depression. They occur to the south and east of a region where there are high contrasts of temperature and also high contrasts of atmospheric humidity, which respectively accompany cool northerly and warm southerly winds.

Now, substituting "dry and north-west" for "northerly" winds, most of these conditions are probably fulfilled in the case of the Bengal tornadoes. Such was certainly the case in the only instance for which the actual data are before me, viz. that of the recent Dacca storm above quoted. This storm occurred in the evening, and the Simla weather chart shows that, on the morning of the day, there was a barometric depression in Northern Behar, to the north-west of Dacca. On the morning of the previous day a cyclonic storm had appeared in the Punjab, and it may be inferred from the subsequent changes of the barometer and winds, that this had passed to the eastward over the Himalaya, and that the depression in Northern Behar was a portion of its southern limb. In any case, on the forenoon of the 7th, winds were northerly over North-western India, north-west to west down the Gangetic plain, and between south and south-west in Bengal. The north-west winds down the Gangetic plain were very dry, the relative humidity at 8 A.M. varying between 20 and 36 per cent of saturation at different stations, while the south-west winds in Bengal had a humidity in most cases of over 80 per cent. The difference of temperatures of the two winds near the ground surface was not indeed very great, and in the afternoon evanescent, but it is well known that in the dry atmosphere of the spring months in Upper India the vertical decrease of temperature is very rapid in the daytime, much more so than in the damp atmosphere of Bengal; and it is probable that at the elevation of the clouds, in which tornadoes originate, there was a large difference of temperature. It seems very probable, then, that the circumstances under which tornadoes occur in Bengal are similar, in all essential respects, to those which hold good in the United States of America as above defined.

No tornado has been traced in Bengal for a greater distance than a few miles. In America, according to Lieut. Finlay, the length of their path has varied from 2 miles to 130 miles, and their rate of progression from 15 to 80 miles an hour. Their average duration at any given spot was only 45 seconds. In the great majority of instances, the direction of the wind's revolution is the same as in cyclones. The exceptional reported instances in which it was said to be "probably" anticyclonic or clockwise, are only 6 per cent of the whole.

PART II

CLIMATES AND WEATHER OF INDIA IN RELATION TO HEALTH AND INDUSTRY

PART II

CLIMATES AND WEATHER OF INDIA

THE CLIMATES OF INDIA

Heterogeneity of Indian Climates.—In the heading of this section and in the title of the work, I use the word *climate* in the plural, in order to emphasise the fact that in this respect, and consequently in all that depends on climate, the different parts of India exhibit very great diversity. Northern or extra-tropical India alone, in its most easterly and most westerly provinces, in Assam on the one hand and in Sind on the other, presents us with the greatest possible contrast of dampness and dryness, a contrast greater than that of the British Isles and Egypt; and when, further, we compare the most northerly province, the Punjab, with the most southerly, such as Travancore or Tenasserim, we have in the former a continental climate of the most pronounced character, extreme summer heat alternating with winter cold that sometimes sinks to the freezing-point, and in the latter that almost unvarying warmth in conjunction with a uniformly moist atmosphere, that is especially characteristic of the shores of a tropical sea. To speak, then, of the climate of India, as we might speak of the climate of Ireland, as if such expression denoted certain definite conditions of heat and moisture, varying only within moderate limits, and nearly uniform in the several provinces, would be as misleading as if we were to speak of its inhabitants in terms implying that they are a homogeneous race, alike in ethnic and social characters, culture and belief.

That they are *Indian natives* certainly implies that they are human beings, denizens of some part or other of Her Majesty's Indian Empire, and little indeed beyond. And in like manner the term *Indian climate* means little more than that, on the general average of the year, the sun is higher in the heavens, and the temperature some degrees greater, than in Europe or other lands of the temperate zone. In all else we must be prepared to find the greatest variation, and even the greatest contrasts.

In respect of this one feature, the mean temperature, notwithstanding that one-half of the area of India lies without the tropical zone, the country as a whole is remarkable as being one of the warmest regions of the earth's surface. It has long been known that what is termed the thermal equator (the line of greatest mean heat of the globe), does not follow the course of the equinoctial line, where, on the average of the year, the sun's action is most direct and intense, but, in the longitude of India, bends northward, running up through Ceylon and the peninsula to beyond the tropic and, passing through Sind, crosses thence to the Arabian peninsula; furthermore, that where it traverses the low plain of Southern India is one of the hottest portions of this line. But it has been shown in the discussion of the temperature in the earlier portion of this work, and it will presently be explained more in detail, that it is not in Southern India that the most intense heat is experienced in the summer season. From May to August the heat of North-western India exceeds that of any part of the peninsula at any time of the year; and probably nowhere on the earth's surface has the thermometer, when properly shaded in the manner explained on page 6, registered higher temperatures than stand recorded in the dry climate of Upper Sind and western Rajputana.

Hill Stations

Before proceeding to describe the varieties of climate to be met with on the plains and plateaux of India, I will notice those of the hill stations, many of which lie around its borders, and which, as the health resorts of Europeans in the military and civil services of India, and as the permanent abodes of many of those who have become domiciled in the country, lay claim to our especial interest from a sanitary point of view. The climates of the Himalayan stations have indeed been described in a special work on the subject, written for the guidance of medical officers by Dr. F. N. Macnamara, and as that work, in addition to the statistics of the climate of each station, gives much other information relative to the prevalent diseases and their treatment, to all such officers it will remain a valued and familiar guide-book, which the present in no way pretends to supersede. But the latest meteorological data available to its author were those of 1876, only 2 years after the Meteorological Department had been established on its present footing; and the records of 10 additional years, which have now been published, enable me to give much fuller and more trustworthy data than were then extant, for illustrating the climatic features of those stations that are provided with observatories. To these and to the similar stations in the peninsula and Ceylon, my own remarks will be restricted. They comprise the following :—

Quetta.	Ranikhet.	Mount Abu.
Leh	Darjiling.	Wellington.
Murree.	Shillong.	Ootacamund.
Simla.	Pachmarhi.	Newara Eliya.
Chakrata.	Chikalda.	

Quetta.—The climate of this important military station, within a few miles of the Afghan frontier, has become known only within the last few years, thanks to the excellent observatory established and superintended by Dr. Fullerton, the residency surgeon. In many respects it is unlike that of any other high-level station, but may be taken as representative of that of the greater part of the western frontier, and especially of Northern Baluchistan. The part of Baluchistan in which Quetta is situated is a tableland, extending from the summit of the Bolan range, and declining gently towards the west. At the *Kotal* or summit of the Bolan pass, it is 6500 feet above the sea, and consists of broad loamy plains, almost bare of vegetation, intersected and surrounded by numerous ridges of bare rocky hills, some peaks of which rise to 11,000 feet above the sea. Quetta is built at an elevation of 5500 feet on one of these valley plains, forming part of the drainage of the Lora river, running north and south for about 30 miles, and varying from 3½ to 5 miles in breadth. To the north of Quetta it bends westwards. The station is thus surrounded by hills, except to the south and north-west, those on the east being about 2 miles distant, and those on the west little more than 1½ miles. To the north-west the vista is open, nearly from north to west, and 30 miles off is seen the Khojak range, at the near foot of which is the frontier military post Pishin.

The whole character of the scenery around Quetta testifies to the extreme dryness of its climate. The naked hill scarps, on which every rocky outcrop stands conspicuous as in a geological diagram, the absence of trees and for the most part of bushy scrub on the pebble-strown slopes at their foot, and the grassless flats of loam that occupy the intervals between the ranges, dotted here and there with mounds of earth excavated in the construction of *karezes* or underground water channels, bear witness, even in the absence of any meteorological record, to the scantiness of the rainfall and the aridity of the prevailing westerly winds. The summer monsoon rains, if indeed they reach Quetta at all, do so only to the extent of a rare shower; the total rainfall of July and August amounting, on an average, to less than an inch and a half, distributed over four rainy days, and in some years failing completely. Indeed the only rain of importance is that of the first four months of the year, when, on an average, it rains or snows on one day in four; and the total average rainfall of the year is barely 10 inches. Even this small amount is very uncertain and precarious. In 1879 it was little more than 4 inches, and in the following year barely 6 inches, whereas in 1885 it exceeded 21 inches; and the number of rainy days in the year has varied from 18 only to 73. January and February are the dampest months, but even then the mean humidity of the air does not exceed two-thirds of saturation, and the driest is September, when it does not much exceed one-third. All through the summer monsoon the average humidity is below 50 per cent.

As is invariably the case in dry regions, the ordinary range of temperature is very great, both in the course of the 24 hours and also between winter and summer. The diurnal range is least in February, when it amounts to 21°, and is greatest in October, when it averages not less than 35° between the early morning and the afternoon. Although the station is not less than 5500 feet above the sea, the mean daily maximum of July and August is 91° or 92° in the shade, and a reading of 103° has been registered within the last 6 years. The mean temperature of these months, however, is only 77° and 75°, and as a rule in the early morning, the thermometer, protected from radiation, sinks to between 63° and 65°, and often lower. In January the plateau is swept by storms of snow and sleet with bitterly cold winds, alternating with frosts, which, however, speedily disappear under an unclouded sun, and from November to February the shaded thermometer, on most days before sunrise, sinks to a temperature below the freezing-point. The nights are always cool throughout the year. As above remarked, the winds are almost always more or less westerly, most commonly north-west, and this is the case throughout the season of the Indian summer monsoon, when in the greater part of Northern India easterly winds prevail.

Leh.—Situated 4° farther north than Quetta, and 6000 feet higher; this station, in the leading characteristics of its climate, represents that of Western Tibet. Its observatory, 11,500 feet above the sea, is the most elevated in Asia, and has now furnished a register for 12 years.

Leh, the chief town of the province of Ladàk, is situated in the upper valley of the Indus, which is here from 6 to 8 miles across for a distance of 40 or 50 miles, and shut in on both sides by portions of the Tibetan plateau, averaging

16,000 feet above sea-level, and bearing some of the loftiest ranges in the Himalaya. The flats and terraces along the banks of the river are all cultivated, and dotted with numerous villages; but for a mile on each side, between the fields and the foot of the hills, is a waste of sand, gravel and large boulders, similar to that which fringes the much lower hill ranges of Baluchistan, but on a larger scale. The town of Leh nestles under the hills north of the valley, at a distance of some 4 miles from the river, up a long, gentle, gravelly slope.

The atmosphere of the valley is remarkably clear and transparent, and the heat of the sun very great. There is generally a difference of more than 60° between the reading of the exposed sun thermometer *in vacuo* and the air temperature in the shade, and this difference has occasionally exceeded 90°. It has been mentioned on a former page that Dr. Cayley succeeded in making water boil by simply exposing it to the sun in a small bottle blackened on the outside, and shielded from the air by inserting it in a larger phial of transparent glass. Owing to the diminished pressure of the atmosphere at the elevation of Leh, this would, however, take place at 191° or 192°, or about 20° below the normal boiling-point at the sea-level.

The mean annual temperature of Leh is 40°, that of the coldest months (January and February) only 18° and 19°; but it rises rapidly from February to July, in which month it reaches 62°, with a mean diurnal maximum of 80°, both in that month and August, and an average difference of 29° or 30° between the early morning and afternoon. The mean highest temperature of the year is 90°, varying between 84° and 93° in the last 12 years. On the other hand, in the winter, the minimum thermometer falls occasionally below 0° Fahrenheit, and in 1878 reached as low as 17° below zero. The extreme range of recorded temperatures is, therefore, not less than 110°.

The air is as dry as at Quetta, and rather more uniformly so. In the driest month (June) the mean humidity is but 37 per cent of saturation, and in the dampest months (January and February) only 61 per cent. But the skies are more cloudy than these circumstances might lead one to expect, and in no month does the mean cloud proportion fall much below four-tenths of the expanse. The amounts of rain and snow are, however, insignificant. The average rain (and snow) fall is only 2·7 inches in the year, and twice that amount is the greatest yet recorded in any one year (1879), while in one year (1876) it was less than half an inch. It snows most frequently in January and February, but the falls are very light in the valley and soon disappear. Rain is most frequent in July and August, but even then it occurs, on an average, only on one day in ten; between one and two-tenths of an inch being the average fall of each rainy day. Agriculture is, therefore, almost entirely dependent on irrigation.

The winds are generally light, and depend on the local direction of the valleys. At Leh, which stands at the entrance of the valley leading up to the Khardong pass, the most common directions are between south and west in the daytime and summer, and from north-east in the night, especially in the later months of the year. In January and February the air is generally calm, and April and May are the most windy months of the year.

Murree.—This is the chief sanitarium of the Western as Simla is of the Eastern Punjab. It is situated on the summit of the ridge that divides the Jhelum valley from the Pàtwar (the tableland above the Salt range), and commands an extensive view of mountain and plain. The hills around are well wooded, except in the direction of the plains. None of those in the vicinity of Murree are of much greater elevation than the station itself, which stands nearly 7500 feet above the sea. The observatory which was established on its present footing in 1875 is at the Lawrence asylum, the military school, and fully 1000 feet lower than the station, on the crest of a spur that runs down towards the plains, and has a somewhat higher mean temperature than the station itself, while it is screened from northerly winds by the main ridge.

Being thus situated on the crest of one of the outermost spurs of the Himalaya, its climate, like that of other Himalayan sanitaria, is of a very different type from that of Leh and Quetta. Although drier than that of stations such as Mussoorie and Naini Tal, similarly situated but farther to the south-east, its atmosphere is much damper than that of the plains immediately below, and subject to smaller variations of temperature both annual and diurnal. The mean temperature of the observatory is 56°, in January and February 39°, and in June 71°. From this time it falls gradually to 65° in September, and then rapidly to the end of the year. The lowest temperature is generally reached in February, when the mean minimum reading is 34°. The lowest yet recorded, in 1886, is 16·7°. Owing to the comparative lightness of the summer rainfall, the temperature, from June to September, is higher than at the more easterly hill stations. Notwithstanding the elevation, the shaded thermometer not infrequently rises above 90° in June, and in 1880, a very dry year, registered as high as 98·7°. The diurnal range varies but little in the course of the year. It is rather smaller in the winter and in August, when it amounts to 14°, than at other times of the year, and is greatest in the driest months, viz. from April to June, when, however, it does not much exceed 17°. These figures, it must be remarked, represent the conditions of the observatory only, and it is probable that the amounts would be found to differ considerably, with differences of position and aspect, in different parts of the station, but those general characters which distinguish the climate from those of other hill stations are doubtless shared by all parts of the site.

From June to November the air is much drier than that of Simla, or any other hill station of the outer Himalaya, but in February it is apparently damper, and in the remaining months of the winter and spring of about the same character. The humidity is highest in August, when it amounts to 80 per cent of saturation. In March the skies are as cloudy as in the rains.

In the spring months, from March to May, it rains on

an average about one day in three, most frequently in April, the rainfall on each rainy day averaging about four-tenths of an inch. Rain is less frequent in June, but increases again in July and August, when about half the days are more or less rainy, and the falls heavier than in the spring, averaging three-fourths of an inch on each day of rain. The number of rainy days in the year has varied from 74 only, in the dry year 1880, to 123 in 1885, and the total quantity from 39·1 inches in the former to 71·8 inches in the latter year.

October and November are the clearest months, but even these, and every month up to May, are more cloudy than at Simla or on the plains. On the other hand, in July and August there is less cloud than at Simla or other more easterly stations, as might be expected from the smaller frequency of rain.

Simla.—The well-known summer capital of the Government of India, distant 58 miles by road from the plains and 23 by direct measurement, is built on a ridge, much of which is rather over 7000 feet, separating the valley of the Sutlej from the drainage of the Tonse and Jumna, and therefore forming part of the main watershed of India. The greater part of the station crowns the crest of an amphitheatre of hills facing to the south-west, but shut off from the plains by the peak of Tara Devi opposite the station, and by a more distant ridge, on which are the military sanitaria Dugshai and Kasaoli. This amphitheatre is formed by the main ridge and a spur which runs southwards from its highest point, the forest-clad hill Jako, which rises above its north-east corner to a height of 8048 feet. A small part of the station occupies the northern face of Jako and the crest of a spur which runs off to the north in the direction of the Sutlej. Trees are rigorously protected within the limits of the station, and many of the houses are embosomed in a thick growth of deodars, blue pines, and oaks; but the lower spurs and hillsides are for the most part bare of timber, presenting steep grassy or rocky slopes, and every available acre of moderately-inclined ground is terraced and cultivated. The descents on all sides are very steep to the stream beds, about 2000 feet below the crest of the ridge. The observatory is attached to the Government offices, on the southern aspect, a hundred feet below the crest of the ridge.

With a higher rainfall and, on the whole, a somewhat damper atmosphere than Murree, the climate of Simla is subject to greater variations of humidity at different seasons, though not of temperature. The mean temperature of Simla is 55°, or nearly the same as that of Murree, and the same as that of Milan; that of January and February 41°, and that of June, the warmest month, 67°. The former of these temperatures is the same as that of Florence in January, and the latter that of Frankfort and Vienna in July. The lowest readings reached in the course of the year vary between the freezing-point and 13° below it, 19·7° being the lowest yet recorded. The highest are generally below 90°, but in 1877 rose to 94·4°. The course of variation during the year is much the same as at Murree. The extent of the diurnal range is, however, more variable. In April and May, when it is greatest, there is an average difference of 20° between the early morning and afternoon—the same as in the south of England in July, while in August this difference is only 11°. During the six winter months it is pretty steady, averaging between 16° and 18°.

The driest month is December, when the mean humidity of the air is only 47. But in May, the mean humidity of which is 49 per cent of saturation, in some years, for short periods, it has been known to fall much below these averages, even as low as 20 per cent. The early part of June is sometimes equally dry, but the rains which, as a rule, set in towards the end of the month, are often preceded by afternoon storms, which mitigate the dryness of the air and reduce the temperature. In July and August the rains are heavy, and the mean humidity as high as 88 and 91. At this time the station is frequently enveloped in cloud and mist, but less continuously than Darjiling and Naini Tal, or even than Mussoorie; and even at the height of the rains the evenings are frequently fine. The rains cease about the middle of September, and their cessation is followed by the clear mild weather that renders Simla at this season one of the most charming places of residence to be found in the Indian empire or elsewhere. During the latter part of September, and all through October and November, the skies are generally cloudless, and only as an exceptional occurrence is the calm serenity of this season interrupted by a day or two of rainy weather. About Christmas the clouds begin to gather, and the first three months of the year are chequered by occasional storms of snow, and, later on, of rain and hail. Thunder and hail storms become more frequent in May and June, though interrupted by long spells of hot dry weather, and it is during these intervals that the exceptionally high temperatures above noticed are experienced.

The mean annual rainfall of Simla is 70·1 inches; higher, therefore, than that of Murree, the excess being entirely due to the greater heaviness of the summer monsoon rains. The average number of rainy days in the year is, however, not much greater, viz. 99. It rains less frequently in the spring months, but more frequently in the rains. In the wettest years there have been 136 rainy days. Snow usually falls several times in January and February, eight being the average number of snowy and rainy days in these two months, on most of which the precipitation takes the form of snow. About once in 8 or 10 years the accumulation may be to a depth of 4 or 5 feet in a single fall, but as a rule it is much less. Hail-storms are common in March, April, and May, and occur occasionally in June, the precipitation sometimes, though rarely, taking the form of soft hail, which is unknown on the plains. Though more frequent than on the plains, the hailstones are generally smaller on the hills and less destructive, the commonest size being that of small marbles.

The winds at Simla are generally light; those from north and north-west are dry; and the houses, on the Elysium spur and the north face of Jako, which are most exposed to these winds, have a more bracing atmosphere than those of Chota Simla and Boileaugunj, which have a southern aspect. The latter also receive more sunshine, and in the rains more cloud and mist. There is thus a very appreciable difference between the northern and southern portions of the station, the latter being much the more extensive.

Chakrata.—This is a small military sanitarium on the crest of a ridge which rises to the west of the Jumna valley, a few miles above Kalsi, where it makes its exit from the Himalaya. It is 50 miles in a direct line south-east of Simla, and 22 to the north-west of Mussoorie. Except the low range of the Sivaliks and the spurs which run down from Deobund to the Doon, no hills intervene between the Chakrata ridge and the plains. The station lies well within the zone of heavy rainfall that borders the outer Himalaya, and at a height of 7000 feet above the sea. The Deobund range, immediately north of Chakrata, is a few hundred feet higher, but to the north-east falls towards the Jumna, and the station commands a magnificent panorama of Jumnotri and the snowy peaks at the head waters of the Jumna and Ganges. The hills around the station, like this part of the Himalaya generally, present bare precipitous slopes to the south, and are more or less wooded on their northern aspect.

Being at almost exactly the same elevation as Simla, but more southerly and nearer to the plains, its mean temperature is about 1° higher, viz. 56°, and as the course of the annual variation at the two stations is almost identical, there is a similar difference in most months of the year. The diurnal range varies in like manner with the seasons, and is 1° or 2° less than at Simla at nearly all times of the year. The mean annual range is also 2° less, and somewhat less variable from year to year. The mean highest temperature of the year is 85°, the lowest 24°; the highest yet recorded 89·6°, and the lowest 18·7°. These differences are not indeed great, but they illustrate the modification that ensues in the climate of the outer Himalaya in the course of only 50 miles, as the chain is followed to the south-east.

Except in March, April, and August, the air is rather damper than at Simla, and most so from October to February. But except in July, August, and September, Chakrata is by no means a damp station, and its situation on the crest of a ridge exposes it freely to the usually light winds of the mountain zone. On the average of 17 years the rainfall is only 61·5 inches, and more than half of this falls in the two months July and August. The winter rainfall is about the same as at Simla, that of April and May little more than half as great. The number of rainy days in the year averages 110, and in the wettest year has not exceeded 127, while in the very dry year, 1877, there were but 95.

Ranikhet.—This, too, is exclusively a military sanitarium, situated in the Kumaon Himalaya, 120 miles south-east of Chakrata, rather more than 6000 feet above the sea, on a little plateau, 25 miles distant from the plains, from which it is shut out by the higher range, in which lie embosomed the station and lake of Naini Tal. This range shields it from the heavier rainfall of the southern face of the hills, and it enjoys a more equable climate than any station yet noticed. The plateau is well wooded with pine groves, amid which clearings have been made for the buildings. To the south and south-west extends a broad valley clear of forest, leading down to the Ramgunga river, and to the north an extensive depression in the hills of similar character affords a wide view over lower ranges and valleys, terminating some 20 miles off against a higher range, above which towers a gigantic group of snow-clad peaks, including Nanda Devi and the Trisùl. To the north-west the view is equally unobstructed, and on clear days the great independent groups of snowy peaks, Badrinath, the Banda Poonch, etc., may be seen rising one beyond the other in diminishing perspective to a distance of more than 100 miles, towering above the lower mountains, and separated by great valleys from the watershed of the Upper Sutlej.

The mean temperature of the station is 60°, and the average annual range from 30° to 86°; the extremes, hitherto recorded, being 26° as the minimum, and 87·8° as the maximum. The absolute range of temperature is therefore 9° less than that of Chakrata, and 13° less than that of Simla. The mean diurnal range is from 1° to 3° less in different months, except from July to October; and in April and May, when greatest, does not exceed 17°. This comparative equability of temperature is not, however, due to greater dampness. On the contrary, the mean humidity of the air, in most months, is appreciably lower than at Chakrata, and even in July and August is only 85 and 86 per cent of saturation. There would appear also to be somewhat less cloud, especially in the winter months. The rainfall, 49 inches, is less by 12 inches, and little more than half that of Naini Tal; and there are on an average but 95 rainy days in the year; 107 is the greatest number yet recorded.

The winds are light and calms frequent at all times of year. Those from the south-west greatly preponderate, owing probably to the situation of the station at the head of the broad valley above mentioned. On the whole, Ranikhet enjoys a more equable, moderate, and at the same time rather drier climate than any other station in the north-west Himalaya from which we have comparable data.

Darjiling.—An interval of 570 miles separates this station from Ranikhet, the greater part of which is occupied by the kingdom of Nepal. For this part of the Himalaya, save Pithoragarh, an unimportant station 50 miles from Ranikhet, and Katmandu the capital of Nepal, in a valley 4000 feet above the sea, any accurate climatic data are wanting; and Katmandu is rarely accessible to Europeans, and not available as a sanitarium. Darjiling is situated 16 miles from the plains, on a branch of the Singalelah spur, running southwards from Kanchinjinga, the gigantic peak which rears itself at a distance of only 40 miles to the north of the station. The station is built, partly on the crest of the ridge, partly on its western and south-western slopes, at elevations between 6500 and 7800 feet; but it is shut off from a direct view of the plains by the Goom Pahàr ridge, running east and west between Sinchal (8608 feet), and Tonglo (10,081 feet). On the north the Darjiling ridge descends rapidly to the valley of the Great Rungit, across which and the ridges beyond, the station commands a magnificent view of the snow-clad monarch which dominates it, and with its attendant peaks affords one of the grandest spectacles to be met with in the Himalaya and probably in the world. The highest part of the Darjiling spur is where it runs up to join the Goom Pahàr, and it is on the crest of this spur, at 7421 feet above the sea, that is situated the present observatory, established in 1881. This house, now occupied by St. Paul's school, is the same as that at which Dr. (now Sir Joseph) Hooker recorded his observations in 1848. For this site we now possess a meteorological register for a little more than 4 years. For 14 years previously the observatory was at the telegraph office, 500 feet lower down, on the west face of the ridge, and more on a level with the greater part of the station. An abstract of the results of both is given in the Appendix at the end of this work. The existing data for the new observatory are hardly sufficient to show its climatic difference with great precision, but it would appear that its mean temperature is not less than 3° lower than that of the old observatory, and its atmosphere about 3 per cent of saturation damper. The diurnal range of temperature is moreover 2° or 3° less.

Taking the registers of the old observatory as representing the climate of the greater part of the station, it appears that, notwithstanding its more southerly position and (excepting Ranikhet) its somewhat lower elevation, the mean temperature of Darjiling is slightly lower than that of any of the north-western sanitaria, viz. 54°. But the most striking point of difference is its much greater dampness, a difference which characterises it in almost every month of the year, and more especially in the spring season. Even in the driest month, March, the mean relative humidity of Darjiling exceeds 70 per cent, and from June to September inclusive is above 90 per cent of saturation. Its cloudiness is not less striking. The clearest and most serene months are November and December, but even in these, on an average, nearly half the sky is clouded. During the rains, from June to October, for many days in succession, the whole station is shrouded in cloud and mist, with almost incessant rain. Even in May more than half the days are rainy, and from June to September the number of rainy days varies from 22 to 28 on an average in each month. As the rains begin early, so also they last late, and it is not till the middle of October that they can be considered fairly at an end, the average rainfall of that month being 6·4 inches. The mean rainfall of the year is 120 inches, distributed over 149 days, and in 1870 there were 171 rainy days.

It is to this dampness and cloudiness, and the frequency of rain throughout the spring and summer, that must be attributed the comparative coolness of the station. The highest temperature reached in the year is, on an average, only 78°, which is 8° lower than the average maximum of Ranikhet and 10° lower than that of Simla. The average minimum of the year is 30°, which is the same as that of Ranikhet; and 26° is the lowest recorded at the old observatory in 14 years. The mean difference of extreme night and day temperatures varies from 11° in June, July, and August to 17° in March, April, November, and December; a range of variation which is the same in amount as at Ranikhet. The winter months are generally fine, and notwithstanding the frequency of cloud, rain and snow are less frequent at Darjiling at this season than at the north-western stations.

The military sanitarium on the Jellapahar is on the same ridge as the new observatory, but 400 feet higher up, and its climate, therefore, differs in the same way from that of the old observatory, and to a greater extent. It is colder, and the site is more frequently enveloped in mist.

The gardens which produce the fine teas of Darjiling are all at lower levels than the station, and in a warmer climate. Some gardens, established in the early days of tea cultivation between 6000 feet and 7000 feet, were found to yield indeed a tea of fine flavour, but the produce was insufficient to be highly remunerative, and tea cultivation is now carried on at all elevations from 5000 feet down to the plains.

Another important product of the Darjiling valleys is cinchona bark, the cultivation of which is carried on at the Government plantation, on a larger scale than anywhere else in India. The gardens are situated below Darjiling, to the east of the station, on the slopes that run down to the Tista valley. A register of temperature and rainfall has been kept here during the last 8 years, under the same conditions as at other Government observatories, at the superintendent's residence at Mongpoo, about 3000 feet above the sea, and affords the following summary of the temperature and rainfall, which will be of much interest to those who are engaged in the cultivation of the cinchona. The rainfall averages are obtained from 13 years' registers.

Temperature and Rainfall of Mongpoo Cinchona Garden

	Temperature.					Rainfall.	
	Mean.	Mean Max.	Mean Min.	Mean Range.		Ins.	Days.
				Daily.	Month.		
January .	54	59	49	10	17	0·8	2
February .	55	60	50	10	23	0·8	3
March .	63	69	57	12	25	2·1	6
April . .	68	73	62	11	21	4·0	9
May . .	69	73	64	9	21	9·2	17
June . .	72	75	68	7	17	29·9	22
July . .	73	78	69	9	14	31·2	25
August .	72	76	68	8	15	26·0	24
September .	72	76	67	9	17	16·5	16
October .	68	72	63	9	19	5·3	7
November .	61	67	55	12	19	0·1	1
December .	55	60	49	11	18	0·2	1
Year .	65					125·9	133

Mean highest temperature of year	83°	Lowest recorded reading .	40·5°
Mean lowest temperature of year	43°	Absolute range . . .	43·1°
Mean range of temperature of year	40°	Rainfall of wettest year	155·8 ins.
Highest recorded reading .	83·6°	Rainfall of driest year .	92 ins.
		Rainy days in wettest year .	156
		Rainy days in driest year .	116

The most striking characteristic of the Mongpoo climate disclosed by these registers is its great equability. With an annual mean temperature of 65°, nearly the same as that of Athens, Algiers, and Madeira, the extreme variation during the year is rather less than that of Moulmein—less than that of any place enumerated in the table on page 18, save only Bombay, Cochin and Galle; and throughout the year the daily range is hardly greater than that of the dampest provinces of India in the rainy season. Moreover, this equability is displayed not only in the average conditions of many years, but perhaps with even greater emphasis in the regularity of their recurrence year after year. Thus, in the course of 8 years, the highest temperature reached in each year has varied only between 81·5° and 83·6°, and the lowest between 40·5° and 46·5°. It has never descended to within 8° of the freezing-point.

Shillong.—This station, which is the seat of Government in the province of Assam, is situated on a gently undulating tableland 4800 feet above the sea-level, immediately north of the culminating ridge of the Khasi hills, and about midway between the valley of Assam and the plains of Silhet. It is only 30 miles north of Cherrapunji, (long notorious as the wettest known place in the world), but the higher plateau that intervenes effectually shields it from the heavy rainfall of the southern escarpment of the hills, and the rainfall of the station averages only 85 inches, or little more than two-thirds of that of Darjiling. The Shillong ridge, immediately to the south of the station, dominates it by more than 1000 feet, but to the north the general level of the plateau falls away gradually towards the Brahmaputra. The surface of the plateau around Shillong, as indeed of the Khasi hills generally, is a rolling green sward, with occasional clumps of pines; but trees are for the most part restricted to the valleys which, in the course of ages, have been eroded by the hill drainage, the higher parts of the surface being kept clear of forest by the fires annually kindled by the natives for the destruction of the coarse grass and the stimulation of a young growth fitted for pasture.

Partly owing to the difficulty of access from other provinces, and partly to the want of spare house accommodation, Shillong has hitherto been but little visited by others than the Government officials and a few residents of the Assam valley. The former obstacle is now in a great measure removed by the completion of the railway from Calcutta to Dhubri, and the establishment of a line of swift steamers daily from that place to Gauhati, from which Shillong may be reached in 10 hours; and when new houses are provided to meet the increasing demand, it may be expected that in the course of a few years the place will become better known; and a station which, in its mild climate and open downs, holds forth to the residents of Bengal many of those attractions which make the Madras Nilgiris so charming a resort, will attract from the overcrowded boarding-houses of Darjiling some of those who seek a temporary refuge from the "branding summers of Bengal." To such persons the following sketch of its climatic features will be welcome.

On the average of the 4 years, 1869-1872, the mean temperature of Shillong was 62°, which is about the same as that of Constantinople, Barcelona, and Oran; in fact of an average Mediterranean climate, and the July temperature of the south of Ireland and the more northern midland counties of England. Shillong itself has this temperature in March and October. In the warmest months, June to August, it is below 70°, and in April and May intermediate between the two, since in the Khasi hills, as in the province of Assam generally, rain is so frequent in the spring months that the temperature does not rise to a maximum in May, as on the plains of Northern India, and suffers no abatement when the monsoon rains set in in June. The highest temperature reached during the 4 years was 84°, and the maxima of the other years were a degree or two lower. The lowest reading recorded was just above the freezing-point; the average minimum of the 4 years 34°, and the mean temperature of December and January 51°. The winter temperature is therefore nearly the same as that of

Lisbon and Palermo, and the July temperature lower than that of either. In December and January, the most serene months of the year, the mean difference of the early morning and afternoon temperatures is 19° or 20°, which is as great as that of Simla in April and May, and is that of the south of England in July. In July it is only half as great.

While, in respect of temperature, the climate of Shillong much resembles that of places in the south of Europe, in the dampness of its atmosphere and its rainfall it is eminently tropical. In the driest month, March, the humidity is indeed only 59 per cent of saturation, but from July to October inclusive it ranges between 86 and 89, and from June to September, on an average, eight-tenths or more of the sky is clouded. Except the winter months, or from November to March, there is no dry season. Even in April it rains on one day in three, in May on two days out of three, and in the four succeeding months even more frequently. On the average of 18 years there have been 150 rainy days in the year. But although it rains more frequently than at Darjiling, the rain is much less heavy, since, as above remarked, the average rainfall of the year hardly exceeds 85 inches, of which 70 inches falls in the five months, May to September. In the remainder of the year the climate equals that of the most favoured countries in Europe.

Pachmarhi.—This is a small station on the Mahadeva or Pachmarhi hills, in the Central Provinces, about 1° south of the Tropic of Cancer. It is the hot weather residence of the Chief Commissioner and some of the higher officials of the Province, and is also used as a sanitarium for a small body of British troops. The Pachmarhi hills form a prominent group in the belt of hilly country that stretches across India to the south of the Sone and Nerbudda valleys, now generally known as the Satpura range. On the south, the Pachmarhi group terminate in a precipitous escarpment of thick-bedded sandstones, and on the north, slope rapidly but less abruptly towards the Nerbudda valley. Their summit forms a little undulating plateau, grassy and dotted with clumps of trees, which give it the general aspect of an English park, diversified with bluffs and higher peaks of sandstone which, in Dhupgarh, rise to 1000 feet above the general level of the tableland. The scenery is peculiar and very beautiful. The elevation of the station being only 3500 feet above sea-level, and 2500 above that of the Nerbudda valley, its climate is of the same general character as that of the Central Provinces around, with the lower temperature and higher rainfall due to its elevation and the form of the ground. It is sufficiently high to afford cool nights even at the hottest season, but in the daytime the sun is very powerful, and in some years the shaded thermometer rises to above 100°. As a sanitarium it cannot be considered as a rival to the Himalayan stations and the Nilgiris of Southern India, but it affords a grateful place of refuge from the intense heat of the lower plains from March to May, and at other times of year, except perhaps sometimes in the rainy season, the climate is charming.

The mean annual temperature of Pachmarhi is 69°; that of December, the coolest month, 56°; and that of May, the warmest, 83°, rising to 94° in the daytime and sinking to 74° at night. The rains, which usually set in about the middle of June, bring down the day temperature nearly 20°, and the night temperature by 5° or 6°. The former rises again slightly at the end of the rainy season, in September and October, but this slight rise is compensated by the gradual fall of the night temperature, and after October the decline is rapid. Even in the coldest season the thermometer rarely sinks to the freezing-point, 35° being the mean minimum of the year.

Except in the four months of the rains, the atmosphere is very dry, especially in the spring or hot season. In April the average humidity is only 26 per cent, and from November to May below 60, the cloud from two to three-tenths of the expanse, and the rainfall from three to seven-tenths of an inch in each month. In July rain is heavy and frequent, amounting to nearly 29 inches on 25 rainy days. In August it is lighter, but almost equally frequent.

In the cyclonic storms, described on a previous page, which frequently traverse this part of India in the rains, falls of 3 or 4 inches in the 24 hours are by no means unusual, and occasionally as much as this is registered on two or more days in succession. The rains come to an end in the latter part of September, but there are usually a few heavy showers in October before the weather finally clears. The natural drainage of the plateau is complete and rapid. A little lake has been formed by damming a small watercourse that intersects the station, and this is considered to have somewhat impaired the salubrity of the place, which is hardly above fever range.

Chikalda.—About 100 miles to the south-west of Pachmarhi, this small sanitarium has been established at 3656 feet above the sea, on the highest part of the Gawilgarh hills, which, like the Pachmarhi group, are a part of the Satpura range. It is little known beyond the province of Berar, in which it is situated, and in elevation, and the general character of its climate, much resembles Pachmarhi. The geological formation of the Gawilgarh hills is, however, different, being a portion of the great Deccan basaltic trap. Chikalda stands on the southern margin of the hills, overlooking the plain of Berar, which has here an elevation of over 1000 feet. A Government observatory, established here in 1875, affords the data summarised in the Appendix, from which are gathered the following particulars of its climate.

The mean temperature of the year is 70° or 1° warmer than Pachmarhi, and the annual course of its variation is of the same general character, except that its range is much smaller, the winter months being 6° warmer, while those of the monsoon (June to September) are 2° or 3° cooler. The highest temperature of the year is, however, about the same as at Pachmarhi, the mean of May being 94°, and the mean highest reading in the year 99°, with an absolute maximum of 103° in the dry year 1876. The lowest yet recorded is 39·5°, which is 9·5° higher than that of Pachmarhi; and therefore the absolute range of recorded temperature is only 63·5°, or 9° less than that of the more northerly station, the elevations of the two being the same within 30 feet.

The rainfall is 11 inches less than at Pachmarhi; that of July alone being 11 inches less, and that of both August and September 2 or 3 inches less, but there is more rain in October, viz. over 4½ inches on an average, whereas Pachmarhi has but 2. In other respects, the two climates are very similar.

Mount Abu.—Mount Abu is a lofty isolated hill or cluster of hills, 7 miles to the west of the Aravali range; the name given to the chain of rocky hills that runs south-west from Delhi through Rajputana to the northern limits of the Nerbudda valley, forming the western margin of the Malwa tableland. Mount Abu is a rocky eminence rising to 5653 feet above the sea-level, and crowned by a Jain temple of great sanctity, a renowned place of pilgrimage to the adherents of that sect, from all parts of India. The station of the same name, which is the hot-weather residence of the chief officials of Rajputana, the seat of a Lawrence asylum for soldiers' children, and is also resorted to by the residents of Guzerat, is built on this hill at an elevation a little below 4000 feet, that of the observatory being 3945 feet above sea-level. Being on the border of the arid region of western Rajputana, it shares the general characteristics of its climate, save that, owing to its elevation, it is much cooler, and has a rainfall between double and treble as great as that of the plains around.

Although it lies more than 2° north of Pachmarhi, and is 400 feet higher, the mean temperature of the year is only 1° lower, viz. 68°, and the January temperature the same; that of the spring months, March to May, being 3° or 4° lower, and that of the three final months of the year higher by the same amount. In January, the coolest month of the year, the temperature is 58°. The lowest reading varies much in different years, not having fallen below 45·3° in 1882, while in 1878 it was as low as 32·8°. On the average of 9 years it was 39°. In May, which is the warmest month, the mean temperature is 79°, but the afternoon average is 89°, and the highest readings have varied from 93° to 101° in different years: the nights, however, are cool, the temperature falling, on an average, to 71° before sunrise. Owing probably to the position of the station, on a hill which descends on all sides to plains 3000 feet or more below the station, the daily range of the thermometer is smaller than at Pachmarhi, not exceeding 18° on the average of any month in the year, and being only 16° or 17° in the winter months, and 8° or 9° in the rains.

As already remarked, the air is very dry during the greater part of the year. The mean humidity of March or April is only 29 or 30 per cent of saturation, and from November to May inclusive, in only one month does it exceed 40. In August it rises to 87. From October to May the skies are very clear, the mean proportion of cloud varying from one to three-tenths of the sky expanse, and the rainfall is equally slight, indeed insignificant, not exceeding two-tenths of an inch in any month, on an average, from November to April. Only in July and August is it at all heavy, the average of each of these two months being rather over 22 inches, and this falls in quantities averaging about 1 inch on each rainy day. The mean rainfall of the year is a little over 63 inches, distributed over 78 rainy days. The greatest in any one year was 123 inches, and the least 19·2 inches.

From May to August or September the wind is steady from the south-west or west, and of moderate strength, averaging 10 or 12 miles an hour; in November and December northerly winds prevail, and in the remaining months they are more or less variable, but with a preponderance of west winds, the dry winds of all North-western India.

Mahableshwar.—Except Matheran, within a few hours' railway journey from Bombay, the only hill station in the Bombay Presidency is Mahableshwar, on an eminence of the Western Ghats, 4540 feet above the sea. For this station I have no recent and trustworthy registers of climate, save only of rainfall. This has long been known to be exceptionally heavy at Mahableshwar, exceptional even in the rainy zone in which it is situated, and is surpassed only by that of Cherrapunji and a few other places in Assam. The residents of Bombay resort to Mahableshwar during the hot season that precedes the rains, but from June to October the favourite place of residence is Poona, which is more accessible from Bombay and, together with a light rainfall, not exceeding 25 inches from May to October, nevertheless enjoys a cool atmosphere throughout the rainy season. The climate of this place will be noticed further on, in connection with that of the Deccan generally.

Ootacamund.—Beside one or two smaller settlements and numerous plantations of coffee and cinchona (chiefly on the western slopes), two sanitaria, one civil and one for military purposes, have been established on the Nilgiri hills, one of the loftiest of the hill groups dotted over the southern part of the peninsula. The first, Ootacamund, is the summer residence of the Madras Government; the second, Wellington, formerly known by its native name, Jakatalla, is exclusively occupied by British troops; but there is, in its vicinity, the civil settlement of Coonoor, the climate of which is very similar.

The Nilgiris are a lofty mass of hills at the southern extremity of the Mysore tableland, where the Eastern and Western Ghats converge. Their summit is a grassy undulating and hilly plateau about 20 miles across, and averaging between 6000 and 7500 feet above the sea. On its western margin it is crowned by a somewhat higher ridge, the western face of which falls away abruptly towards the plains of Malabar, as a continuation of the Western Ghats of the Wynaad and Mysore; and about midway in its extension it is crossed from north to south by another and higher range of hills, the culminating point of which, Dodabetta, is 8640 feet above the sea. Ootacamund and Wellington are on the opposite sides of this latter range, the former to the west, the latter on its eastern flank, and thus, while Ootacamund is more exposed to the westerly monsoon which blows in the summer months, Wellington receives more rain, when, in October, the wind has changed to east and north-east; a change which ushers in the north-east monsoon of the Bay of Bengal.

One could hardly desire, and it would certainly be hard to find within the limits of the Indian empire, or perhaps elsewhere, a more charming climate or one more fitted to the European constitution than that of the Nilgiri hills. The Laureate has sung of the "sweet, half-English Nilgiri air;" but it is the air of the English spring and summer, without Atlantic storms or the bitter east winds of March; a climate where one may inhale the fresh breeze that blows over rolling downs, and brave with impunity the ardour of a tropical sun, and even enjoy the cheerful companionship of an evening fire through the greater part of the year. The flowers of Southern Europe and well-flavoured English vegetables flourish with a luxuriance unknown in the torrid zone, save on such as this and its sister hill groups, the Shevaroy and Pulni hills. Shielded by the Kùndah and Makùrti ranges from the heavy rainfall of the Western Ghats, fine and cool cloudy weather is the rule rather than the exception, even in the summer monsoon; and in mid-winter the clear frosty atmosphere of the early morning is warmed up daily by a genial unclouded sun, and day follows day untroubled by snow-storms, and free from their unwelcome sequelæ, thaws, slush and catarrh. Without the grandeur of the Himalayan gorges, or the majesty of eternal snows, the Nilgiris have a soft beauty of their own, recalling to the Englishman the undulating contours of his own western hills; and the pedestrian can strike a bee-line over hill and dale, whithersoever his fancy leads him, unimpeded by any obstacle more serious than an occasional peat-bog or a brawling hill stream. Among all the pleasant memories of more than 30 years of Indian life, and an experience of all parts of the empire from Peshawar to Sibsagar and Point de Galle, I can recall no more charming scene and climate than those of the Nilgiri hills. Many an old Anglo-Indian, whom choice or necessity has led to fix his home in India, has found in these hills scenery as beautiful and a climate as enjoyable as any in the most favoured lands of the Mediterranean shores.

One striking characteristic of the Ootacamund climate, due to its position low down in the tropics, is the comparative uniformity of its temperature throughout the year. While the mean temperature of the year is the same as that of Simla (viz. 55°), that of May, the warmest month, is only 4° above this mean, and that of January, the coldest, only 7° below it. Even this difference is due much more to the variation of the night temperature than to that of the day, and the latter is affected more by the amount of cloud and rain at different seasons than by the sun's declination and the length of the day. The afternoon temperatures are lowest in July, but vary very little from that month to December, after which they rise steadily till April. From April to July they fall 10°. The night temperatures are lowest in January, and rise 18° between that month and May; from which time they vary only 1° or 2° till October. The highest thermometer reading recorded in 1880, the only year for which we have a trustworthy register, was 77·3° and the lowest 25·3°. The daily range is very great in clear weather, as is usually the case on tablelands. In January, February, and March it varied, on an average, between 27° and 31° in the 24 hours, but only from 10° to 14° in the cloudy months, June to October.

The atmosphere is not very dry, except in March, the only month of 1880 in which the mean humidity ranged below 50 per cent of saturation. From June to November inclusive, the average of any month fell little below 80 per cent, and in November was 90. From June to October the skies were very cloudy; but as the rainfall register of Wellington shows that 1880 was a remarkably wet year on the Nilgiris, there is reason to believe that Ootacamund is in general a less damp and cloudy station than might be inferred from the table in the Appendix. In 1857, the geological survey of the Nilgiri hills was carried out during the summer monsoon with no more interruption from bad weather than would have been experienced in an English climate. Notwithstanding the prevailing cloudiness at this season, the rainfall is by no means heavy. From May to August the weather is showery, but the total amount of rain is only 4 or 5 inches in the month, and out-of-door life is as little checked by weather as in an ordinary English summer. The heaviest rain is in October.

Wellington.—At a level 1000 feet lower, and with an additional screen interposed between it and the westerly monsoon, in the Dodabetta range, which dominates it by 2500 feet, Wellington is 6° warmer than Ootacamund, and has less cloud and rain; while in some respects it enjoys even greater equability of temperature. During the summer monsoon it is frequently fine and clear to the east of Dodabetta, while completely overcast and raining at Ootacamund; a contrast which we meet with in a much higher degree on the opposite sides of the culminating range of the Ceylon hills. But, after October, the relations of the two stations to the rainy winds are reversed; and in the final months of the year Wellington is the rainier station of the two.

The mean temperature of Wellington, as deduced from between 12 and 14 years' registers, is 61°. That of May, the warmest month, is only 5° higher; and that of January, the coldest, only 6° lower. The average afternoon temperature of May is 76°, and the highest readings recorded in each of 6 years have varied only between 79·5° and 80·7°, or little more than 1°. The early morning temperature in January averages 45°, and the lowest has varied between 34·2° and 37·5°, or little more than 3° in different years. In the winter months the range during the 24 hours is considerably less than at Ootacamund. It amounts to 24° in February, and 21° only in January and March; and from June to November does not exceed between 13° and 15° on an average.

The air is drier than at Ootacamund. The humidity is above 80 only in October and November, but it is below 70 only from February to May. There is also appreciably less cloud than at Ootacamund, but clear skies predominate only from January to April. The rainfall is moderate—about 48 inches only in the year. From April to August there falls from 3 to 4 inches in each month on an average, and it rains about every second day. In October and November it is about three times as heavy, but even then there average 13 or 14 fine days in each month. In January and February rain is rare, and in March and April it rains on an average one day in five, chiefly in afternoon thunderstorms.

Newara Eliya.—The mountain region of Ceylon is an isolated mass in the south of the island, surrounded by a broad undulating plain on the south-east and east, and a more extensive plain on the north, while, to the south-west, a generally hilly country of greatly inferior elevation and intersected with alluvial flats extends between it and the sea. Most of the highest elevations are on the south, where Adam's Peak and some other eminences rise to over 7000 feet; and from this range a spur of equal height is given off to the north, running north-north-west across the middle of the mountain tract and dividing it into two nearly equal portions. Midway in this spur is Pedro Talla-galla, the highest peak in the island, rising to 8296 feet above sea-level; and on its south-western flank, 2000 feet below the summit, lies the little sanitarium of Newara Eliya. The station is built on and around a small alluvial flat, formed by a small stream that drains the western slopes of the mountain, and then precipitates itself through the Rangbodde Pass, by which the road descends in the direction of Kandy. The ridge, of which Pedro Talla-galla is the culminating peak, acts as a barrier to the rainfall of the summer monsoon, in the same way as the Dodabetta range on the Nilgiris, but in a much more emphatic manner, producing a very striking contrast of weather on its eastern and western slopes. The following description was written many years ago, after a visit to Newara Eliya in the month of June, and its accuracy has been confirmed by persons of much longer experience of the station. In the summer monsoon, day after day and week after week, Newara Eliya lies under a dense canopy of cloud that shrouds all the higher peaks and pours down almost incessant rain. But let the traveller leave the station by the Badulla road, and crossing the *col* of the main range at a distance of 5 miles from Newara Eliya begin the descent to Wilson's bungalow, and he emerges on a panorama of the grassy downs of the lower hills bathed in dazzling sunshine, while on the ridge above, he sees the cloud masses ever rolling across from the west, and apparently dissolving away in the drier air to leeward of the hills. The disappearance of the cloud is doubtless due to the descent of the cloud-laden current that has swept over the crest of the range, and to the warming which it necessarily undergoes in such descent; which, according, to theory, must be 1° in about every 450 feet, until the cloud is dissipated. A similar phenomenon has been described by Sir John Herschell on the Table mountain at the Cape of Good Hope, and on a smaller scale, and less continuously, may be frequently witnessed on the ridges of the Himalaya.

Newara Eliya is a much rainier though not cloudier station than Ootacamund, being unprotected (unlike the latter) by a higher range to windward. The table in the Appendix shows that it rains on an average 195 days in the year, and the annual rainfall is nearly double that of Ootacamund. Indeed, the only season that can be called fine is restricted to the first four or four and a half months of the year, and even in these it rains on one day in three or four. October, which is the first month of the winter monsoon rainfall, is as rainy as June, July, and August, and even in November it rains on an average on two days out of three.

The mean temperature of the station is 59°, or 4° higher than that of Ootacamund, a difference due in a great measure to its lower elevation. In no month does the mean vary more than 2° above or below this; and the chief variation at different seasons is that the extreme temperatures of day and night differ more widely in February, March, and April than in other months of the year. January and February are the coolest months, and it is warmest in May, just before the setting in of the monsoon. But the highest temperature recorded has not exceeded 79° during the last 11 years, and it has varied only between that and 74° in different years. The lowest trustworthy reading on record is 32°, and the average minimum of the year 35°. Much lower readings were registered in 1875 and 1876, but there can be little doubt that for two and a half years an untrustworthy instrument was in use, and these accordingly I have rejected.

The diurnal range of the thermometer is not so great as at Ootacamund, but rather approaches that of Wellington. It varies from 12° in June and July to 23° or 24° in February, March, and April. It is a very damp station, the humidity of the air being between 80 and 90 per cent of saturation from May to the following January, and not falling to 70 on the mean of any month. A damp and cool but not cold atmosphere, varying but little throughout the year, is therefore the leading characteristic of the climate of Newara Eliya.

General Characters of Hill Climates.—Herewith I conclude my notice of the climates of the hill stations of India and Ceylon, of such of them at least as are represented in the Indian Meteorological Reports. We have seen that they present almost as great variety as do the plains below them. In all cases their atmosphere is cooler and damper than that of the neighbouring plains; but while those in the North-west Himalaya are subject to great vicissitudes of heat and cold, dryness and dampness in the course of the year, those of Southern India and Ceylon are comparatively uniform in these respects, and their fine clear season is shorter than at the northern stations, and by no means so dry. Of all the stations enumerated, Ootacamund is that which most frequently reminds one of England; but the resemblance is not very close, and consists chiefly in its coolness, its cloudy skies and moderate rainfall. The climate of Simla and the Kumaon hill stations is rather South European in type; and to Darjiling hardly any parallel can be found, except perhaps, and that imperfectly and in respect of dampness, in Cornwall and the south-west of Ireland. As to the stations at lower elevations, Shillong, Pachmarhi, Chikalda, and Mount Abu, they retain too much of a tropical character to be put in comparison with any European analogue. They are pleasant places of refuge from the scorching or suffocating heat of the plains, and seem cool by comparison; but, except in the cold season, when they are charming, they are scarcely such as an European, free to select his place of residence, would probably choose as an abode.

Climates of the Plains

I now pass to the description of the climates of the plains of India, beginning with the Northern and Western provinces, then tracing out the changes that gradually ensue as we proceed eastwards to Assam and southwards to Ceylon, and concluding with what is at present known of the climate of the Burman peninsula. The tabular summaries of temperature, humidity, etc., at a number of the more important stations given in the Appendix, will serve to illustrate these descriptions, and to afford exact data for many of the military cantonments and larger civil stations, in which the European population chiefly centres.

Punjab.—Of all the provinces of India, the Punjab, which is the most northerly, is also that in which the vicissitudes of climate are greatest. It has the coldest winters, and, together with Sind, the hottest summers; and it has been shown at page 12, in the illustrative instance of its capital, Lahore, that the irregular changes of temperature from day to day, which accompany changes of weather, are greater and much more frequent than in the damper province of Bengal, or, it might be added, in any other part of India. Next to Sind and Western Rajputana, which border it on the south, it is in great part also the driest province; but in this respect there are great differences in different parts of its area. Much of its southern and western districts is virtually desert, not owing to any want of natural fertility in the soil, for, wherever it can be brought under irrigation, no soil is more productive, but solely to the scantiness and uncertainty of the rainfall. But the tract, from 50 to 100 miles broad, that extends along the foot of the Himalaya, receives in most years a rainfall sufficient for successful agriculture; and as this occurs at two different seasons of the year, viz. from January to April and from July to September, two series of crops are raised annually, the spring crops or *rabi* consisting chiefly of wheat, barley, and gram (*cicer arietinum*), the autumn crops of millet, maize, various pulses, oil seeds, and sugar cane. Cotton, indigo, and some rice are also among the agricultural produce of the Punjab.

How the climate impresses the European is well shown in the following extracts translated from a lecture by the Rev. J. M. Merk, for 16 years a resident in the province. With some slight modifications, to be noticed in due place, the description will apply to the greater part of North-western India.

"Like the rest of India, the Punjab has really but three seasons: the summer or hot season, the rains, and the winter, which, in India, we speak of simply as the cold season. The hot season begins in April, but in March it is already so warm that barley and wheat ripen and are harvested. From April to June, as a rule, there is no rain. The west wind holds sway, and, blowing from the sandy wastes of the Indus region, is a veritable hot wind (Glutwind). A denizen of the temperate zone can hardly realise to himself the desiccating, truly scorching heat of this wind. When exposed to it, one may imagine one is facing an open furnace. The thermometer rises in the shade to over 50° C. (122° Fahr.) In order to enjoy fresh air at this season one must take exercise in the early dawn, between 4 and 5 in the morning; for no sooner has the sun risen than the heat sets in again. After 7 A.M., save of necessity, no European leaves his house, and should business oblige him to do so, he must protect himself from the sun with an umbrella and a thick head-covering. . . . At sunrise, or soon after 5 A.M., houses must be closed, only a small door being left open for communication with the outside. Thus the house of a European is more like a gloomy prison than an ordinary dwelling-house. So long as the hot winds blow strongly and steadily, rooms may still be kept in some measure cool by means of *tatties* or grass screens set up in front of the doorway, and continually sprinkled with water, or by the fan vanes of the so-called 'thermantidote,' which a servant keeps revolving and sprinkles with water; and at night the punkah is worked. Whoever cannot provide himself with these artificial cooling appliances must suffer the daily torment of insupportable exhausting heat. Man and beast languish and gasp for air, while, even in the house, the thermometer stands day and night between 35° and 45° C. (95° and 113° Fahr.) Little by little the European loses appetite and sleep; all power and energy forsake him. Vegetation suffers equally; almost all green things wither; the grass seems burnt up to the roots; bushes and trees seem moribund; the earth is as hard as a paved highway; the ground is seamed with cracks; and the whole landscape wears an aspect of barrenness and sadness. At length, in June, the hot winds cease to blow, and are followed by a calm; and now indeed the heat is truly fearful; tatties and thermantidotes avail nought; all things pine for the rains; but no rain, not even a shower, can one hope for, till the south and east winds shall have set in. And even then, the rains do not extend to the whole of the Punjab; Lahore has but little rain, Mooltan scarcely any; and the peasant of the Western Punjab is dependent entirely on artificial irrigation for the watering of his crops.

"The southerly and easterly winds bring first clouds and violent storms with heavy rain showers, which are repeated daily, or, at all events, every 2 or 3 days; and, finally, the rains which, in the Himalaya, set in at the beginning of July and cease at the end of August or in the middle of September. In July the trees begin a second time to burst into leaf; grass springs up once more, and soon a vegetation is developed that, fostered by warmth and moisture, is scarce to be kept within due bounds. The peasant now works hard at ploughing, sowing, and weeding his fields. Rice is sown in June, during the great heat; in September it is reaped; and within 2 months, maize is sown and harvested. . . .

"After from 4 to 6 weeks of heavy rain, often falling uninterruptedly for 2 or 3 days in succession, it clears up, and sometimes some weeks pass without further rain; after which, a week or two more of rainy weather bring the season to a close. Grateful as is the coolness brought by these showers, the more oppressively hot and sultry is it, when the rain ceases and holds off, if only for half a day. The atmosphere weighs on one like a heavy coverlet; and then comes the daily and nightly plague of mosquitoes. Insect and reptilian life is now active; of evenings it hums and buzzes and croaks all around; frogs make their way into the house, and with them more serious and unwelcome visitors, scorpions and snakes; for which reason it is unwise at this time of year to go about in the dark.

"One can hardly picture to oneself in our European climate how serious and disagreeable are the effects of excessive moisture, as experienced towards the end of the rains. Woodwork swells, and doors and windows can be fastened only with much difficulty. Shoes and all articles of leather become thickly coated with fungus, books become mouldy and worm-eaten, paper perishes, linen becomes damp in the presses, and despite the oppressive heat, one must often light a fire on the hearth, only to neutralise in some degree the influence of the damp.

"The period which immediately follows the rains up to October is the most unhealthy season in the year. Decaying vegetation under an ardent sun generates miasma, the consequences being fever, dysentery, and not unfrequently cholera. Towards the end of the rains one rejoices indeed to see the heavy dark clouds disappear, but the heat soon becomes once more so great that one longs for the cold season, and more than ever turns an anxious eye to the wind vane, watching for some sign of the cool westerly and northerly winds. With the beginning of October these winds set in steadily, clearing the skies, and now the blue firmament appears in all its splendour, so glorious in the torrid zone. . . . From October to Christmas, as a rule, the weather is clear and fine, the air is pure and most delicious, and one can hardly imagine a more charming climate ; but it must never be forgotten that an Indian sun shines overhead, and that even in the cold season one must never expose the unprotected head to its rays. The European now once more breathes freely, and it is a delight with the head well covered to move about in the open air. Fruit certainly is over, but the European now bethinks himself of his own garden, for it furnishes him with most of the vegetables of Europe, and our loveliest garden flowers bloom and delight his eye with their well-remembered forms. . . . For 5 or 6 weeks Europeans can work vigorously and with pleasure.

"In December and January the fire burns all day long on the hearth, and in the morning and evening is especially grateful. The nights are positively cold ; even on the plains, ice and hoar frost form, and near the ground the thermometer sometimes sinks to 5° C. (23° Fahr.) During the second half of the cold season we have in the Punjab a good deal of rain, without which indeed the barley and wheat harvest is but poor ; the pulses also require the winter rains. In February we have a short spring ; many trees unfold their leaves, and every bush furnishes its quota of flowery adornment. But this spring is of short duration, and in March it is already warm on the plains and the hot summer is at hand ; an occasional dust-storm, however, for a while keeps off the summer heat. A dust-storm is indeed in itself unpleasant, the air being so charged with dust as to bring about an Egyptian darkness, no matter what may be the hour of the day."

The above description gives a more vivid picture of the chief features of the climate and will better serve to bring them home to the mind of the English reader than any mere summary of the statistical data. Except, perhaps, that the extreme temperatures are slightly overstated,[1] it represents fairly enough the experience of any resident of Delhi or one of the large military cantonments along the northern border of the province, where, during the two months, July and August, the rainfall is such as the writer describes. But in the west and south rain is much more scanty and uncertain, and although the air is sufficiently damp to be very oppressive to the system and to render nugatory the usual artificial appliances for cooling the atmosphere of houses, even the temporary respite of heavy cooling rain is unfrequent throughout the summer monsoon. At Mooltan it rains on an average on but five days altogether in the two months, at Dera Ishmail Khan on seven only, and at Peshawar there are but six rainy days at this season. In 1877 no rain fell at Peshawar from the end of April till the end of September, and at Mooltan there were but four rainy days in the six months from May to November, and thirteen only in the whole year.

It now remains to notice more particularly the variations at some of the chief stations in different parts of the province. In its extreme north-western corner, on a plain surrounded by mountains and watered by the Cabul river above its junction with the Indus, is the important frontier station Peshawar. The military cantonment of Nowshera is near the eastern extremity of the same plain. This part of the province enjoys nearly seven months of fairly cool pleasant weather, for it is not until the latter part of April that it begins to be unpleasantly warm in the daytime, and even in that month the average maximum is only 85° in the shade, and the night minimum 60°. The most cloudy and rainy season here is not that of the summer monsoon, but from January to April. The rainfall of the former averages only about 4 inches, and, as in 1877, sometimes fails almost completely, while that of the latter is almost double as great. The winter is cloudy and cold, the temperature not infrequently falling to the freezing-point before sunrise ; while there is nothing to mitigate the ardour

[1] This is probably due to no inaccuracy on the part of the writer. Before 1876 the instruments at the Government observatories were so exposed as to register higher day and summer temperatures, and lower night and winter temperatures than really represented those of the air. See, on this point, the remark on page 15. In 1876 the system was altered, and the data quoted in the present work are taken from the subsequent registers.

of the summer sun, and during the three months, June to August, the average afternoon temperature is 100° and upwards. The highest temperatures reached in June are not less than in the more southerly parts of the Punjab, averaging 115°, and having been as high as 119° in the shade. Notwithstanding the proximity of the mountains, the atmosphere of Peshawar is as dry as that of Mooltan, in July and August indeed drier, and, needless to remark, the power of the sun is very great in these months. The average rainfall is only 13·5 inches in the year, and cultivation is therefore dependent on irrigation from hill streams and the Cabul river; from which latter source the city is well supplied.

Of the same general character and only less extreme is the climate of Rawalpindi, 90 miles to the east of Peshawar, on the plain above the Salt Range, 1700 feet above the sea, and within a few miles of the outermost hills of the Himalaya. The atmosphere of Rawalpindi is less dry than that of Peshawar at all times of the year, and the rainfall between twice and three times as great; that of July and August alone averaging more than 14 inches. There is more rain also in the winter and spring, and on rare occasions snow has been known to fall and accumulate to a depth of 4 inches in and around the station. This part of the Punjab, lying to the west of the Jhelum and north of the Salt Range, is a small plateau, elevated in places as much as 2000 feet above the sea, and there is a marked difference in its climate and vegetation as well as in the general aspect of the country, as compared with the plains to the east of that river.

About 110 miles to the south-east of Rawalpindi lies the important military station Sialkot. It is built on high ground to the east of the Chenab, about 15 miles from the foot of the hills. On the mean of the year this station is about 4° warmer than Rawalpindi, owing to the higher temperature of the winter, spring, and autumn, for in the monsoon months there is but little difference. The monsoon rainfall is about 6 inches heavier, but that of the winter and spring months about 2½ inches less, and also less frequent. Notwithstanding the somewhat heavier rainfall, the number of rainy days is little more than two-thirds of that at Rawalpindi. This cantonment is one of the healthiest in the Punjab.

Lahore, the capital of the province, lies 64 miles almost due south of Sialkot, at a greater distance from the hills and in a drier zone; indeed, on the border of the desert tract that occupies the high ground of the Bari Doab between the Ravi and the Sutlej. It is 2° or 3° warmer than Sialkot in almost every month of the year, a difference partly due to its more southern position, partly to its drier atmosphere. On the average of 10 years, the highest temperature registered in the year is 117°, and this varies but a few degrees from year to year, 120·3° being the highest yet recorded. In the winter, the shaded thermometer but seldom sinks to the freezing-point, 34° being the mean minimum of the year. The air is dry, the mean humidity of the year being only 50 per cent of saturation, and that of May, the driest month, 33. Both the summer and autumn rainfall is less than that of either Sialkot or Rawalpindi, the annual total being 22 inches, and on an average it rains on but 37 days in the year. The station is well planted with trees and gardens artificially irrigated, but a few miles to the south begins the desert tract occupied by Capparis bush with an occasional *babûl* tree (*Acacia Arabica*), which extends to Mooltan ; save where irrigated from the Bari Doab canal. Such is the case, for instance, in the Changa Manga forest, a tract of 32 square miles in extent, about midway between Lahore and Mooltan, where the annual rainfall is under 16 inches. In the course of 20 years this forest, partly consisting of *sissoo* (*Dalbergia sissoo*), has attained an average height of 40 feet or more, under the influence of a good annual watering from the canal.

90 miles to the east of Lahore, and somewhat farther south, is Ludhiana, 3 miles from the left bank of the Sutlej and 50 miles from the foot of the Himalaya. Although so much farther from the hills than Sialkot, the climate is very similar, the temperature of the two places scarcely differing more than a degree in any month, and the mean temperature of the year being the same. The winter is rather drier and less cloudy, and in most months of the year the rain is rather lighter ; in other respects there is little difference.

Delhi, in the extreme south-east of the Punjab, is not less than 130 miles distant from the Himalaya, but the climate, although dry, is much less so than at stations equally distant from the hills in the more westerly parts of the province. The winter is less cool, and the shaded thermometer, during the last 11 years, has never sunk to within 4° of the freezing-point. But the summer temperatures in May and June are nearly as high as in the drier parts of the province, the mean maximum of the year being 116°. In July and August, however, the heat is less intense (though scarcely less oppressive), the average afternoon maximum being 4° lower than at Lahore in these months. The winter rains are less frequent and heavy than at stations nearer the hills, while the summer rainfall is nearly the same as at Ludhiana. Although, therefore, the climate of Delhi is somewhat less extreme than that of places farther to the north-west, it is of the same general type.

Returning now westwards to the drier parts of the Punjab remote from the hills, we first come to Sirsa, 150 miles west-north-west from Delhi, and about the same distance south by east from Lahore, almost midway between the Jumna and Sutlej, and on the border of the Bickanir desert. Although the mean temperature of this place is only 1° lower than that of Delhi, yet the winter temperature is 3° lower, and the average minimum of the year 6° lower, while, owing to the scantiness and uncertainty of the monsoon rainfall, the day temperatures of July and August are 5° higher than at Delhi. The diurnal range of temperature throughout the year is also from 3° to 9° greater. When at its greatest, in November and December, there is a mean difference of 36° and 34° between sunrise and the afternoon. The humidity of the air is lower than at any station yet noticed, the annual mean being only 47 per cent of saturation, that of April and May 36 and 37, and that of July and August only 60. The winter as well as the summer rains are infrequent and precarious, and the annual average is barely 15 inches, of which about half falls in July and August.

The climate of Dera Ishmail Khan, in the extreme west of the province, on the right bank of the Indus, is, in some respects, still drier than that of Sirsa. Owing perhaps to the proximity of the river and the irrigation derived from it, the air of the observatory is slightly damper, but this difference is probably purely local. The rainfall is little more than half as great, viz. little over 8 inches in the year, of which no more than $3\frac{1}{2}$ inches fall in July and August. The mean winter temperature is 3° or 4° lower, the average minimum of the year being 31°, and in 1878 it fell to 26°. On the other hand, the intense summer heat is more prolonged, the afternoon maximum ranging above 100° from May to September, and having once reached 121·5°, which is nearly as high a reading as any yet recorded by a well-shaded verified thermometer in India.[1]

Lastly, Mooltan, 115 miles south of Dera Ishmail Khan, and a little above the junction of the Sutlej and Indus, has a still lower and much less frequent rainfall. Rain falls, on an average, on only 15 days in the year, a difference as compared with Dera Ishmail Khan, principally due to the infrequency of winter and spring rain. Both in 1875 and 1883 it rained on only 9 days in the whole year. The mean temperature of Mooltan is about 2° higher, but in the winter it has fallen to the freezing-point in 4 out of the last 11 years. In the early months of the year the rise is very rapid. Already in April the thermometer often rises to 100°, and in May the average maximum is 104°. June is the hottest month, as in other parts of the Punjab, but July is nearly as hot, and not till the end of October does the afternoon temperature range below 90°. Mooltan and Sirsa may be taken as representative of the climate of the Southern Punjab generally. In dryness and rainlessness they are surpassed, and that but little, only by Upper Sind.

Sind.—The great plain traversed by the Lower Indus from the Punjab frontier to the sea, is at once the driest and, as a whole, the hottest of all the provinces of India. What the Nile is to Egypt that is the Indus to Sind. Wherever the arid soil can be brought under the influence

[1] It was exceeded in 1886 at Pachpadra in Rajputana, where 123·1 was registered on the 25th May, and at Jacobabad, where 122·2 was recorded on the same day.

of its vivifying waters, there agriculture flourishes, and even extensive forests of *babùl* and tamarisk afford ample supplies of small timber, fuel and fodder. But beyond such limits all is a sandy or stony waste, thinly tenanted by a leafless bush of *Capparis aphylla* and the gray-leaved *Prosopis spicigera*, or, as in the Cutchee desert, a broad plain of hard bare clay, void of vegetable life. The aridity which, in a high degree, characterises the whole province, reaches its highest expression in Upper Sind, represented in our registers by the climate of Jacobabad. The atmosphere of this place shows a lower degree of humidity and a lower rainfall than even Sirsa, Dera Ishmail Khan, and Mooltan, and it is probable that even this would be found surpassed by that of the Pat or Cutchee desert to the west, since the country around Jacobabad, extending to about 4 miles beyond it on the west, is under field crops, and the station is shaded by magnificent trees fed by irrigation and subsoil infiltration from canals supplied by the river. As in other dry regions the vicissitudes of temperature are very great. Notwithstanding its more southerly latitude, the extreme winter cold is as great as at Mooltan and Sirsa; indeed rather greater; and the extreme summer temperature surpasses that of both. Nevertheless, even Upper Sind is not entirely uninfluenced by the dampness of the summer monsoon. May is still the driest month of the year, and although June is virtually rainless, the south and south-east winds that prevail from June to September, alternating with oppressive calms, bring a slight increase of moisture, lowering the temperature a few degrees in July and August. Rain is, of course, very rare. Indeed, the only occasions on which it falls are when, in the summer monsoon, a cyclonic storm travels from the eastward, and reaches Sind before it is broken up; or when, in the winter, a barometric depression forms in Sind or passes across it from Baluchistan. On such occasions the rainfall is sometimes very heavy, and owing to the stony impervious character of the soil, and the rapid drainage of the hills on its western frontier, heavy floods sometimes sweep over the low country, breaching the railway and causing serious damage. The climate of Lower Sind is slightly damper and less extreme, and the strong west-south-west winds that blow steadily from April to September, although they bring little or no rain, afford the means of ventilating dwelling-houses, which, as may be seen in the native city of Hyderabad, are surmounted by permanent wind-sails of masonry, to catch the current, and divert it through the apartments below.

On the coast and for a long distance inland, this south-west and west wind is remarkably strong and steady, blowing throughout June, July, and August with an average rate of from 20 to 25 miles an hour. The same directions prevail, more or less, through the greater part of the year. On the coast, north-east winds preponderate in December, but in no other month; and even at Hyderabad, where north winds prevail from November to January and north-east winds in February, from the middle of March to the end of October the south-west wind is steadily dominant. Its dryness shows, however, that it is only in part a sea wind; it is doubtless fed to some extent from the dry atmosphere of the Baluchistan highlands, and indeed it is to this source that the climate of Sind, and more particularly Upper Sind, owes its extreme aridity.

As already pointed out, this characteristic feature is displayed in the strongest degree by Jacobabad, which stands on the flat alluvial plain to the west of the Indus, 40 miles from its bank, and 65 miles from the Khirthar range of hills on the west, through which the Bolan Pass leads up to the highlands of Baluchistan. The mean temperature of Jacobabad is 2° higher than that of Mooltan, and that of the winter months from 2° to 4° higher; but this is due to the greater power of the sun in the daytime, since the night temperatures are quite as low on an average, and in extreme cases somewhat lower than at Mooltan. Consequently, the daily range of the thermometer is greater, and averages between 30° and 34° in 7 months of the year. There is but little cloud at any time of year; more from February to April than in any other month, but rain is very rare in the winter and spring, and the small quantity that falls in July and August, averaging 3 inches, makes up two-thirds of the annual total. Even in these months it rains on an average on only one day in ten, and this is very uncertain and irregular.

The climate of Hyderabad is more moderate. The city stands on an elevated platform of rock, 3 miles to the east of the Indus, commanding an extensive view over the alluvial flat of the river. It is thus well exposed to the south-west and west winds that sweep across the river, and the air is somewhat less dry than that of Jacobabad in every month except December. The rainfall of July and August is also double as great, but it amounts to only 6 inches on an average, and is equally precarious. In 1880, it rained on only five days during the whole year at both these places. From September to the following June, both inclusive, the rainfall is insignificant. The winter temperatures are from 4° to 6° higher than at Jacobabad, and the summer temperatures 5° or 6° lower. The difference of the summer day temperatures alone is even greater, amounting to not less than 8° in July and August.

Lastly, Kurrachee, within 3 miles of the sea, is a cooler station than either of the foregoing. Owing to its more southerly position, as well perhaps as its proximity to the sea, the winter mean temperature is 2° or 3° higher than at Hyderabad, and the night temperature, even in January, has never fallen to 41° in the last 11 years. In the summer a maximum reading of 117·6° has been registered, which is nearly as high as the highest at Hyderabad, but this was quite exceptional, and in most years, 102° to 106° is the highest reached; temperatures which are about the same as the highest of Calcutta. The mean afternoon maximum of the hottest months, May and June, is only 93°, those of Hyderabad being 106° and 103°, and those of Jacobabad 108° and 111°. The strong sea wind which, during the greater part of the year, blows night and day at Kurrachee, much mitigates the heat, and although somewhat relaxing, makes it far superior as a place of residence to other stations in Sind. This wind, however, brings but little rain; the annual average of the place, 7·8 inches, being slightly less than that of Hyderabad; and, as in other parts of Sind, trees will not grow unless artificially watered.

Rajputana.—Both geographically and in point of climate Rajputana is transitional between the very dry, almost rainless, valley of the Lower Indus and the plateau of Central India, where agriculture can be carried on extensively and regularly, at least in the summer months, without the aid of artificial irrigation. It consists of a congeries of native (and one small British) states, the more westerly of which, Bickanir, Jeysulmere, and Jodhpur, are included in what is known as the great Indian desert. The rainfall of this region varies from only 5 or 6 inches in the extreme west to a little under 20 inches in the year along the Aravali hills, and, being very precarious, is insufficient to ensure successful cultivation. For water for domestic purposes and the supply of their cattle, the population is in a great measure dependent on deep wells. In describing this region, Mr. W. T. Blanford remarks[1] that "the name 'Great Desert' usually conferred on it, conveys an imperfect idea, because the tract of country is neither barren nor uninhabited. It is covered with shrubs and bushes in general, and in places small trees are found. Moreover, though the population is thin, villages are scattered throughout, and immense herds of camels, cattle, sheep, and goats are kept and pastured. The desert is, in fact, a great sandy tract entirely destitute of streams of water and with but few hills of rock, and a large portion of its surface consists of sandhills of considerable height. When rain falls, crops of *bajri* (*Holeus spica*) are raised. When rain fails, the population lives principally on the milk of cattle and on imported grain." The boundary between this and the eastern states, Jeypore, Udaipur, and a number of small states, is the Aravali range, a series of rocky ridges rather than a continuous range, which begins in the historic ridge of Delhi and runs to the south-west past Jeypore and Ajmere to Mount Abu. The country to the east of this is the north-western portion of the tableland extending between the Jumna and the Nerbudda, the remainder of which will be noticed under the heading of Central India. Everywhere to the east of the Aravalis the rainfall is generally over 20 inches in the year; in the extreme south, in the little state of Banswarra, over 40 inches; and in the hills, wherever protected from the jungle fires, so destructive of all arborial vegetation, exist extensive tracts of forest, described by Sir D. Brandis as consisting of such trees as *Anogeissus pendula*, *Butea frondosa*, *Bombax malabaricum*, *Acacia catechu*, *Acacia Senegal*, *Dendrocalamus strictus* and *Tecoma undulata*, with a number of bushy shrubs, such as *Grewia populifolia*, two species of *Balsamodendron*, *Balanites Roxburghii*, *Salvadora Persica*, *Cordia Rothii*, *Calotropis procera*, and *Reptonia buxifolia*. Indeed, much of the preserved forest around

[1] *Journ. As. Soc. Bengal*, Vol. xlv., 1876, Part ii. p. 89.

Ajmere consists of bushes rather than trees. As in the Punjab, two series of crops are sown and reaped yearly; the summer or rain crops, consisting, as also there, of millets, cotton, and beans; the winter crops of wheat, barley, gram (*Cicer arietinum*), vetches, lentils, and peas. In the Indore State the opium poppy is also extensively grown. All through Rajputana the winter rainfall is lighter and less regular than in the Punjab, and in general the crops of that season require some artificial irrigation.

The division of the year into a hot season, a rainy season, and a cold season, holds good in Rajputana as in the Punjab; but the first and last are less strongly contrasted. The extreme summer temperatures are indeed but little below the highest registered in the Punjab, but the winter cold is less; the latter season is less cloudy and rainy, and even on the high ground of Eastern Rajputana, in but few places does the shaded thermometer ever sink to the freezing-point. In this part of the province, the rains, which set in in the latter part of June, bring about a rapid and considerable reduction of the heat, and the westerly wind that blows through the summer months from the coasts of Sind and Cutch affords an atmosphere less intolerable to the human system than the close, oppressive, and almost stagnant air then prevailing in the Punjab.

The climate is illustrated in the Appendix by the meteorological statistics of four stations, viz. Bickanir, which represents the northern part of the desert tract; Jeypore and Ajmere just to the east of the Aravalis, and Deesa, which, though administratively in the Bombay Presidency, is close to the southern border of Western Rajputana and, except that the rainfall is rather higher, may be taken as representative of that portion of the province. Neemuch also, in Central India, has a climate very similar to that of Southern Rajputana lying east of the Aravalis.

The climate of Bickanir is drier, and on the average of the year apparently 4° warmer than that of Sirsa, and the rainfall is 2 inches less; while the average number of rainy days is rather greater. In the dry season, from October to May, the mean temperature in many months is as much as 6° higher, and the mean night temperature from 7° to 11° higher. Such at least are the results shown by the registers; but the station is one of the least accessible in India, and it is possible, though there are no independent grounds for such an assumption, that the record may be affected by some undetected cause of error.

Ajmere and Jeypore are respectively 6° and 3° cooler than Bickanir, there being a mean difference of 3° between them, although they are less than 90 miles apart, and their difference of elevation is less than 200 feet. It is not clear to what circumstance this is due, unless indeed to the proximity of the large Anasagar tank and also of considerable hills, from which air cooled by radiation descends at night on the station of Ajmere; for it is noteworthy that the greatest differences between the two stations is in the night temperatures. The difference of the rainfall is little more than 1 inch. The mean humidity of the air is about the same at both; but while Jeypore is much the drier station in October and November, and still more so in April and May, it is very much damper in July. During the height of the monsoon, in July and August, it rains on about two days in five, on an average, at Jeypore and Ajmere; at Bickanir only on about one day in four. In the south-east, in Banswarra, the rainfall of this season is heavier and more frequent.

Except in the rains, the climate of Deesa is much drier than either of the above stations, as dry as that of Western Rajputana, the average humidity of the year being only 46 per cent of saturation; but the monsoon rains are rather heavier than at either Jeypore or Ajmere. The winter months are much less cool, owing partly to its lower elevation, partly to its more southern latitude; the mean temperature of January being 67°, while it is only 58° at Ajmere and 61° at Jeypore. The difference of their summer temperatures is not more than 1° or 2° in any month.

The North-West Provinces and Oudh. The Gangetic Plain.—These provinces include the whole of the alluvial plain of the Ganges and Jumna, eastwards from the latter river to the boundary of the Bengal Province of Behar. Together with Lower Bengal and Behar, they constitute the most fertile and highly-cultivated and also the most densely-populated provinces of India. Traversed by the great tributaries of the Ganges, two of which, the Gogra and the Gandak, rival in magnitude the parent stream and, like it, bring down the drainage of the Himalaya, depositing from their waters the silt that constitutes the soil of these plains, there would be no portion of them unfitted for agriculture, were it not that, in certain of the drier tracts, salts left by the constant evaporation of the subsoil waters have accumulated to such an extent as to be destructive of all vegetation, save only a coarse grass and a few worthless plants that thrive on saline soil. This class of soils, known by the local names of *reh* and *usar*, exists only where the ground is too high to be reached by the flooded rivers, and where, at the same time, the rainfall is too scanty to dissolve the saline efflorescence and carry it away in the surface drainage. With the exception of such tracts and the strip along the foot of the Himalaya, where the more highly sloping marginal zone consists of pebbles and coarse gravel with a filling of sand and earth, the whole of the vast plain is highly productive arable land. The sub-montane zone, known as the *bhàbar*, is essentially a forest tract, the chief home of the sàl tree (*Shorea robusta*) which furnishes one of the most valuable timbers of Northern India.

Although, in the opinion of Indian geologists, there is no reason to question that the whole of the Gangetic plain has been formed by the rivers in past ages with materials washed down from the hills, very much of it is now far above the reach of the highest floods, and in its turn is being eroded and its waste carried down to lower levels. Such portions of the plain are known as *bhàngar* land, and except where they can be irrigated from canals or local wells, are suited only for dry crops. The lower grounds, termed *Khàdar* lands, which flank the present rivers, lying in the depressions cut by the latter in the more ancient *bhàngar*, are more or less subject to inundation, and are especially, though by no means exclusively, the rice lands of the province. As in the Punjab and Rajputana, so also in the North-west Provinces, the summer or *Kharif* crops consist of millets, rice, oil seeds, sugar cane, and cotton, the winter or *rabi* crops of wheat, barley, pulse, gram, and, especially in the eastern districts, of the opium poppy, which is largely cultivated both here and in the adjacent province of Behar. Tobacco, safflower, mustard, and indigo, together with many kinds of vegetables, among which cabbages and cauliflowers rival or surpass in size those ordinarily raised in their European home, are also among the field or garden produce. But in fact there is scarcely any agricultural product of the summer of the temperate zone that cannot be raised on this fertile plain, which, besides being in great part irrigable from wells or from canals led off from the upper waters of the great rivers, has for the most part and in most seasons a rainfall varying from 30 to 50 inches. Only in parts of the *doab*, the tongue of land that separates the Jumna from the Ganges, and in a portion of Oudh to the north of the latter river, does the average rainfall fall below 30 inches in the year ; and it is in this portion of the province that the *reh* soil mentioned above is chiefly found.

It is, however, a meteorological peculiarity of the North-west Provinces, that, more than any other part of India, with the sole exception of the arid tract farther west, it is peculiarly subject to vicissitudes of the rainfall. Not to speak of the great famines of the last century, in 1804, 1834, 1838, 1861, 1868, and 1877, these provinces, ordinarily so fertile, have been repeatedly devastated by famine, resulting from the failure of the seasonal rains ; and whatever may have been the fate of other portions of the plain, the Gangetic doab has invariably been involved in the disaster in its most severe form. It was as a protection to this precarious tract that, after the destructive famine of 1834, the Ganges canal, the greatest work of the kind in India, and indeed in the world, was planned and executed under the direction of Sir Proby Cantley. The trunk canal commenced in 1848 was completed and opened in 1854; and although the districts which it supplies have since suffered from seasons of drought, almost if not quite as severe as that of 1834, in no year has the resulting suffering approached in intensity that of that memorable year.

The climate of the North-west Provinces differs from that of the Punjab in many important particulars. The cool season is less cold, less rainy and cloudy, and comes to an end in March, when strong hot winds set in from the west with great persistence, lasting well into May. These winds are diurnal, beginning 2 or 3 hours before noon, blowing through the heat of the day and lulling towards the evening or after sundown. They are intensely dry, a humidity as low as 6 per cent having sometimes been indicated by the hygrometer; but they are very healthy, and indeed the season when they prevail is, to acclimatised Europeans as well as to natives, the healthiest in the year. The heat is greatest at the end of May or the beginning of June; when, on certain days, the thermometer reaches as high a point as at most of the Punjab stations. As the hot winds flag, the clouds gather more frequently, and occasionally the heat is temporarily relieved by an afternoon dust-storm. Sometimes, and especially in the eastern districts and near the hills, the dust-storm terminates in rain; and in some seasons, after a light easterly wind for a day or two has held possession of the plain, one or two days of steady rain follow. This is known as the *chhoti barsàt* or "little rains." It has, however, by no means the regularity frequently attributed to it in the popular belief.

The rains set in some time in the latter half of June; rarely earlier. In some years they begin simultaneously all over the province; in others, first of all in the eastern districts. While they last, it rains more often than in the Punjab; but rain is by no means continuous for more than a day or two at a time, and these rainy spells are separated by intervals of finer but steamy and oppressive weather. At Benares there are, on an average, 26 rainless days in the two months July and August; and in the drier parts of the province, as at Agra, rain falls on only about one day in three. In most seasons there occur intervals of a week or more during which rain is almost or quite suspended, such interruptions being termed "breaks" in the rains, and in certain years these breaks last for several weeks, and westerly winds set in, less hot and dry but more oppressive than the hot winds of the spring. In 1877 such was the case almost throughout the summer monsoon. Save, however, in such exceptional seasons of drought, the temperature of July and August is several degrees lower than in the Punjab.

The rains cease, as a rule, in September, earlier or later in different years. Generally they last a week or more longer in the eastern than in the western districts. A few weeks of close and warmer weather follow, but under the clear skies of the lengthening nights the temperature gradually falls; and if, as sometimes happens, a late and final fall of rain comes at the end of the month or in October, its cooling effect is rapid and permanent. Light airs begin to move from the west and gradually strengthen till they become the steady cool wind of the winter months. As a rule, the weather remains fine and clear, with cloudless skies, till near Christmas, when the heavens become covered with a sheet of filmy cloud, and after some days of rather close weather, and either a still atmosphere or light southerly and easterly winds, the first rain of the cold weather falls, followed by a great fall of temperature and a stronger and cooler wind from the west. These falls may be repeated once or twice or oftener in January and February, but they are not lasting, and the rain is lighter than that of the summer months. As already mentioned, the rainfall of this season is less frequent than in the Punjab, except on the hills of Kumaon and Garhwal.

The meteorological statistics of seven stations are given in the Appendix, illustrative of the climate of the North-west Provinces and Oudh.

The first of these, Dehra, lies close to the foot of the Himalaya, in the north-west corner of the province, or rather between the Himalaya and a range of low hills, the Siwaliks, which run parallel to it at a distance of 8 miles, constituting a sort of outwork to the main chain. The intervening strip of plain, 2000 feet or more above the sea, is known as the Dehra Dùn, and enjoys a more temperate climate than the great plain to the south of the Siwaliks. Tea is grown here in tracts cleared from the forest, and some Europeans have chosen it as a permanent place of residence. The annual rainfall is as much as 76 inches, two-thirds of which falls in the months of July and August. The mean annual temperature is 71°, that of June 84°, and the average afternoon temperature in May and June 93° and 94°, and at the annual maximum 105°, which is fully 10° lower than on most parts of the western half of the Gangetic plain. From December to February the mean of the day is from 55° to 57°.

Roorkee lies 15 miles south of the Siwaliks, and nearly 900 feet above the sea. It is the headquarters of the engineering staff of the Ganges canal. The mean annual temperature is 75°, and the average rainfall 42 inches, which is higher than on the plains farther south.

Meerut is a large military and civil station on the Gangetic doab, 40 miles north-east of Delhi, and 75 miles from the hills; with a rainfall of only 28 inches in the year, and a mean annual temperature of 76°; and Agra, 160 miles farther south, on the high southern bank of the Jumna, has again a slightly lower rainfall, 26 inches, and a mean temperature of 79°. These two stations represent the climate of the drier western districts.

The next station, Lucknow, stands in the very heart of the Gangetic plain, about 100 miles from the Himalaya, and at an equal distance from the high ground south of the Jumna. It is an important military and civil station, and the former capital of Oudh. It is warmer than any of the foregoing, except Agra, the mean annual temperature being 78°, but the rainfall is higher, viz. 37 inches. Allahabad, the seat of the local government, is again farther east, at the confluence of the Ganges and Jumna. The European part of the station is on high *bhàngar* land, and the climate is but little affected by the proximity of the two great rivers. The mean temperature is the same as that of Lucknow, 78°, but the highest temperature reached in the year appears to be a little higher, viz. 116°. In the winter it never sinks to the freezing-point. The average rainfall is between 37 and 38 inches.

Finally, Benares, 75 miles farther east, and not far from the eastern boundary of the province, also situated on the bank of the Ganges, has a very similar climate, and a slightly higher rainfall, viz. 39·6 inches.

The Central Indian Plateau.—The high ground that extends to the south of the Jumna and Ganges, almost from the banks of these rivers to the valleys of the Nerbudda and Sone, includes, besides the native states of the Central Indian Agency, the districts of Jhansi and Lalitpur, under the government of the North-west Provinces, and that part of the Central Provinces formerly known as the Saugor and Nerbudda territories. It forms a low tableland, sloping from south to north, and terminating in the former direction, abruptly, within a few miles of the Nerbudda, in the escarpment which, in modern maps, bears the name of the Vindhyan range. The eastern half consists of sandstones and other sedimentary rocks, the successive outcrops of which form a series of parallel ranges running north-east and south-west; while the western portion is covered with a sheet of volcanic rock, the northern extension of the great trap formation of the Deccan. Between them, the northern portion of the plateau, in Bandelkhand, consists of an extensive exposure of crystalline rocks. The highest part of the plateau is the south-western, in Malwa, where it is more than 2000 feet above the sea, and where the productive black soil that covers the basaltic rocks yields rich crops of wheat and the opium poppy. Of the drug manufactured from this last, Malwa is one of the most productive provinces in India.

The climate of the Central Indian tableland is drier than that of the Gangetic plain, and except from the beginning of April to the setting in of the rainy season in June, the more elevated tracts are comparatively cool. In these months its surface is swept very persistently by dry hot west winds; but, so soon as the rains set in, the temperature is greatly reduced, and although the rainfall is moderate (between 30 and 50 inches) the atmosphere is by no means so still and oppressive as in most parts of the Gangetic plain. To acclimatised Europeans, such stations as Saugor, Jubbulpore, and Indore are not unpleasant places of residence during the greater part of the year. The cool season lasts about four months, from the beginning of November to the beginning of March. Less cold than in the Punjab, it is also less damp and cloudy; and rain is infrequent, there being not more than one or two rainy days on an average in the month. As a rule, bright, clear, pleasant weather lasts uninterruptedly for weeks together at this season. In these cool months, the eastern part of the plateau, the tract between Jubbulpore and Benares, is cooler than the central and western districts, owing, as it would seem, to the greater prevalence of north-east winds. From November to March the difference of temperature between Jubbulpore, to the north of the Satpura range, and Nagpur, to the south, amounts to as much as 6° or 7°, the distance between them being not more than 160 miles, and the difference of elevation little more than 300 feet.

The meteorological statistics of six stations on the Central

Indian plateau are given in the Appendix, in illustration of its climate, viz. Neemuch and Indore in the west, Jhansi and Saugor in the central, and Sutna and Jubbulpore in the eastern division of the area. Politically, Neemuch, Indore, and Sutna are in the native states of Gwalior, Indore, and Rewah, Jhansi in the North-west Provinces, and Saugor and Jubbulpore in the northern districts of the Central Provinces.

Neemuch and Indore are respectively situated 1600 and 1800 feet above the sea, and have mean annual temperatures of 75° and 74°, and an annual range of 73° and 68°. The temperature of the coolest month is 62°, that of the warmest, May, 87° or 88°. In the winter, the temperature rarely falls below 40°, and 108° and 111° are the average highest readings of the year. These are from 4° to 7° lower than those reached at stations on the Gangetic plain, farther north. In the rains, the mean morning temperatures range from 71 to 74 in different months, and those of the afternoon are 83° or 84° at Indore, 85° to 87° at Neemuch. The average annual rainfall is 33 inches at Neemuch, 36 inches at Indore, and the number of rainy days 57 at the former and 76 at the latter station.

Jhansi, although more northerly than either of the above, but at a lower elevation (850 feet), being surrounded by numerous rocky hills which absorb the sun's heat by day and radiate it at night, is a much warmer station, especially in the hot season. The mean temperature of the year is 79°, that of January 63°, and that of May, the warmest month, 95°. In June, the morning minimum averages 84°, and the afternoon maximum 104°. In May they are respectively 81° and 107°. The highest reading of the year averages 115°, the lowest 43°, and the highest yet recorded is 117·9°. The annual rainfall is about 36 inches.

Saugor is a more elevated station, 1750 feet above the sea, but in a slight depression, on the margin of a large artificial tank, which probably aids in mitigating the temperature in the hot season. The mean temperature of the year is 76°; that of January 63°, and that of May 89°; therefore 6° lower than that of the same month at Jhansi. The morning minimum in June is 7° lower. The rainfall is 46 inches in the year and the average number of rainy days 70. The mean humidity of the air is only 50 per cent of saturation, and that of the driest months, April and May, only 28.

Sutna is a small station on the railway between Allahabad and Jubbulpore, and except as the headquarters of the Political Agency of the eastern states of Central India, has little importance. It is warmer than Saugor from May to October, being at a lower elevation (1040 feet), but cooler in the winter and early spring, owing to the circumstances already noticed as affecting the eastern portion of the plateau.

Finally, Jubbulpore is an important station at the junction of the East India and Great Peninsula railways, 1340 feet above the sea, and with a considerable European population. Although there are some low rocky hills round about it, it is open to the winds from all quarters, and is comparatively a cool station. The mean temperature of the year is 75°, that of December 61°, and that of May 90°. The highest in the year varies between 106° and 113°; the lowest between 32° and 38°. The rainfall is higher than at stations to the north and west, viz. nearly 54 inches in the year; and in July, the wettest month, it rains on an average on 22 days. In the whole year there are 80 rainy days. Here, as in most parts of Central India, the rains last from the middle of June to the middle or latter part of September, but there is little rain from November to May inclusive. In the latter month and June an occasional thunder-storm precedes the rains. Jubbulpore has a damper atmosphere than any of the places previously noticed, the mean humidity being 57 per cent of saturation. This difference manifests itself at all times of year, but especially in the cool season. It is not due to any peculiarity in the position of the observatory, which stands on an open grassy plain, not in the neighbourhood of any swamp or water expanse, and has been, ever since its establishment, under the competent and careful superintendence of Dr. Rice, now Deputy Surgeon-General of the North-west Provinces.

Behar and Chutia Nagpur.—Behar includes that portion of the Gangetic plain which lies between the North-west Provinces and Bengal; including a considerable alluvial tract to the south of the river, which is crossed by its great southern tributary, the Sone, and a tract of high land between this last and the northern half of the Gangetic delta. Towards the south, this rises gradually to the Chutia Nagpur plateau, which is virtually an extension of the belt of hilly country that stretches across India south of the Sone and Nerbudda, and is generally known as the Satpura range. At Hazaribagh and Ranchi, the level of the plateau is about 2000 feet above the sea, and has a few hills of much greater elevation, the most prominent of which is Parasnath to the east of Hazaribagh.

Behar is a rich and densely-peopled province, highly cultivated and enjoying a higher average rainfall than the more westerly extension of the Gangetic plain, the part already described under the North-west Provinces and Oudh. In addition to the ordinary crops, such as wheat, rice, millet, pulses, oil seeds, etc., it is the chief seat of the indigo cultivation, and, in Bengal, of the opium poppy, from which the drug is manufactured at the Government manufactory at Patna. Besides many smaller streams, it is traversed by three great Himalayan tributaries of the Ganges, the Gandak, the Gogra, and the Kosi, and to the south of the main river by its great Central Indian tributary the Sone; and in most years the water supplies from these rivers and the rainfall render Behar one of the richest agricultural provinces in India. But, like the North-west Provinces, it is subject to the occasional failure of the rains, and in such seasons it has frequently been the seat of disastrous dearth and famine; the more disastrous, owing to the density of the population which its very fertility has fostered.

The climate of Behar is more akin to that of the North-west Provinces than to that of Lower Bengal, but damper, especially in the eastern districts, and to a certain extent transitional between the two. The winter months are cool and bracing, and the rainfall of this season is less than on the plain farther west. Hot west winds set in in March, and last into April and sometimes into May; but in April and May light damp easterly winds blow intermittently to the north of the Ganges, more particularly in the neighbourhood of the hills; and thunder-storms accompanied with heavy rain, and sometimes with hail, take the place of the rainless or almost rainless dust-storms of the Upper Provinces at this season. The rains set in in June, about the same time as in Lower Bengal, and last to the end of September or the beginning of October, in which month the rainfall of the province amounts, on an average, to between 2 and 3 inches, and to upwards of 4 inches in the most easterly districts and in the neighbourhood of the northern hills.

The Chutia Nagpur plateau bears much the same relation to Northern Behar as does that of Central India to the North-west Provinces and Oudh. Less than half a century ago it was a wild forest-clad tract thinly peopled by aboriginal tribes; but it includes all the valuable coalfields of Bengal, and the opening up of these, together with the extension of the railway, and the spread of cultivation which has followed in the track of the pioneer of commerce, has now effected the denudation of all but the more remote and hilly parts of the province, and all over the eastern and northern parts of the plateau, all large timber has long since disappeared. In many parts there remains little but an occasional sacred grove of large trees, and patches of sàl coppice, intermingled with trees of less value as timber, but some of which yield other valuable products, such as the flowers of the *Mahowa* (*Bassia latifolia*), largely used by the natives as food and for the distillation of a weak and nauseous spirit; also lac gum and dye, and the cocoons of the *tussch* silk moth. The field produce of Chutia Nagpur is similar to that of the highlands farther west, wheat being largely grown in the cold season.

Chutia Nagpur has a more copious rainfall than most parts of Central India; in general, but little short of 50 inches, and in some places exceeding that amount. But it is swept by dry west winds in the spring months, and hence the attempts that have been made to cultivate the tea plant in the neighbourhood of Hazaribagh have not been attended with success. The higher parts of the plateau enjoy a dry and pleasant climate; and, in the rainy season, Hazaribagh, were it more accessible and provided with the requisite accommodation, would afford a grateful place of temporary refuge from the saturated and oppressive atmosphere of Lower Bengal.

In the Appendix are given the statistics of temperature, humidity and rainfall of two stations in Northern Behar, two in Southern Behar, and one in Chutia Nagpur, viz. Durbhanga and Purneah on the alluvial plain north of the Ganges, Patna on the southern bank of the river, Gya on somewhat higher and rocky ground between the Sone and the uplands of Chutia Nagpur, and Hazaribagh on one of the highest parts of the Chutia Nagpur plateau.

In point of temperature, Durbhanga and Purneah are very similar. They have a mean annual temperature of 77°; in January of 62°, and from April to September a nearly uniform mean temperature between

83° and 85°, but with a greater range in April and May than in the subsequent months. At Purneah, but not at Durbhanga, April is on the whole the warmest month, the heat of May being slightly mitigated by the damp easterly winds that are felt intermittently several weeks before the rains, and bring an occasional thunder-storm. But the highest afternoon temperatures are generally reached in May, during intervals of fine weather. The mean maximum of the year is 105°, and no reading so high as 110° has been recorded at either station during the last 9 or 10 years. The mean lowest of the year is 45° at Durbhanga and 40° at Purneah ; a difference probably depending on some local peculiarity of position, possibly on the fact that the Durbhanga observatory is in the immediate neighbourhood of a large tank, while that of Purneah is far from any water expanse. It may be owing to the same fact that the diurnal range of temperature at the Durbhanga observatory, at all times of the year, is several degrees less than at Purneah. The mean annual humidity, however, is the same at both places, viz. 71 per cent of saturation. The rainfall of Purneah, owing to its more easterly position, is one-fourth greater than that of Durbhanga, the former being $64\frac{1}{2}$ inches, the latter $47\frac{1}{2}$. As already remarked, the rainy season begins earlier and ends some weeks later than in the Upper Provinces, and the September rainfall is but little if at all less than that of July and August. From the middle of October to the end of May, at Durbhanga, the average monthly rainfall varies from 0·1 to 0·6 inch only ; but at Purneah the fall of April already exceeds 1 inch, and that of May averages between 3 and 4 inches, being distributed over 6 rainy days. On the mean of the year it rains on 66 days at Durbhanga, on 88 at Purneah, and in the wettest of the last 15 years there were 86 rainy days at the former and 112 at the latter. In respect of rainfall, therefore, there is a considerable difference between the eastern and western districts of Behar.

Patna, which lies somewhat farther west than Durbhanga and more to the south, has a drier climate than either of the above stations and a greater diurnal range of temperature. The mean temperature of the year is 78°; that of January 61°, and that of May 89°. In both April and May the highest afternoon readings average 100° or more, and the highest of the year has varied between 107° and 113° during the last 11 years. In the coldest season it has not fallen below 36°. The rainfall averages nearly 43 inches in the year, which is somewhat lower than that of the plain north of the Ganges. July is the wettest month, and November and December the finest. The September rainfall is under 8 inches, that of October between 2 and 3 inches. On an average it rains on 71 days in the year.

Gya, owing perhaps to its position on high ground in the neighbourhood of some rocky hills, is a very warm station. Its mean annual temperature is 79°; that of January 64°, and that of May 92°, or 3° higher than that of Patna, only 60 miles farther north. The average afternoon temperature of this latter month is 104°, and the highest of the year has varied from 109° to 116°. The lowest reading recorded has been but little under 40°. The average rainfall is only 41 inches, and its distribution throughout the year is nearly the same as at Patna, five-sixths of the whole falling between June and September and between 2 and 3 inches in October; after which follow two and a half months of almost rainless weather. In January and February the winter rains of Northern India contribute about $1\frac{1}{4}$ inches, after which follow March and April with dry winds and a smaller rainfall. In May the average fall is rather over 1 inch, in June over 6 inches, and the heaviest (nearly 12 inches) is in July. The average number of rainy days is 75 in the year.

Hazaribagh, at 2000 feet above the sea, is much cooler, the mean temperature of the year being only 74°, or 5° lower than at Gya. In December and January it is 61°, the mean of the minima 50° and 51°, and the lowest reading in the year varies between 39° and 46°. The mean temperature of May, the hottest month, is only 85°, the same as at Durbhanga, and the highest in the year varies from 102° to 109°. During the spring months, hot west winds blow at Hazaribagh as they do at stations at lower levels, but, so soon as the rains set in, the afternoon temperature falls 13° or 14°, and from that time forth the mean temperature of the day averages no more than 78°, while the movement of the air across the plateau renders the station cool and pleasant, affording a great contrast to the comparatively stagnant and oppressive atmosphere of the lower plains, and especially of Bengal. The rainfall is moderate, amounting to about 13 inches in each of the months, July and August, and about 8 inches in June and September. In October the temperature falls to 74° and the morning minimum to 66°, and from this time to March the climate is very pleasant and healthy.

Bengal and Orissa.—These provinces include the whole of the alluvial plain that skirts the northern shore of the Bay of Bengal, that of Bengal being the combined deltas of the Ganges and Brahmaputra, together with the extension of the plain up to the foot of the northern mountains and the gates of Assam; that of Orissa the deltas of the Mahanadi and of a number of smaller rivers that drain the highlands of Chutia Nagpur. Intersected by innumerable river channels and abounding in swamps, while open to the damp winds from the bay, which begin to blow on the coast as early as February and gradually penetrate farther inland with the increasing heat, the climate of Bengal is as characteristically damp and relaxing as that of North-Western

India is the reverse. The dry westerly winds that play so great a part in the meteorology of the Upper Provinces are felt only occasionally and intermittently in the province of Bengal, during the spring months and chiefly in the warmest hours of the day, and even then with a reduced temperature and of a less parching character, owing to the moisture taken up from the surface over which they blow.

The customary division of the year into three seasons, the cool season, the hot season, and the rains, holds good in Bengal as in the more westerly provinces, but the first is shorter and less bracing, and the heat of the second, if less intense, owing to the greater dampness of the air, is on this account, perhaps, more trying to the European constitution. The rains are also longer and more copious.

At Calcutta, the cool weather scarcely sets in before the second week of November, and lasts only to the middle or latter part of February. For ten or twelve weeks the weather is delightful, pleasantly cool in the daytime, and in the evening sufficiently cold to make a fire agreeable, though perhaps hardly indispensable to comfort. But the dampness of the climate manifests itself at night in frequent fogs on the river and the low grounds around; and in the native part of the town and the crowded suburbs, the smoke from the huts condenses the moisture of the air and enshrouds the habitations in a veil of fog that remains suspended in the still atmosphere. About the end of February the days begin to be appreciably warmer, the sun's power greater and his glare more intense, and in March it is sufficiently warm in the house to require that the punkah be set in movement once more after its three months' holiday. But the nights, or at least the hours after midnight, are still comparatively cool, and as the air is as yet not highly charged with moisture, to an acclimatised European the heat is by no means oppressive. During this month and April the rise of temperature is very rapid, and by the end of the latter month the afternoon readings of the thermometer are often as high as any recorded during the year.

In December any rain is exceptional. In January and

February it occurs less regularly and frequently than in Upper India, but there are generally two or three rainy days in these months, preceded by some days of close cloudy weather with light southerly winds; and when the weather clears a great fall of temperature follows and a renewal of the cold season. In March and April rain becomes more frequent, but in the shape of the thunder squalls known as nor'-westers which, as a rule, come on towards the evening, and are often preceded by a dust-storm. Their general character has been described in a previous part of this work. For a graphic but perhaps somewhat exaggerated account of one of these little storms the reader may be referred to M. Rousselet's well-known work on India, where, however, the writer appears to be under the mistaken impression that he had experienced a cyclone. These squalls are more frequent in May, and sometimes usher in a day or two of continuous rain; but, more frequently, an hour or two expends their force and they are followed by a refreshing night and a somewhat cooler day. Now and then, at intervals of some years, a cyclone may pass over Bengal in May or the early part of June, but these more formidable visitors are perhaps rather to be expected at the end than at the beginning of the monsoon, and happily, at any season, are to be reckoned as meteorological rarities in Bengal.

An advantage which Calcutta enjoys in the hot season, and which is denied to places much farther inland, is the southerly breeze which, at the close of a hot day, often blows up from the wide estuary of the Hooghly, bringing some hours of agreeable relief. It does not last beyond midnight, but it mitigates the heat of the dinner hour, and renders the southern verandahs of the large Chowringhee houses pleasant places for the post-prandial lounge.

At length, in the early part of June, the clouds gather more thickly, while the barometer falls to a lower point than it has reached since the beginning of the year; and in the first or second week, heavy and continuous rain ushers in the monsoon. This first burst of the rains usually accompanies a cyclonic storm, formed either at the head of the bay or over the delta itself. As has been explained in a previous part of this work, such storms are not attended with very strong winds, at least on land, though the weather may be stormy at sea; and the barometric depression at their centre does not exceed two or three-tenths of an inch. The first onset generally carries the rains to the greater part of Bengal, and sometimes, but not always, to Behar. As a rule, the rainfall does not penetrate to the Upper Provinces till some days or even weeks later. Its immediate effect is a great fall of the day temperature; and the comparative coolness, supervening on many weeks of close oppressive weather, brings a sense of relief. Bursts of rain of a similar character, alternating with sporadic showers and an occasional rainless interval, rarely lasting more than a day or two, follow in succession through July and August. The air is saturated with vapour; vegetation grows apace, and indoors and out of doors every absorbent material reeks with moisture; but so long as the rainfall is abundant, and the intervals of its suspension short, the climate, if not exactly pleasant, is not very oppressive nor notably unhealthy. When, however, in September, the rainless intervals become longer, and the day temperature begins to rise, while the air, still highly charged with moisture, is almost motionless, the relaxed energy of the human system fairly rebels against this further trial of its endurance, and all who are not compelled by their avocations to remain at their post hasten to escape to the temporary refuge of a hill station. September and October are thus the most trying and unhealthy season of the year; and in Bengal it is not until the end of October or the early part of November that an appreciable fall of temperature brings relief.

Such, according to the author's experience of many years, is the climate of Calcutta, and the description may serve as fairly representative of that of the greater part of Bengal. Farther north indeed, and on the higher ground to the west of the delta, the hot season is somewhat drier and its temperature rather higher; and in Eastern Bengal the differences are of the opposite character. In Orissa the temperature is higher throughout the year, and in other respects the climate differs in the same manner as in Western Bengal.

The agriculture of Bengal is determined by the watery character of its surface. Rice, which in the Upper Provinces is of subordinate importance and an article of luxury, in Bengal is the universal crop and the food of the people, and two sowings are reaped during the year, viz. the *Aūs* crop, sown in the early spring and harvested in July and August, and the *Amun* crop, which is sown in April, transplanted in August, and reaped in the cold season. Oil seeds and dàl are also largely grown, and in Northern Bengal tobacco and jute-fibre are important crops, widely cultivated for export. Among the minor but still important articles of produce may be mentioned sugar-cane, betel pepper, and locally, the mulberry for the culture of silkworms, hemp, and cocoanuts, which last never grow at any great distance from the sea. The only important forests on the plains of Bengal are those of the Sunderbuns, the marshy islands along the seaward face of the delta, of which the most important tree is the *Sūndri*, producing a tough wood, much used in carriage and cart building.

In the Appendix are given the tabular statistics of Burdwan and Berhampore, representing the western margin of the alluvial tract, the former being the more southerly. These lie in the drier portion of the province; Calcutta and Jessore, which are more central; and Dacca and Chittagong in the eastern and damper districts. Northern Bengal is not represented by any station, but its climate may be regarded as intermediate between that of Purneah in Behar and that of Dhubri, near the entrance of Assam. The province of Orissa is illustrated by Cuttack.

The mean temperature of Bengal is nearly uniform between 77° and 78°; that of January 65° or 66°, Chittagong being a degree warmer. April and May are about equally hot, viz. 85° or 86° in the western and drier districts, and between 81° and 83° at Dacca and Chittagong. In the former, the mean temperature falls 1° or 2° on the advent of the rainy season; at the latter it remains nearly uniform from April to the end of September, the night temperature rising in the same measure as the day temperature falls, until the daily range is reduced by one-half, more or less; and not until the middle or latter part of October does a general fall set in with the light northerly winds that succeed the rains. The highest temperature of the year occurs in April, or, more frequently, in May; and varies between 106° and 111° at Burdwan, 105° and 113° at Berhampore, 97° and 106° at Calcutta, 99° and 108° at Jessore, 94° and 106° at Dacca, and 91° and 99° at Chittagong. In the cool season the lowest temperature occurs as a rule in January, sometimes in February, and varies between 44° and 51° at Burdwan and Berhampore, 45° and 52° at Calcutta, Dacca, and Chittagong, and 39° and 48° at Jessore. The diurnal range of temperature in the drier seasons of the year is less in Bengal than in the Upper Provinces, not exceeding 22° in any month at Calcutta, 23° at Chittagong, and 25° or 26° at the other stations, except Berhampore, where, in March, it averages 28°.

The humidity of the air averages 69 per cent of saturation at the driest and most westerly station, Burdwan; 74 at Berhampore, 75 at Jessore, 77 at Dacca, 78 at Calcutta, and 80 at Chittagong; and in the driest month, either February or March, is as high as 55 at Burdwan, 57 at Berhampore, 62 at Jessore, 65 at Dacca, 69 at Calcutta, and 70 at Chittagong. Calcutta, owing probably to the low level of the site, and the proximity of a large sheet of water, the Salt Lake, which skirts its eastern suburbs, is therefore damper than most other stations in the western and central part of the delta, but the dampest climate is that of Chittagong. At the height of the rains the humidity of the air averages 87 per cent and upwards at all stations except Burdwan; at Calcutta as much as 89 in August.

The rainfall has a similar distribution, being heavier in the eastern districts and the marginal portion of the delta, and heaviest at Chittagong. The extremes are Berhampore and Burdwan with 55 and 58 inches respectively, and Chittagong with 106 inches. Calcutta has an annual rainfall of 65½ inches, Jessore of 68 inches, and Dacca of 74 inches. In the wettest year on record (1871) the rainfall of Calcutta amounted to over 93 inches, and in the driest (1837) to only 43½ inches. The number of rainy days is less variable than the amount of the fall in different parts of the province. It amounts on an average to 91 at Berhampore, 105 at Burdwan, 110 at Dacca, 118 at Calcutta, 121 at Jessore, and 122 at Chittagong.

The mean temperature of Cuttack is 81°, 3° or 4° warmer than Bengal; that of May, here the warmest month, 89°; and the highest temperature of the year has varied during the last 10 years between 106° and 118°. Owing to the lightness of the spring rainfall, and the prevalence of dry westerly winds, the hot season is here, therefore, for the time, as intense as in the Upper Provinces. December is the coolest month, and the lowest winter temperatures have never fallen below 49°. The mean humidity of the year is about the same as at Burdwan, but in the driest month does not average below 62, and in the dampest does not exceed 82. It is therefore more uniform at different seasons than in the province of Bengal. This peculiarity is probably local, owing to the position of the station, at a low level, between two rivers. The mean annual rainfall is $57\frac{1}{2}$ inches, distributed over 106 rainy days. The rains are rather more prolonged in Orissa than in most parts of Bengal, as this province receives more of the October rainfall, when the winds become easterly in the north of the bay and the season of heavy rain is setting in on the Madras coast. On the other hand, the rainfall of the spring season is lighter than in Bengal.

Assam and Cachar.—This province, which, including Sylhet, lies to the east of Bengal, and to the north of Burmah, Manipur, and Tipperah, consists of two alluvial plains, that of the Brahmaputra and its affluents, and that of the Baràk; together with the intervening hills, different sections of which, named from the several indigenous tribes that occupy them, are distinguished on our maps as the Garo, Khasi, and Naga hills. The valley of the Brahmaputra, which constituted the former kingdom of Assam, is long and narrow, and is nearly divided into two parts, Upper and Lower Assam, by the Mekhir hills, an isolated and independent group which stands out midway in the valley between the Naga hills and the Himalaya. The valley of the Baràk, forming the province of Cachar and the plain of Sylhet, which opens out to the west of the former and is traversed by its two branches, the Soorma and Kushiyari, is much shorter, and occupies the interval between the Khasi hills on the north and the low hills of Tipperah and the Looshaie country on the south.

The plain of Sylhet has long been cleared and cultivated. Many parts of it are low, and in the rainy season are flooded to a considerable depth by the torrential drainage from the hills around. Nevertheless large tracts are cultivated with rice, and the long low hill spurs that run out from the southern hills, the low hillocks that dot the plain along the bases of these and the northern hills, and the higher parts of the plain itself, together with similar tracts in Cachar, are occupied by tea gardens, established for the most part on clearings made in the forest during the last 40 years.

In the Assam valley tea cultivation is still more extensive, but the cleared area is far exceeded by that of the virgin forest. Writing in 1872, Mr. Peal describes the aspect of the valley, as seen from the slopes of the Naga hills, as "literally a sea of jungle forest, an enormous dead level. The smallness of the area under cultivation surprised us more than anything; it did not look 1 per cent. The Potars I could easily recognise; yet they were but little green streaks, hardly noticed in the general view. The amount of waste land is enormous."[1] And Col. Godwin Austen, describing the valley of the Dunsiri in the following year, says, "At 5 miles from Golaghat the forest is entered, and this is continuous to the foot of the hills for a distance of 44 miles."[2] Around the rivers which intersect the plain, tracts that are too swampy for forest are covered with tall grass. In these, rice cultivation is gradually extending, but the native population is at present too sparse to deal with more than some small portions of the available area, and thus the greater part of Assam is still in the condition of forest and swamp.

As the principal seat of tea cultivation in India, and the country in which the tea plant is indigenous, still growing wild on the hills around, the climate of Assam is one of peculiar interest. Its most characteristic feature is its dampness at all seasons, in conjunction with the moderately high and comparatively equable temperature, due to the sub-tropical position of the province and to its being protected from all desiccating winds by the hills which enclose and seclude it. In the cool season thick fogs cover the low grounds, often remaining undissipated till the sun has attained its noonday power. In the spring, with the increasing warmth, thick clouds gather daily over the valleys, and frequent showers and thunder-storms drench the dense foliage of the evergreen forest while moderating the heat; so that, instead of a hot season preceding the rains, such as we have seen to be an universal characteristic of India

[1] *Journ. As. Soc. Bengal*, Vol. xli. Part i. p. 10.

[2] Report on Survey Operations in the Naga hills and Manipur, 1872-73.

proper, the temperature rises gradually and uniformly as the sun becomes higher and the days longer, and reaches its highest point in July, in the middle of the rainy season. At this season, the rainfall of the Assam valley is not more copious than in Bengal, though torrential on the southern face of the Garo and Khasi hills. But while it begins much earlier it lasts till October; and up to the end of that month, and even in November an occasional day's rain continues to stimulate vegetative growth. December is the only month in which the average rainfall is below 1 inch, and it is not till the beginning of December that the plucking of the tea bushes comes to an end for the season, and that the plant after pruning is allowed to rest till the showers of February and March bring out the first flushes, and afford materials for the manufacture of the following season.

A climate so uniformly moist in a country of such a character can hardly be described as eminently favourable to health and comfort. But it has been proved by half a century of experience, that when the jungle has been cleared and due regard is had to efficient drainage and the selection of the dwelling site, Europeans of good constitution who observe the precautions indispensable in all tropical climates and lead an active life, may preserve their health in Assam as well as in other parts of India. The cool season, if less bracing than in the Punjab and the drier provinces of Northern India, is still such as an Englishman may enjoy, and if from May to October he has to endure the oppressive heat of a vapour-laden atmosphere, it is questionable whether even this is more trying than the torrid fervour of an Indian hot season with some months of damp heat to follow. And to those whose natural taste and training enable them to appreciate the beauties of nature and to feel an interest in her works, Assam presents many attractions. No province in India affords a more exquisite presentment of nature in her native glory of hill, river, and forest. To the sportsman, indeed, the very impenetrability of the forests is an obstacle which may deter any but the most persevering from penetrating to the yet unvisited haunts of the rhinoceros and the wild buffalo. But to the naturalist the field is one of extraordinary richness. The province stands on the borderland of the Indian, Malay, and Indo-Chinese regions, and is tenanted by numerous forms of animal and vegetable life unknown in India. In describing the hills that border the north-east frontier, Col. Godwin Austen writes: [1]—

"We have a mingling in this direction of the Indian with the Indo-Chinese forms. Many birds, extremely rare in collections, and only represented by a few solitary examples, described by Hodgson and Blyth from Nepal, etc., were obtained here again, and a large number have yet to be identified or compared with specimens from other distant quarters. Some 280 species were collected. Other families of natural history are equally rich, none more so than the insects, and interesting forms of land mollusca are numerous. The great forest of the Dunsiri swarms with insect life, the Lepidoptera conspicuous by their numbers and coloration. Leaving this and ascending the spurs of the Burrail, with the changing flora new forms are constantly appearing, and I could point to few areas where a naturalist can find more rare objects for his search than on the slopes of this range."

The botany of the province is equally rich and varied.

"The mountains display a rich vegetation of the most tropical forms which India produces. *Anonaceæ* are numerous, several species of *Myristiceæ* occur, and the India-rubber fig forms large forests in some places. *Calami* and *Plectocomia* abound in the dense jungles, as well as other rare and interesting palms belonging to the genera *Livistonia*, *Licuala*, *Arenga*, *Areca*, *Wallichia*, etc. Oaks and chestnuts are also characteristic types, as are *Guttiferæ*, *Ternstræmiaceæ*, *Magnoliaceæ*, *Saurauja*, and tree ferns." [2]

Besides tea, which is the chief agricultural product of Assam, rice is largely grown, but, as regards the Assam valley, insufficient even for local consumption. From Sylhet, however, there is a considerable export, and among minor articles of produce the *supari* or betel nut, oranges, and India-rubber, the latter collected in the forests by the hill tribes around.

Of the observatories now or formerly existing in the province, the climatological data afforded by those of four stations are given in the Appendix. One of these, Shillong,

[1] Report on Survey Operations in the Naga hills and Manipur, 1872-73.

[2] Hooker and Thomson, Introductory Essay to the Flora of India.

on the Khasi hills, has already been noticed among the hill stations of India. The others are, Sibsagar in Upper Assam, Dhubri on the confines of Bengal opposite the entrance of the Assam valley, and Silchar, the chief town of Cachar.

The mean temperature of Sibsagar is 73°, the same as that of Sialkot and Ludhiana in the Punjab: but while that of the coolest month, January, is 59°, or 7° higher, that of the warmest month, here July, is 84°, or 7° lower than the June temperature of the latter stations. A still greater difference is shown by the extreme temperatures of the year. At Sibsagar the thermometer very rarely reaches 100°; this has occurred in only 2 years out of 10; and the highest reading yet recorded is 102°, while the lowest vary between 40° and 46°. The highest averages 17° or 18° lower, and the lowest 7° or 8° higher than the corresponding extremes at the Punjab stations. In like manner the daily range of temperature is much less, in April little more than half that of Ludhiana. All these differences are due to the dampness of the climate. This is very high. It averages 83 per cent of saturation, and only in one month of the year is it below 80. In the cold season the air is relatively damper than in most months of the rains, and hence the frequency of fog. Cloud is common at all seasons. The rainfall of Sibsagar averages 93 inches. It is least in December, but even then, on an average, it rains one day in ten; and in all other months, except November, more frequently. Even in March there are 14 rainy days, and on the average of the whole year 104. In 1874, a very wet year, it rained on 195 days. The rainfall of the valley above Sibsagar is still higher.

Dhubri, which may be taken as in some measure representative of the climate of Lower Assam, and also of that of North-eastern Bengal, has a higher mean temperature than Sibsagar, viz. 75°, and even a smaller annual and diurnal range of variation, except from February to April. January is 4° warmer, while July, which here as in Upper Assam is the warmest month, is 4° cooler. There are as yet only 4 years' registers for this station, and the highest and lowest temperatures yet recorded, both in the same year, are 101° and 45° respectively. In the driest months there is a mean difference of 22° between the morning minimum and the afternoon maximum in the shade, but in the 4 months of the rains only 8° or 9°. The mean humidity is not so high as in Upper Assam, being only 78; but while it is much lower in the cold season, it is rather higher in the rains. The rainfall is about the same, viz. 94 inches, but there is very little from the end of October to the end of February, less in March and April than at Sibsagar, and much more in May and June. With an equal total annual fall, the rainy days are only two-thirds as numerous as at Sibsagar.

The greater part of Lower Assam being under the lee of the Garo and Khasi hills, has a lower rainfall than either Dhubri or Sibsagar. Thus Gauhati has but 69 inches, Nowgong 79½, and Tezpur 76 inches; but it is no doubt higher along the northern margin of the valley, at the foot of the Himalaya. There are numerous tea gardens in Lower as well as in Upper Assam, even where the rainfall is least, but they are less productive.

Silchar is in the narrower part of the Baràk valley, on an alluvial plain, but little above the flood level of the river. Ridges of low forest-clad hills run southwards from the river bank at intervals of 6 or 8 miles, and these intervals are occupied by plains, a great part of which are under water in the rains, and form swamps in the dry season. 20 or 30 miles to the south, these ridges run up into the hill country of South Cachar and the Looshaie hills; and 10 miles to the north of the river are the Naga hills. About 20 miles farther east, the valley is nearly closed by the Burrail range, which rises to 5000 feet, and other higher ranges intervene between it and Manipur.

Thus situated, Silchar has a climate somewhat warmer than that of Upper Assam, in virtue of its lower latitude, but quite as rainy. Indeed the rainfall is heavier than that of Sibsagar, viz. 120 inches, and it rains nearly as often. But in the cool months fogs are less frequent. The mean temperature is 76°; that of January 64°, and that of the three months, June to August, 82°. A temperature exceeding 100° has been registered but once in the last 11 years; but the average maximum of the year, which is remarkably uniform, is 99°, and the minimum 45°. In point of humidity the atmosphere rather resembles that of Dhubri than that of Upper Assam; but Silchar is much damper from February to April.

Central Provinces (Nagpur) and Berar.—These provinces lie entirely south of the Tropic of Cancer, and occupy the northern central portion of the Indian peninsula. To the south of the Sone and Nerbudda valleys a tract of hill country, in places 100 miles in breadth, known in ancient annals as the Vindhya, and on modern maps bearing the name of the Satpura range, stretches across the peninsula from the high plateau of Amarkantak to the fortress of Asirgarh, beyond which, after a short interruption, it is continued in the Rajpipla hills to the Gulf of Cambay. It separates the drainage of the Ganges and Nerbudda from that of the Tapti, the Godavery and Mahanadi, and extra-tropical from inter-tropical India. From a geological point of view, it is not a definite mountain range, presenting no uniformity of structure or rock character, but is rather a belt of broken highland, a series of tablelands varying from 2000 to 4000 feet in elevation, which have remained while the country to the north and south has been worn down to lower levels by long ages of denudation. To the south of this hill-belt, the former kingdom of Nagpur, now included in the Central Provinces, and the province of Berar, formerly a portion of Hyderabad, consist of three great stretches of plain, in great part cultivated and, at the foot of the Satpuras, elevated more than 1000 feet above the sea, together with several ranges of hills which divide and surround them; some bare and rocky, but more generally covered with forest, or with the scrub jungle that remains from the destruction of the former forest.

The three plains are that of Berar, drained by the Poorna, a tributary of the Tapti; that of Nagpur, traversed by the Waingunga, a tributary of the Pranhita and Godavery; and that of Raipur or Chhatisgarh, on the upper course of the Mahanadi. The black soil which covers these plains is very fertile, and yields abundant crops of cotton and wheat, which grain is here at its southernmost limit as a staple crop. Other produce is of the same kinds as are raised generally in Northern India. But there remain large stretches of culturable land, now or formerly covered with forest, which, in due course of time, with the increase of population, will doubtless be brought under the plough. Besides these, however, the provinces include hilly tracts of enormous extent, formerly bearing forests of valuable timber, which have been terribly devastated and, for the time being, rendered of little value, by the wasteful temporary cultivation of the nomad hill tribes, and by the destructive fires which the villagers annually kindle to burn off the coarse grass, and provide a young growth for pasturage. Some three-fourths of the Central Provinces are thus nominally forest; and of this an area of 20,000 square miles is the property of Government. At present not more than 3500 miles are rigorously preserved, but this area may and doubtless will be largely extended; and it may be expected that, in the course of some years, the forests of the Central

Provinces will become one of the most valuable and productive sources of teak and sàl timber in the peninsula. Other valuable timber trees are the Sàj (*Terminalia tomentosa*), Bìjesàl (*Pterocarpus marsupium*), Shishum or Sissoo (*Dalbergia latifolia*), Kàwa (*Pentaptera arjuna*), and Anjun (*Hardwickia binata*).

In describing the climate of Nagpur, Mr. (now Sir Charles) Grant says :—

"The mean temperature is higher than in many parts of India, at the same height above the sea-level. But the absence of the really bracing air in the cold season of Upper India is in some measure compensated for by the fresh cool weather during the greater part of the monsoon; and by tolerably cool nights in the summer months.

"As in other parts of India, there are three seasons, the hot, the cold, and the rainy. The positively hot weather ordinarily commences about the first of April and lasts till the first week in June. The monsoon lasts throughout June, July, and August. At this season the climate, though full of moisture, is fresh and pleasant to the feelings. In September there are long breaks between each fall of rain, when the weather is often close and sultry, though never so much so as in the plains of the north of India at this time. October is generally sultry and unpleasant, but diversified occasionally by refreshing showers. The cold weather does not fairly set in till the middle of November. From the 15th of November to the end of February the air is generally cool and pleasant. Often, however, with the appearance of clouds, the thermometer rises as much as 7° or 8°, and the climate becomes disagreeable and close. From the 15th of February the weather gets warmer, and the hot winds blow from the beginning of April till the monsoon. Rain falls during every month in the year, usually during the hot and cold season only in showers, but sometimes accompanied with violent storms. Hail falls occasionally in January, February, and the early part of March, sometimes in very large stones, doing much damage to the spring crops.

"The climate during the rains is considered by the poorer inhabitants, who are exposed to it, as more trying than the cold of the real cold weather. In July and August it is not unusual to see people sitting round a fire in the very early morning, before going out for their day's labour. The climate is certainly not unhealthy. . . . Fever is the most frequent among the epidemic diseases. The most unhealthy season is from the second week in September to the second week in December. The jungle tracts are certainly not free from malaria until the cold weather has well set in."

This description applies to the plain. On the Satpuras it is much cooler. Thus, of Betùl, Mr. Grant writes :—

"It is almost out of the reach of hot winds, and would no doubt be an agreeable residence in the hot season. The climate of Betùl generally, at least to Europeans, is fairly salubrious; its height above the plains (2000 feet) and the neighbourhood of extensive forests moderate the heat of the sun and render the temperature pleasant throughout the greater part of the year. During the cold season the thermometer at night continually falls to several degrees below the freezing-point.[1] Little or no hot wind is felt before the end of April, and even then it ceases after sunset. The nights in the hot season are invariably cool and pleasant. During the monsoon the climate is very damp, and at times even cold and raw, thick clouds and mists enveloping the sky for many days together."

A similar account is given of the climate of Chhindwara; that of Pachmarhi, at 3500 feet, has already been noticed among the hill stations; and that of Seoni may be gathered from the statistics given in the Appendix.

In the Appendix are given the tabulated data of four stations; three of which, Akola, Nagpur, and Raipur, represent the three southern plains; and the fourth, Seoni, the Satpura tableland.

Beginning with the most easterly, Raipur, which is situated on the undulating plain of Chhatisgarh, at an elevation of 960 feet above the sea. The mean temperature is 78°, the same as that of Calcutta and Burdwan in Bengal, fully 2° farther north, and 3° cooler than Cuttack in nearly the same latitude. The earlier months of the year have a higher temperature than at some of the Bengal stations, but this is compensated by the greater coolness of the rains and later months. December is the coolest month, with a mean temperature of 66° and a mean minimum night temperature of 54°. The lowest reading of the year has varied between 41° and 51° in the last 10 years. May is the hottest month, with a mean temperature of 92°, and of 105° in the afternoon, the absolute highest in the year being between 108° and 116°. In the rains the mean temperature falls below 80°, and that of the afternoon to 87°, and the temporary rise on the cessation of the rains is only 1°. The diurnal range amounts to 25° or upwards from January to May, decreases to 12° in the rains, and then increases again gradually till the end of the year. The atmosphere is moderately dry, the mean humidity being 59 per cent. April and May are the driest months, when the hot winds blow and the mean humidity sinks to 36 per cent, and much lower in the daytime. The rainfall is nearly 52 inches, but of this but little falls from the end of October to the end of May, the cool months, Decem-

[1] This must refer to an exposed, hardly to a well-shaded, thermometer.

ber to February, being especially rainless. The average number of rainy days in the year is 76, of which 61 occur in the four months, June to September.

Nagpur, although but little farther south and at a somewhat higher elevation (1025 feet), is rather warmer, the mean temperature of the year being 79°. From November to May it is 1° or 2° warmer, a difference due solely to the higher temperature of the daytime, since the morning minima are as low or lower than those of Raipur. The extreme readings of the year have varied between 112° and 118° for the maximum, and between 43° and 51° for the minimum. The climate is drier than that of Raipur, the mean humidity being 53 per cent of saturation, and that of April and May but 28 and 30 per cent. The rainfall also is lower, viz. 45 inches, and its distribution in the different seasons is similar. But it would seem to rain rather oftener, as the registers show an average of 84 rainy days in the year.

Akola, the westernmost station, situated on the plain of Berar, is again drier. Its mean temperature is 78°; that of January 68°, and that of May 93°. In all months the mean temperature differs but little from that of Nagpur, but the winter minima are 2° or 3° lower and the annual minimum some 5° lower. The mean humidity is but 50 per cent, and that of April, the driest month, 22 per cent; while that of the rainy season does not average more than 74 or 75 per cent. The rainfall is much less than at the more easterly stations, not exceeding 30 inches, and the number of rainy days is 66, of which 9 only occur in the seven months, November to May.

Seoni, on the tableland, almost due north of Nagpur, and 2030 feet above the sea, has a mean temperature 5° lower, viz. 74°; a difference which holds good all through the dry season pretty equably, but is reduced to 3° in the rains. The afternoon temperature of May averages 103°, the morning minimum in December 50°; and while the highest readings of the year have varied between 105° and 111°, the lowest have ranged between 36° and 47°. The average humidity is rather higher than that of Nagpur, except in the last three months of the year; and the rainfall is rather greater, viz. 51 inches, distributed over 98 rainy days, of which, on an average, 12 occur in the six months, November to April, and 6 in May.

The West Coast of India. The Konkan and Malabar.—The west coast of the peninsula and the strip of hilly and undulating country that extends below the Ghats from the Gulf of Cambay to Cape Comorin has the dampest and most uniform climate of any part of the peninsula. While open to the westerly winds from the ocean, which mitigate the more intense effect of the tropical sun and maintain the verdure of the land surface and of the evergreen forests that clothe the slopes of the Ghats, the country is shielded by this range from the desiccating winds of the Deccan tableland, and its surface is watered by the numerous small streams that bring down its drainage and open out into the tidal estuaries and backwaters that intersect the seaward margin of the tract.

Although extending through 13° of latitude (from 8° to 21° N. lat.), the annual mean temperature of the western coast is nearly the same throughout, viz. 79° or 80°. But in some other respects there are considerable differences between its northern and southern extremities. To the north of Bombay the climate of January and February, if not comparable with the cold season of Northern India or even of Bengal, is still such as to afford cool nights, while the ordinary damp heat of the daytime is tempered by northerly winds; and as far down as the southern limits of the Bombay Presidency (in N. lat. 14°) the season from the latter part of October to the end of May is almost rainless. But in South Canara, Malabar, and Travancore, the day temperatures of January and February are little if at all lower than those of the subsequent months, and in the early morning the thermometer does not fall much below 70° at places near the sea; while spring showers precede the rains, and the autumnal rainfall lasts to the end of October. At Cochin, January and February are the only months in which the average rainfall does not exceed 1 inch, and July is cooler than January.

The rainfall of the summer monsoon is heavy all along the west coast, and still heavier on the Ghats, which force the saturated current to ascend to a height of from 2000 to 7000 feet before pursuing its course across the Deccan tableland and the loftier hill groups of the Nilgiris, the Anamalais, and the Pulnis. The cooling which it undergoes in this ascent, amounting to about 1° in each 400 feet, causes the enormous precipitation on the face of the Ghats, of which Mahableshwar with 255 inches, Baura Fort with 250 inches, and Matheran with 244 inches, are examples.

On the coast the fall is less. From June to October it amounts to 119 inches at Mangalore and 132 inches at Honawar, where it is heaviest; decreasing thence northwards to 100 inches at Rutnagiri, 73 inches at Bombay, and to only 42 inches at Surat. In the south, on the coast of Travancore, the rains set in somewhat earlier than at Bombay, generally in the latter part of May, and are heaviest in June. In Bombay they may be expected in the first or second week of June, are heaviest in July, and come to an end in October; but in the south they last quite to the end of October, and the rainfall of this month is rather heavier than that of September.

It is almost needless to remark that in such a climate the vegetation has all the luxuriance that is commonly associated with our ideas of the Tropics. The strip of low plain that borders the greater part of the coast is covered with cocoa-nuts and rice fields, and the villages of Malabar and Travancore are embowered in groves of betel nut palms and Talipots. "Cassia, pepper, and cardamoms flourish wild in the jungles, and form staple products for export. The fact that the pepper is cultivated without the screens used in other parts of India to preserve the humid atmosphere about it, is the best proof of the dampness and equability of the climate. The low valleys are richly clothed with rice fields, and the hillsides with millets and other dry crops, whilst the gorges and slopes of the loftier mountains are covered with dense and luxuriant forest." [1]

Of the forests Sir D. Brandis writes:—

"The richness of the vegetation . . . may be seen in the forest tracts at the foot of the Ghats, north-east of Bombay, stretching to the neighbourhood of the Tapti river. These forests, in which the Teak is the most important tree, consist chiefly of species that shed their leaves in the dry season. The dense evergreen forests, characteristic of the wet zone, begin to the south of the Rutnagiri district. In Canara, Malabar, and Travancore they occupy large tracts at the foot of the Ghats and also on their summits and the higher hill groups, which, like the Nilgiris and Anamalais, branch off to the east of the chain. Mahableshwar and the surrounding slopes are clothed with

[1] Hooker and Thomson, Introductory Essay to the *Flora Indica.*

evergreen forest which, some miles farther east, where the climate becomes drier, yield place to a mixed forest of deciduous trees. On the windy plateau of Mahableshwar, as well as elsewhere on the crest of the Ghats, the trees are indeed small and stunted ; but in protected depressions and valleys the thickest evergreen forest of the wet zone is luxuriantly developed, and besides Laurels, *Eugenia* and the wild Mango, consists of *Anonaceæ* and other characteristic tropical families. . . . The variety and richness of this evergreen forest increases to the south, and reaches its highest development on the western slopes of Coorg, the Wynaad, and the Nilgiris, opposite that part of the coast where the rainfall is highest. Extensive forests with a mean height of 200 feet are here not rare. Between the tree-boles the ground is covered not only with dense bush and herbage, in which species of *Strobilanthes*, an Acanthaceous genus with large beautiful flowers, plays a conspicuous part, but also with young trees whose parents protect them with their dense canopy. The trees of these forests as well as the undergrowth are all shade-loving plants, which can thrive in the deep gloom of the forest. When, on the fall of some forest giant, a gap is made in the leafy canopy, the young trees shoot up, mastering the underwood, and, striving towards the light, contend with each other for pre-eminence of growth, till the weaker yield place to the stronger. Besides the families already mentioned, the southern tract especially abounds in *Guttiferæ*, *Dipterocarpæ*, *Meliaceæ*, *Leguminosæ*, *Rubiaceæ*, *Euphorbiaceæ*, and *Urticaceæ*. In many places tree-ferns, palms, and bamboos give a characteristic aspect to the forest."[1]

The climate of the west coast is illustrated in the Appendix by four stations, viz. Surat in the extreme north, Bombay, Mangalore and Cochin ; and that of the summit of the Ghats by Mercara in Coorg. The gradations of temperature, humidity, and rainfall, experienced in passing from north to south, will be best shown by the following comparison of the four coast stations :—

The mean annual temperatures of all four stations are within a degree of each other, viz. 79° or 80°. But whereas that of Surat in January is 70° and in May 86°, the corresponding temperatures of Bombay are 74° and 85°, of Mangalore 76° and 83°, and of Cochin 79° and 82° (that of April being 84°). The extreme temperatures of the year show a much greater variation. At Surat they are respectively 109° and 48°, at Bombay 95° and 61°, at Mangalore 94° and 63°, and at Cochin 95° and 67°. The Bombay observatory being situated on a narrow point of land, surrounded on three sides by many miles of sea, has doubtless more equable temperatures than places on the coast of the mainland in the same latitude.

[1] Die Beziehungen zwischen Regenfall und Wald in Indien (translation).

The mean humidity of the atmosphere is comparatively high, and equable at all the stations except Surat. At this place it averages but 62 per cent of saturation, and remains between 50 and 60 from November to May. At the remaining stations it averages between 77 and 80, and never falls to 60 in any single month. At Cochin the lowest mean humidity of any month is 70 per cent. The humidity of the air, therefore, like its temperature, becomes more constant in passing from north to south, and probably this circumstance, quite as much as the rainfall and the rise of the winter temperature, determines the change in the character of the vegetation noticed in the foregoing description. The average cloudiness of the sky is about the same at Bombay, Mangalore, and Cochin, viz. between four and five-tenths of the sky expanse, but only three-tenths at Surat.

The distribution of the rainfall has already been noticed. The average number of rainy days in the year is 66 at Surat, 108 at Bombay, 142 at Mangalore, and 164 at Cochin.

Mercara, which is situated in Coorg, only a few miles from the crest of the Ghats at an elevation 3700 feet above the sea, has a cool, equable, and pleasant climate at all times of the year. The mean temperature is only 67°, and that of the warmest month, April, 72°; the average afternoon maximum of this month being 85°, and the highest reading recorded in each of the last 5 years within a degree or so of 90°. The mean of December and January is 64°, that of the early morning 55° or 57°, and the mean minimum of the year 49°. In the rainy season the air is almost saturated with vapour, and the station enveloped in cloud, and during the four months, June to September, it rains almost daily, and often day and night long without intermission. From December to the end of March, however, there is but little rain, and in the first three months of the year the humidity is between 60 and 70 per cent of saturation. The annual rainfall averages 127 inches, of which about 106 inches falls from June to September.

The Wynaad, which is the chief coffee-growing district of India, lies on the Ghat range, and at its foot, immediately to the south of Mercara, and its climate is similar but somewhat warmer, owing to its lower elevation. A high rainfall, a moderately high and equable temperature, and a generally damp atmosphere, are the conditions most suitable to the plant, as will be further illustrated by the climatic features of Kandy, in Ceylon. Cinchona is also largely cultivated on the slopes and on the plateau of the Nilgiris. Indeed, the plantations are, altogether, much more extensive than those of Darjiling, and certain varieties, especially the crown-barks (*C. officinalis*), also *C. Ledgeriana* and *C. pubescens*, are grown with greater success than in the Sikkim Himalaya.

Khandesh, the Deccan, and Mysore.—In India without the Tropic, the most arid province, Sind, is separated from the dampest, Assam, by the whole breadth of Northern India, a distance of 22° of longitude; and while the former lies in the extreme west the latter occupies the extreme east of the region. In the peninsula these relations are in a measure reversed, and whereas the seat of the heaviest rainfall and the dampest atmosphere is the west coast of the peninsula, the traveller has but to ascend the Ghats and strike eastward across the tableland beyond their crest and, in the course of only 30 or 40 miles, he will have passed from the evergreen forest and torrential rainfall of the Sahyàdri range to the rolling plains of black soil and the flat-topped terraced hills of the Deccan, where a shady tree is among the rarest of nature's gifts, and where, notwithstanding the natural fertility of the soil, all agriculture is more or less precarious owing to the uncertainty of the scanty rainfall.

The interior of the peninsula, above the Western Ghats, is a plateau, the greater part of which is elevated between one and two thousand feet above the sea. Its general slope is eastward, and the great rivers that carry off its drainage, the Godavery, the Bhima, the Kistna, the Tungabhadra, and the Cauvery, and many of their tributaries, take their rise on the very crest of the Sahyàdri range and flow eastwards to the Bay of Bengal. The northern portion, comprising nearly all that lies within the Bombay Presidency and the greater part of Hyderabad, consists of flat sheets of volcanic rock, and is thus described by the authors of the *Geological Manual of India*:—

"The volcanic region of Central and Western India is distinguished by marked peculiarities of scenery, and the characters of the surface are widely different from those found in other parts of the Indian peninsula. Great undulating plains, divided from each other by flat-topped ranges of hills, occupy the greater part of the country, and the hillsides are marked by conspicuous terraces which may often be traced to great distances, and are due to the outcrop of the harder basaltic strata or of those beds that resist best the disintegrating influences of exposure. . . . The vegetation of the trap area differs no less conspicuously from that which is found on other formations; the distinction in the dry season being so marked that, especially when taken in connection with the form of the surface, it enables hills and ranges of trap to be distinguished at a distance from those composed of other rocks. The peculiarity consists in the prevalence of long grass and the paucity of large trees, and in the circumstance that almost all bushes and trees, except in the damp districts near the sea, are deciduous. The result is that the whole country, except where it is cultivated, presents during the cold season, from November to March, a uniform straw-coloured surface with but few spots of green to break the monotony; whilst from March, when the grass is burnt, until the commencement of the rains in June, the black soil, black rocks, and blackened tree stems present a most remarkable aspect of desolation. During the rainy season, however, the country is covered with verdure, and in many parts it is very beautiful, the contrast afforded by the black rocks only serving to bring into relief the bright green tints of the foliage."

The southern half of the plateau, in the Madras districts of Bellary and Anuntapur and the kingdom of Mysore, has a different aspect. This, too, consists of rolling plains, but the formation being that of the older crystalline rocks, the hills are of rounded whale-backed forms, and frequently the eminences are crowned with great tors of highly picturesque and bizarre shapes or rounded hummocks of bare rock, the result of ages of atmospheric action on rocks of varied internal structure.

As a general rule, the climate of the Deccan and Mysore, beyond 30 or 40 miles from the crest of the Ghats, is very dry. The driest portion of the Deccan is a strip running north and south, parallel with the Ghats, and from 50 to 80 miles to the east of them, from the foot of the Satpura range in the Tapti valley to almost the foot of the Nilgiris. As far south as the latitude of Poona, the zone of country with a rainfall below 30 inches averages not more than 100 miles in width; but to the south of this it extends right across the plateau to the Eastern Ghats, and even beyond to the sea-coast. In the centre of this part of the plateau, to the south and south-east of Bellary, is an area of 6000 or 7000 square miles, within which the annual rainfall is below 20 inches; a tract which has repeatedly suffered from famine in its severest form. In this dry tract the waste lands are covered with coarse grass or a thin bushy scrub, but in the hilly country of Sandùr to the west, where the rainfall is higher, there is a flourishing forest. In the forests of Mysore, especially those of the Eastern Ghats around Kollegal, the most valuable tree is the Sandal wood (*Santalum album*), which is restricted to the Deccan and Mysore plateau, and to the south of latitude 16° or 17°. The Teak is limited to the zone fringing the Ghats, where the rainfall exceeds 30 inches; and it occurs also in the Sandùr hills, and in the hilly country to the east of the dry zone which extends far into Hyderabad, and where the rainfall is also in excess of 30 inches. The red sanders (*Pterocarpus santalinus*), which yields a valuable dye-wood, is restricted to the eastern margin of the plateau, chiefly in the districts of Cuddapah and North Arcot.

In Khandesh and the greater part of the Deccan districts of Bombay, north of the Kistna river, the winter and spring months are almost rainless. Land winds from the east in the earlier, and from north-west in the later months, prevail almost up to the setting in of the summer monsoon. The winter rainfall of Northern India does not as a rule penetrate so far south, and the thunder-storms of the spring months, fed by vapour from the Bay of Bengal, rarely occur in the northern part of the peninsula much farther west than the meridian of Nagpur. But farther south in Dharwar, in Hyderabad, and also in Mysore and the Madras districts of Bellary, Anuntapur, and Cuddapah, showers occur occasionally in March and April, and more frequently in May. During the monsoon the plateau is swept by a strong steady west wind, which only occasionally slackens and gives a light rainfall; but the rains last longer than in most parts of Northern India, and the October rainfall generally amounts to between 3 and 4 inches in the Deccan, and in Bellary and Mysore is quite as copious as that of September. The rainy season, if much less rainy than in Northern India, is cloudy, cool, and pleasant.

Poona is the chosen seat of the Bombay Government all through the rains; and the great military stations Secunderabad and Bangalore enjoy at this season a climate which is by no means oppressive, either on account of heat or excessive moisture.

In the Appendix, the stations that represent this tract are Malegaon in Khandesh, 1430 feet above the sea; Poona, at 1850 feet, distant only 30 miles from the summit of the Ghats, and on the border of the dry tract; Sholapur, near its eastern limit; Belgaum in Dharwar, opposite a depression in the Ghats that admits the monsoon current less obstructed than in most parts of the range; Secunderabad, Bellary, and Bangalore; the last on a very high part of the plateau, nearly 3000 feet above the sea.

The mean annual temperature of Malegaon is 76°; that of December, the coolest month, 66°, and that of May, the warmest, 88°; the extreme temperatures of the year being between 36° and 43° as a minimum, and between 107° and 111° as a maximum. In the first three months of the year, when the atmosphere is very dry, the diurnal range of temperature averages not less than 34° or 35°, and even in the rains is as much as 14°. The mean humidity is only 51 per cent of saturation, and in April, the driest month, 28 per cent. The average rainfall is about 25 inches, and of this, less than 2½ inches falls in the seven months from November to May. The average number of rainy days is 66, of which but 7 occur in these months.

Poona, which is rather more than 2° farther south, and 400 feet higher, has a mean temperature of 78°; in December and January of 72°, and in April of 86°. Since 1878, when for the first time the instruments were properly exposed, the lowest readings in the year have varied between 40° and 50°, and the highest between 101° and 110°. The former average 4° higher than at Malegaon, the latter 3° lower. In the rainy season, from July to September, the mean temperature is 75°, and the afternoon maximum 81° in July and 83° in the two following months. In the daytime, therefore, Poona is as cool in the rainy season as in December; but the night temperatures are but little below 70° in the former, while they average but 54° in the latter. In the spring months the climate is as dry as that of Malegaon, and the mean humidity of the year is but little greater, viz. 52 per cent. The average rainfall is 28 inches, of which but 2 inches fall from November to April.

Belgaum is again 2½° farther south, and 700 feet higher than Poona. Its mean annual temperature is 74°; the December temperature 71°, and that of April 81°. During three months of the rains it averages only 70° or 71°. The winter temperature, therefore, differs but little from that of Poona, but that of the spring and rainy season is from 3° to 5° lower. The climate is much damper and the rainfall heavier (49 inches). Showers fall in April and May on 6 or 7 days in each month, and this rainfall greatly mitigates the temperature. Moreover, all through the monsoon, a very strong west wind blows constantly through the depression in the Ghat range, opposite Belgaum; the mean rate of movement, as shown by the anemometer, being between 20 and 30 miles an hour, night and day, in June, July, and August. These months and September are very rainy, the average number of rainy days being from 21 to 28 in each month, and even in October it rains nearly every second day. The climate is therefore much damper than in most other parts of the Deccan, especially those to the north and east.

Sholapur is on the eastern margin of the dry tract, and 1600 feet above the sea. Its mean annual temperature is 79°, or 5° higher than that of Belgaum; but the winter temperature is about the same as that of Poona and Belgaum, viz. 70° in December and 72° in January. In the hottest month, May, it is 89°, and in the rainy season from 77° to 79°. The lowest winter readings have ranged between 42° and 49° in the last 10 years, and the highest summer readings between 108° and 112°. The atmosphere is very dry, the mean humidity of April being only 26 per cent of saturation; and even in the rains it averages only between 60 and 70 per cent. The westerly monsoon, though less strong than at Belgaum, blows very steadily, with an average movement of 10 or 12 miles an hour, and brings comparatively little rain. The mean monthly fall of each of the three months, June to August, is between 4 and 6 inches only, but that of September is higher, when, the westerly current having slackened, easterly winds begin to set in from the Bay of Bengal; and in October the rainfall amounts to $3\frac{3}{4}$ inches. The total fall of the year is barely 30 inches, and the number of rainy days 83.

Secunderabad, which lies 160 miles farther east, at about 1800 feet above the sea, has a very similar climate, but slightly cooler. The mean temperature of the year is 78°; that of December and January 69° and 70° respectively, and that of May 89°. The extreme temperatures of the year are nearly the same, the minimum varying between 41° and 53°, the maximum between 106° and 111° in different years. The atmosphere is somewhat damper; the mean humidity of the air being 56; that of the driest month 36, and that of the three months, July to September, from 72 to 75 per cent. The spring months are more showery, there being, on an average, 12 rainy days from March to May inclusive. But the monsoon rainfall is equally light, and that of September somewhat less than at Sholapur. In October it is between 3 and 4 inches. It rains on an average on 89 days in the year.

Bellary is a very dry station as regards rainfall, indeed one of the driest in the Deccan, and is rather warmer than any of the foregoing, though elevated 1450 feet above the sea; its mean annual temperature being 80°. The mean of December and January is 73°, that of April and May 89° and 88°, and the lowest and highest readings recorded in each of the last 11 years have varied between 50° and 57° in the former case, and from 106° to 111° in the latter. Owing to the lightness of the monsoon rainfall, the mean temperature from June to September ranges a little over 80°, and in October is 79°. The air is much drier than at Secunderabad, except in May and June and in the last two months of the year, and especially so from July to September; the highest mean humidity of any single month being 64 per cent. Occasional showers fall in April and May, but the average rainfall of both July and August is below 2½ inches; and most rain falls in October, the average of this month being 4 inches. The total mean fall of the year is barely 18 inches, and it rains on an average on only 55 days.

Lastly, Bangalore, a favourite climatic resort of the residents of Madras, and one of the largest military stations in Southern India, has a climate only second in attractiveness to that of the Nilgiri hills, and especially suited for raising such fruits and vegetables as thrive in the warm temperate zone. The mean temperature of the year is 73°; that of December and January 67°, and that of the warmest month, April, 80°. The highest temperature yet recorded is below 100°, and the lowest in the winter have varied between 46° and 56°. The atmosphere is neither very damp nor very dry at any season. The mean humidity of the year is 66; that of the driest month 49, and that of the dampest 77. The rainfall is moderate and well distributed through eight months of the year. January and February are the most rainless months; but in March there are usually one or two showers, and in April more; and from May to October from 3 to 6½ inches fall on an average in each month. The mean rainfall of the year is 35½ inches, distributed over 96 rainy days.

The Carnatic.—The Ghats, which form the eastern boundary of the great central tableland of the peninsula, are by no means so well defined or continuous as the Sahyàdri range on the western margin of the Deccan and Mysore. To the north of the Godavery they rise from the plain of the Northern Circars to the level of the tableland of Jaipùr, and are here well defined. But to the south of the Kistna they are represented by a crescentic system of parallel ranges, the successive outcrops of an ancient series of stratified rocks, which advance nearly to the sea-coast, some miles north of Madras; and from this point an ill-defined escarpment, bounding the tableland of Mysore, trends away to the south-west, leaving a number of more or less isolated and independent hill groups, from 2000 to 4500 feet in height, between it and the coast plain of the Carnatic. These hills, which bear various names, occupy the centre of the peninsula as far south as the Cauvery valley, in latitude 11°. In this latitude, which also coincides with the southern limit of the outlying ridges of the Nilgiri hills, the hill country of the peninsula is broadly interrupted. A plain, traversed by the Cauvery river in its lower course, leads up by a gentle ascent to the foot of the Nilgiris and the outlying ranges, and at Coimbatore has an elevation of 2000 feet above the sea; and thence descends by a steeper incline to the west coast, separating these latter hills from the Anamalai and Pulni groups, which rise at a distance of from 20 to 25 miles to the south, and rival the Nilgiris in height. These link on to the Travancore hills farther south, and carry on the line of the western highlands to the southern extremity of the peninsula. The interruption of the Ghats, opposite Coimbatore, known as the Palghat gap, exercises an important influence on the weather of the Central Carnatic on the one hand, and of the Malabar coast, south of Calicut, on the other, affording an unobstructed passage to the easterly and westerly monsoons. In the introduction to their report on the geological survey of this part of the Carnatic, Messrs. King and Foote remark:—

"The south-west wind blows with great force into the Palghat Pass, which has a width of about 25 miles, diminishes a little as the mountains recede and form a funnel, out of which the wind issues. Afterwards it meets with little or nothing in the Coimbatore district to oppose it, till, east of the Cauvery, the Shevaroys and Kolamalais break its force, and receive in heavy showers a large share of its moisture then remaining. . . . Southward of the Kolamalais no obstruction is offered to the westerly wind, owing to the absence of mountains, the nearest being the Dindigul mountains at a distance of some 40 miles; and, in consequence, it blows very strongly across the delta of the Cauvery to the east coast. This (westerly) wind, added to the evaporation from the vast area of country laid under water in the delta, during the prevalence of freshets in the Cauvery, tends to cool very greatly the intensely hot climate of the Tanjore district. The coast climate, generally, is much more damp and relaxing than that of the more inland parts of the low country, but the thermometer generally has a lower range near the sea."

Of the hill groups, which occupy so much of the interior, the same authors write :—

"The climate of the mountain plateaux is one much more agreeable and suitable to the European constitution in point of temperature, but unfortunately all the ranges are, during the hottest season, to a greater or less degree the seat of malarious fever, giving rise to very dangerous jungle fever. None of the mountains, even the Shevaroys, are above the so-called fever range.[1] . . . April and May are generally the most unsafe months, and during these the jungle regions, especially those among the lower hills, should be carefully avoided. During the cold season, however, especially in December and January, the malaria seems to be quite in abeyance; it is also said to be so for some weeks after the heavy rains of the south-west monsoon."

The width of the coast plain is very variable. To the north of the Pulicat lake the hills advance to within 30 miles of the sea. But to the south of Madras it has an average breadth of about 80 miles, of which the marginal portion, and certain tracts stretching inland along the rivers, are alluvial flats, and the remainder gently undulating and dotted with occasional little rocky hills. Very much of the area is uncultivated, especially the higher grounds. Agriculture largely depends on artificial irrigation, and except in the deltas of the Godavery and the Cauvery, this is obtained chiefly from tanks, formed by embanking natural depressions in the surface, and fed either by the drainage of the surrounding country, or, in some cases, by canals taken off from the larger rivers, which bring into them a portion of the flood waters. Some of these tanks are 4 or 5 miles across, and form a very characteristic feature of the Carnatic. The waste lands are for the most part covered with bush and scrub, with small trees. Except on the hills, there is little that can properly be termed forest; but the coast is fringed with groves of cocoa-nut palms, and, in the neighbourhood of Madras, extensive plantations of *Casuarinas*

[1] This remark does not refer to the Nilgiris, the Pulnis, or Anamalais, all of which are well above fever range.

have been made for the supply of firewood. There is also a protected natural forest on the island, 33 miles in length, that separates the Pulicat lake from the sea. The principal roads are bordered with avenues of pipal, banyan, tamarind, mango, and other trees, and many of the camping grounds near villages are shaded by *topes* of large and well-grown trees of these and other species. Except wheat and barley, the chief articles of agricultural produce are similar to those raised in other parts of India. Rice is extensively cultivated, and enters largely into the food of the people; but, as elsewhere, the millets and some inferior grains also contribute an important share.

The climate of the Carnatic differs in many important respects from that of other parts of India. The dry season lasts from the middle of December to the latter part of June: but in April and May there are usually little storms accompanied with heavy showers, locally known as the "mango showers," heavier on the hills than on the plains, but even on the latter giving an average of from 3 to 5 inches of rain in the two months. The rainfall of June is lighter than that of May. That of the whole summer monsoon, from June to October, is comparatively light, amounting to 4 or 5 inches a month in July, August, and September on the coast, but to not more than one-third that amount in the dry tract to the north of Combatore, and in the district of Tinnevelly in the extreme south. Heavier rain sets in in October, and lasts till the middle of December; and it is on this that the cultivators chiefly depend for filling the tanks, excepting such as are supplied by feeders from the larger rivers that come down in flood during the summer monsoon. There is in the Carnatic no season that would approve itself as cool to European feelings, otherwise than in a comparative sense; but, for about 6 weeks after the cessation of the late autumnal rains in December, the winds are northerly, the nights pleasant, and the heat of the day not excessive. In the spring months hot land winds blow in the interior, and sometimes down to the coast; but at Madras and other places on the coast, April is the season of the longshore winds, winds which, as their name implies, blow from the south parallel with the coast, and though less hot than the land winds, are damp and very relaxing.

The climate of the province is generally dry as compared with that of the west coast or of Bengal and Arakan; but that of the interior, and of the northernmost and southernmost districts, much drier than that of the coast region from the northern extremity of the Pulicat lake to Point Calimere. The average annual rainfall of this latter tract, to a distance of from 10 to 20 miles inland, exceeds 40 inches. But in the northern portion of the Nellore district, the greater part of Coimbatore, and on the coast of Tinnevelly it is below 20 inches in the year, and over much larger areas around these below 30 inches.

Some of these phases of the climate are illustrated by the five stations, the meteorological statistics of which are given in the Appendix. These are Masulipatam and Madras on the coast; the first in the extreme north, on the sea face of the Kistna delta, the second on a projecting portion of the coast south of the Pulicat lake; Coimbatore, on the dry plain that extends between the Nilgiri and Shevaroy hills, not far from the Palghat gap; Trichinopoly, at the head of the Cauvery delta, almost due east of Coimbatore; and Madura, 70 miles south of Trichinopoly, on the plain to the east of the Travancore hills.

The mean temperature of Masulipatam is 81°; that of December and January 74°, and that of May, the hottest month, 88°. In no month does the afternoon maximum average less than 83°, or the night minimum less than 66°; and in May the former is as high as 99°. The highest reached in the course of the year varies between 101° and 116°. The lowest readings are less variable, being between 58° and 62°. Owing to the situation on the sea-coast of a great river delta, the air is never very dry, nor the range of temperature during the day very great. The latter averages 13° in the dampest months and 19° in the spring. The air is driest in June, when the humidity is 67 per cent, and dampest in October and November, when it is 78; the average of the year is 74. The mean rainfall of the year is about 38 inches, of which only $1\frac{1}{2}$ falls from December to April. In the first three months of the monsoon the monthly fall does not exceed 5 or 6 inches; but it is somewhat greater in September, and heaviest in

October (8½ inches). In November it amounts to about 4 inches, and ceases a full month earlier than in the central and southern districts of the Carnatic. The number of rainy days has varied in the last 11 years from 61 to 101 in the year.

Madras, although also situated on the sea-coast, has a drier atmosphere and a somewhat higher temperature than Masulipatam. Its mean temperature is 82°; that of December and January 76°, and that of June 88°. The afternoon temperatures are lowest in December (83°), but the early morning readings are somewhat lower in January, when they average 68°. The lowest reached in the course of the year, since the thermometers have been properly exposed, have been between 57·6° and 62°. In former years they were many degrees higher. The highest readings, which are recorded with the hot land winds generally in May, have been between 102° and 113° in different years. The daily range of the thermometer is much the same as at Masulipatam. At the observatory, which is nearly 3 miles from the sea, the mean humidity is 71. The driest month is June, when it averages 61, and the dampest November, when it is 79 per cent of saturation. The mean annual rainfall is only a little under 50 inches. In the wettest year on record (1827) it amounted to 88·4 inches, and in the driest (1832) it was only 18·5 inches. From June to September there is less rain than at Masulipatam, the average of each of these three months not exceeding 4 or 5 inches. But that of October is nearly 11 inches, and that of November, which is the wettest month of the year, nearly 14 inches. There is usually a spell of heavy rain in the early part of December; it is in this and the two preceding months that little cyclones frequently form off the coast, whence they pass across the Carnatic bringing a deluge of rain. From January to the end of April there is but little rain, and in May only a few thunderstorms, unless, as sometimes happens, the Carnatic is visited by a cyclone from the bay.

Trichinopoly has the same mean temperature as Madras, viz. 82°, but while it is 1° or 2° warmer from February to May, it is 1° or 2° cooler from June to September. Still less than at Madras is there any cool season, properly so called. Even in December and January the afternoon readings of the thermometer average 85° and 87°, but the nights are comparatively cool at this season, though the thermometer rarely sinks below 60° on the coldest nights in the year. May is the hottest month, when the mean temperature is 88°; but although the average maximum of the afternoon is 102°, no reading higher than 108° has been recorded, at all events during the last 12 years. In June the temperature falls with the setting in of the strong west wind through the Palghat gap, as already described, and this brings cloudy skies; but the atmosphere remains dry and there is but little rain. It is very dry, indeed, at all times, except in the last three months of the year. The mean humidity is 63 per cent, and in April, the driest month, only 54 per cent. Even in June and July it is only 57 per cent, but in November it rises to 76 per cent. The mean annual rainfall is 37 inches. Up to the latter part of April but little falls, but at the end of this month and in May thunder-storms are frequent, and the average rainfall of the latter month is nearly 4 inches. June and July again are dry months, there being on an average only six rainy days in the two months. But rain becomes more frequent in August and September, and reaches its maximum in October, in which month there are about 11 rainy days. November and the first half of December are also more or less rainy, but the rainfall is lighter than at Madras.

Coimbatore also has a very dry climate, but being situated at a high level in the neighbourhood of lofty hills and within the influence of the Palghat gap, it is by no means a very hot station for the latitude. The mean temperature is 78°, the same as that of Calcutta, and 4° lower than that of Trichinopoly and Madras, and in no month of the year does it deviate more than 5° above or 4° below this mean. In December and January it is 74°, in April 83°, and from June to September it fluctuates between 76° and 78°. But the diurnal range is as much as 27° in February and March, and 17° in the dampest months of the year. The lowest readings of the year have varied between 54° and 61° in different years, and the highest between 99° and 104° in the shade. The mean humidity is 66 per cent, and that of February, the driest month, 52 per cent. From this time it rises gradually to August, when it is 73 per cent, and in October, the dampest month, it does not exceed 75. There is less cloud than at Trichinopoly, and the rainfall is much lighter, viz. little above 21 inches in the year. Of this, showers in April and May contribute rather more than 4 inches. Then, from June to September, although there are from 6 to 10 rainy days in the month, the average monthly rainfall is barely 1½ inches; in October and November the total fall amounts to 9 or 10 inches, and after the middle of December there is but little rain till the April showers. Although, as already stated, the rainfall is only about 21 inches, it rains on an average on 85 days in the year.

Madura is a warmer station, having the same mean temperature as Trichinopoly and Madras (82°). In December and January it is 77°, and in April and May 86°. On the coolest night of the year it rarely falls to 60°, and the mean minimum of the three months, December to February, is between 68° and 70°. In April and May the afternoon maximum readings average 101°, but the highest recorded temperatures of the last 12 years have not been higher than between 104° and 107°. From February to May inclusive the daily range averages 25°. The mean humidity of the year is 65, and in the driest months, April, June, and July, it falls to 59. In the last three months of the year it averages from 73 to 75. The rainfall is much higher than that of Coimbatore or than that of the plains to the south-east, viz. 35 inches in the year; and of this, nearly 5 inches fall in April and May, about 9 inches altogether in August and September, 9 inches in October, and 7½ inches in the two final months of the year. Although the rainfall is thus 60 per cent heavier than at Coimbatore, it rains less frequently, the average number of rainy days being only 59 in the year.

Ceylon.—In the course of the foregoing pages we have seen that, in passing from north to south, the different seasons of the year become less and less contrasted, the temperatures of the winter and summer months more and more equal, and at length, in the south-west of the peninsula, on the coast of Travancore, beyond the alternation of a wet and a dry season, there is little to distinguish the different months of the year. In Ceylon, and especially in its south-west province, we have these essentially tropical, or rather equatorial, features of climate in their strongest development. At Point de Galle the difference of the highest and lowest readings of the thermometer in the course of the year amounts, on an average, to no more than 19°, and in some years has not exceeded 17°, a difference less than half that of a single day and night at certain seasons in the Punjab. But, despite the moderate dimensions of the island (less than 300 miles in length and 140 in width), the climate of Ceylon is by no means the same in its several parts. Its east and west coasts, if not presenting so great a contrast as the dry plains of Tinnevelly and the evergreen forests of Travancore on the Indian mainland, exhibit differences of the same kind. The heavy rainfall of the summer monsoon is restricted to the south-west of the island, while the eastern provinces receive their most copious rain in November and December. At this latter season the rain is not restricted to the eastern half of the island; it extends not only to the hills, but also to the south-west province; and in such measure that, both at Galle and Colombo, the average fall of October and November is as heavy as that of May, and considerably heavier than that of June and those months when the summer monsoon blows most strongly on the coasts of India. The south-west province, comprising the generally rugged country between the central hills and the coast from Galle to Colombo, is that in which frequent and copious rain and constant damp heat foster the luxuriant vegetation that has made Ceylon proverbial for its beauty among the beautiful islands of the tropics. But on the north-west coast, opposite to India, and also on the south-east coast, the rainfall is comparatively light, and a jungle of thorny acacias, euphorbias, and deciduous trees, such as characterise the drier parts of the Carnatic, replace the dense evergreen forests of the more humid tract.

The climate of Ceylon has been frequently described, and by no one more fully than Sir Emerson Tennant, from whose work the following passages are extracted in a much abbreviated form :—

"The island is seldom visited by hurricanes, and the breeze, unlike the hot and arid winds of Coromandel and the Deccan, is always more or less refreshing. The range of the thermometer exhibits no violent changes, and never indicates a temperature insupportably high. The mean, on an annual average, scarcely exceeds 80° at Colombo, though in exceptional years it has risen to 86°. But at no period of the day are dangerous results to be apprehended from exposure to the sun, and except during parts of the months of March and April, there is no season when moderate exercise is not practicable and agreeable. For half the year, from October to May, the prevailing winds are from the north-east, and during the remaining months the south-west monsoon blows steadily from the great Indian Ocean. The former is subject to many local variations and intervals of calm. But the latter, after the first violence of its onset is abated, becomes more nearly uniform throughout the period of its prevalence, and presents the character of an on-shore breeze, extending over a prodigious expanse of sea and land.

"There are, of course, abnormal seasons with higher ranges of temperature, heavier rains, or droughts of long continuance, but such extremes are exceptional and rare. Great atmospheric changes occur only at two opposite periods of the year, and so gradual is their approach that the climate is monotonous, and one longs to see again the falling of the leaf to diversify the sameness of perennial verdure. The line is faint that divides the seasons. No period of the year is divested of its seedtime and its harvest in some part of the island, and the ripe fruit hangs on the same branches that are garlanded with opening buds."

The following more detailed description is given of the climate of Colombo :—

"At the opening of the year the north-east monsoon, which sets in two months previously, is nearly in mid-career. It reaches Ceylon comparatively dry, and its general effects are parching and disagreeable. This character is increased as the sun recedes towards its most southern declination, and the wind acquires a more direct draught from the north; passing over the Indian peninsula and almost totally divested of humidity, it blows down the western coast of the island and is known there by the name of the 'along-shore wind.' For a time its influence is uncomfortable and its effects injurious to both health and vegetation. It warps and rends furniture, dries up the surface of the earth, and withers the delicate verdure. These characteristics, however, subside towards the end of January, when the wind becomes somewhat variable with a westerly tendency and occasional showers, and the heat of the day is then partially compensated by the greater freshness of the night.

"February is dry and hot during the day, but the nights are cloudless and cool. Rain is rare, and when it occurs it falls in dashes, succeeded by damp and sultry calms. The wind is unsteady, and shifts from north-east to north-west, sometimes failing entirely between noon and twilight. The difference of temperature between day and night is frequently as great as 15° or 20°.

"In March the heat continues to increase. The day becomes oppressive, the nights unrefreshing, the grass is withered and brown, the earth hard and cleft, the lakes shrunk to shallows, and the rivers evaporated to dryness. Europeans now escape from the low country and betake themselves to the shade of the forests adjoining the coffee plantations on the hills, or to the still higher sanitarium of Newara Eliya. The winds, when any are perceptible, are faint and unsteady with a still increasing westerly tendency; partial showers sometimes fall, and thunder begins to mutter towards sunset.

"April is by far the most oppressive portion of the year for those who remain at the sea-level of the island. A mirage fills the hollows with mimic water. The heat in close apartments becomes extreme, and every living creature flies to the shade from the suffocating glare of mid-day. At length the sea exhibits symptoms of an approaching change; a ground swell sets in from the west, and the breeze towards sunset brings clouds and grateful showers.

"May is signalised by the great event of the change of the monsoon and all the grand phenomena which accompany its approach. About the middle of the month, but frequently earlier, the sultry suspense is broken by the arrival of the wished-for change. As the monsoon draws near the days become more overcast and hot, banks of clouds rise over the ocean to the west. At last the sudden lightnings flash among the hills and sheet through the clouds that overhang the sea, and with a crash of thunder the monsoon bursts over the thirsty land, not in showers or partial torrents, but in a wide deluge that in the course of a few hours overtops the river banks and spreads in inundations over every level plain.

"The extreme heat of the previous month becomes modified in June; the winds continue steadily to blow from the south-west, and frequent showers, accompanied with lightning and thunder, serve still further to diffuse coolness throughout the atmosphere and verdure over the earth.

"July resembles to a great extent the month that precedes it, except that in all particulars the season is more moderate, showers are less frequent, less wind and less absolute heat. In August the weather is charming, notwithstanding a slight increase of heat. The same atmospheric condition continues throughout September, but towards its close the sea breeze becomes unsteady and clouds begin to collect, symptomatic of the approaching change to the north-east monsoon. The nights are always clear and delightfully cool. Rain is sometimes abundant.

"October is more unsettled, the wind veering towards the north with rain pretty frequent. November sees the close of the south-west monsoon and the arrival of the north-eastern. In the early part of the month the wind visits nearly every point of the compass, but shows a marked predilection for the north. The great change is heralded as before by oppressive calms, lurid skies, vivid lightning, bursts of thunder and tumultuous rain. But at this change of the monsoon the atmospheric disturbance is less striking than in May, the previous temperature is lower, the moisture of the air is more reduced, and the change is less agreeably perceptible from the southern breeze to the dry and parching wind from the north.

"In December the wind setting in steadily from the north-east brings with it light but frequent rains from the Bay of Bengal. The morning and the afternoon are again enjoyable, but at night every lattice that faces the north is cautiously closed against the treacherous 'along-shore wind.'"

With respect to the climate of Kandy, Sir Emerson Tennant writes:—

"Kandy from its position shares in the climate of the western coast, but from the frequency of the mountain showers, and its situation at an elevation of upwards of 1600 feet above the level of the sea, it enjoys a much cooler temperature. It differs from the low country in one particular which is very striking—the early period of the day at which the maximum heat is attained. This, at Colombo, is generally between two and three o'clock in the afternoon, whereas at Kandy the thermometer shows the highest temperature of the day between ten and eleven o'clock in the morning. In Kandy the nights are so cool that it is seldom that warm covering can be altogether dispensed with."

And of Trincomalee:—

"At Trincomalee the climate bears a general resemblance to that of the Indian peninsula south of Madras; showers are frequent but light, and the rain throughout the year does not exceed forty inches.[1] With moist winds and plentiful dew this sustains a vigorous vegetation near the coast, but in the interior it would be insufficient for the culture of grain were not the water husbanded in tanks; and the bulk of the population are, for this reason, settled along the banks of the great rivers.

"In the extreme north of the island, the peninsula of Jaffna and the vast plains of Neura Kalawa and the Wanny form a third climatic division, which, from the geological structure and the peculiar configuration of the district, differs essentially from the rest of Ceylon. This region, which is destitute of mountains, is undulating in a very slight degree; the dry and parching north-east wind desiccates the soil in its passage, and the sandy plains are covered with a low and scanty vegetation, chiefly fed by the night dews and whatever moisture is brought by the on-shore wind. The total rain of the year does not exceed thirty inches,[2] and the inhabitants live in the frequent apprehension of droughts and famines. These conditions attain their utmost manifestation in the extreme north, and in the Jaffna peninsula; there, the temperature is the highest in the island, and owing to the humidity of the situation and the entire absence of hills it is but little affected by the changes of the monsoons. The soil, except in particular spots, is porous and sandy, formed from the detritus of the coral rocks which it overlays. It is subject to droughts, sometimes of a whole year's continuance, and rain, when it falls, is so speedily absorbed, that it renders but slight service to cultivation, which is entirely carried on by means of tanks and artificial irrigation."

The agricultural and horticultural products of Ceylon are even more varied than those of most parts of India. There is perhaps no plant native to the tropics, and but few sub-tropical forms of vegetation, that may not be raised either on the low country or the hills; and the actual exports, in addition to tea, coffee, and cinchona, which are (or were) the chief produce of the plantations, include the cocoa-bean, cocoa-nuts, india-rubber, cardamoms, cinnamon, cloves, nutmegs, pepper, and many other spices, vanilla, betel nuts, and most of the agricultural productions of India. Coffee, which, twenty years ago, was the great staple article of ex-

[1] Recent registers show that this is much underestimated. The rainfall of Trincomalee is over 60 inches. But it holds good of the west coast in the latitude of Trincomalee.

[2] This again is underestimated. The average rainfall of Jaffna is 49 inches, and that of Manaar 40 inches, as the result of 16 years' registers.

port, has suffered so severely from the attacks of a fungus, known as the leaf-disease, that from 1877 to 1883 the exports sank from 850,000 to 250,000 hundredweights, and the average produce from 4½ to less than 1½ hundredweights per acre. Many of the estates, formerly devoted exclusively to coffee, have, in consequence, been either abandoned to jungle, or partly planted with tea, cinchona, or cocoa; and at the present time the cultivation of tea, more especially, is advancing by leaps and bounds, in such measure that Ceylon is already a formidable rival of Assam in the production of this important staple. Those who may desire to obtain detailed information on these and kindred subjects may be referred to Ferguson's *Ceylon Handbook and Directory*, a most valuable compendium of general information relating to the colony.

The detailed climatic statistics of three stations are given in the Appendix, viz. Trincomalee, on the north-east coast, representing the drier, but not the driest portion of the island; and that which, like the Carnatic, receives its chief rainfall during the transition from the south-west to the north-east monsoon. Kandy, which may be taken as representative of the climate of the lower hill region, the chief seat of tea and coffee culture, for which it enjoys the great advantage of almost constant humidity, a very equable temperature, and a copious rainfall at two opposite seasons. And finally, Colombo, which is the seat of government, the commercial capital and chief port of the island.

The mean temperature of Trincomalee is the same as that of Madras and the southern Carnatic, viz. 82°, but the annual range is less; the temperature of December and January being only 4° below the mean, and that of the four months, April to July, 3° above it. The minimum night temperatures are even more uniform, averaging between 74° and 77° throughout the year, and never falling below 65°. The day maximum averages 83° in the coolest and 93° or 94° in the warmest months, and reaches to between 97° and 102° only on the hottest days of the year. The mean humidity of the observatory is 72 per cent, in November 82 per cent of saturation, and in July, here the driest month of the year, 65. From May to September, that is, all through the south-west monsoon, it averages below 70, owing to the position of the station on the lee coast of the island, so that the south-west monsoon here blows as a land wind. The mean annual rainfall is 62 inches; of which two-thirds fall in the four months, October to January, and only one-seventh from March to July inclusive. There are, on an average, 104 rainy days in the year, of which also two-thirds occur in the four months, October to January.

Kandy has a still more uniform climate, with a mean temperature of only 76°. That of January is 74°, and that of the three warmest months, March to May, 79°. The minimum night temperatures range from 68° to 73° in different months, and the maximum day temperatures from 80° to 87°. The highest temperature yet recorded is below 95°, and the lowest 54°. The mean humidity is 77 per cent of saturation, varying from 69 in April, the driest month, to 81 in June and 80 in July and also from October to December. The average rainfall is 84½ inches, and this is more evenly distributed through the year than is the case on either coast of the island. October and November have on an average more than any other months, and February and March the least. But in no month does the average fall exceed 12 inches or fall below 2½ inches, and excepting in February and March, it rains on an average on from 10 to upwards of 20 days in each month. The average number of rainy days in the year is 185.

Colombo is 5° warmer than Kandy, but the annual range of temperature is even less. The mean of the year is 81°, but that of the warmest months is only 2° higher, and that of the coolest 2° lower than this mean. On the average of 12 years, the difference of the highest and lowest readings in the year is but 25°, and that of the highest and lowest yet recorded only 5° greater. In February alone, the range of the thermometer is nearly as great as in the whole course of the year. The humidity of the atmosphere is high, viz. 78 per cent of saturation, and even in the driest month, February, it averages 73 per cent; but it occasionally falls much below this during the heat of the day, with a north wind. The average rainfall is 87 inches, of which between 12 and 13 inches falls in each of the three months May, October, and November, and less than 2 inches in February. In most other months except January, the fall is little if at all below 5 inches. There are fewer rainy days than at Kandy, viz. 159 in the year on an average.

Burmah.—Of the climate of Burmah I am able to give a detailed account only in so far as relates to the provinces which have been under British rule since 1825 and 1852 respectively, viz. Arakan, Tenasserim and Pegu. In the case of the recently-annexed province of Ava, it is known generally that the climate is drier and the annual range of temperature greater than in the older maritime provinces, but the only statistical data at present available are the registers of nine months of 1879 at Mandalay, recorded by the late Mr. R. Shaw, and those of two months of 1886. An abstract of most of these is given in the Appendix. The observatories which have now been established at Bhamo and Kindat, in addition to that of Mandalay, will, in the course of a few years, afford a better idea of the characteristic climatic features of the newly-acquired territory.

The position and configuration of Arakan and Tenasserim, on the west coast of the peninsula, with hill ranges running parallel with the coast, expose them to the influence of the south-west monsoon of the Bay of Bengal, in the same manner and as fully as are the Konkan and Malabar to that of the Arabian Sea; and with a similar result, viz. an excessive rainfall from June to September. In Arakan, however, this rainfall is more prolonged than on the west coast of the Konkan in the same latitudes. The southerly monsoon continues to blow in the eastern half of the bay for some weeks after it has ceased in the western half and in India generally, and the stormy weather which frequently prevails on the bay in October and the beginning of November is always accompanied with southerly winds, which discharge a good deal of rain on the hills of Arakan, as well as on those of the more southerly portion of the Malay peninsula.

In Pegu, which lies wholly east of Arakan, and to leeward of the coast range, the sea face of the Irawadi delta shares the heavy rainfall of the coast generally; and the hill ranges to the east of the delta, the Pegu Yoma, and that which separates the Sitang from the Salwin, also have a high rainfall; but the Irawadi valley above Prome is much drier, and this character becomes more and more marked as one ascends the river to Mandalay.

The valley of the Irawadi consists of plains intersected by low isolated ranges of hills, running north and south; and these, at certain places, hem in the main stream of the river, forcing it to flow in a narrow but deep channel. On the west, the valley is bounded by the Arakan Yoma, which commences at Cape Negrais, and consists of a belt of parallel ridges, one of which rises to 4000 feet in about lat. 18°, but sinks again to about 1000 feet at the Pegu frontier. Farther north, it gradually increases in elevation as well as in width; and in latitude 22° 30′ rises in the Blue Mountain to a height of 7100 feet. The extensive hill tract covered by this range and its off-shoots between the provinces of Ava on the east and Arakan and Chittagong on the coast, is occupied by native hill tribes, and covered with forest. But few parts of it have ever been visited by Europeans.

To the east of the Irawadi the Pegu Yoma, a range not more than 2000 feet in height, intervenes between that river and the Sitang, and contains some valuable teak forests. Another and much loftier range separates the latter from the Salwin, and its northern extension becomes a broad series of parallel ridges occupied by the Karens, and east of Bhamo by the Kakhyens. The forests of these hills and those of the independent Shan states east of the Salwin have for many years past furnished the chief supplies of teak timber to the ports of Rangoon and Moulmein. The forests of Burmah and the neighbouring states contain many other valuable timber trees, among which Padouk is already in some demand both in India and Europe. The chief agricultural export of Burmah is rice.

The whole of Burmah is swept by the summer monsoon from the Bay of Bengal, which impinges obliquely on the coast and, as already observed, discharges a very heavy rainfall on the seaward face of the coast ranges and on the marginal portion of the Irawadi delta. On the plains of this river it prevails as a south wind, its southerly direction being probably determined by that of the numerous hill ranges; but it is also probable that it passes over the crest of the higher ranges with its normal south-west direction, blowing towards the plains of China and the slopes of the eastern Himalaya. Ava, which lies under the lee of the Arakan coast range, has a comparatively dry climate and a vegetation which, according to Hooker and Thomson, resembles that of the Carnatic; but the northern part of the province beyond Bhamo is damper, owing to the loftier mountains and the more irregular surface of the country, and the vegetation is very similar to that of Assam. In the winter months the winds are light and northerly; up to the end of January, chiefly north-east at Thyet Myo and in the Irawadi delta, and north-west at Tounghoo. At the end of January they become north-westerly and gradually back to south-west with the increasing heat of the land. From December to April but little rain falls in any part of Burmah, with the partial exception of the most southern districts of Tenasserim; but in May rain becomes frequent, and during the succeeding four months a rainless day occurs but rarely, save only on the plains of the Irawadi north of the delta. During this season the atmosphere is saturated with damp, and at Rangoon and Moulmein the temperature is moderate, not exceeding an average of 83° or 85° at the warmest time of day, and sinking to 74° or 76° at night. In Ava, however, where the rainfall is light, the temperature of this season is much higher, sometimes reaching to nearly 100° in the shade, to judge from the Mandalay registers of 1879.

A very marked feature of the climate of Pegu, and one which probably goes far to explain its notorious unhealthiness, is the great change of temperature during the 24 hours, combined with the prevailing dampness of the atmosphere. At stations in the interior, such as Thyet Myo and Tounghoo, the difference between the thermometer readings at sunrise and in the afternoon, in the earlier months of the year, is as great as in the driest parts of the Punjab, and some 10° or 12° greater than in Bengal, although the atmosphere is not drier than in the latter province. As all the meteorological observations have been recorded under exactly similar conditions at all stations, there is no reason whatever to question the validity of this comparison.

Abstracts of the chief climatological data of seven stations are given in the Appendix, viz. Akyab, on the coast of Arakan; Mandalay (for 11 months only), as representing Ava; Thyet Myo, in the driest portion of Pegu, 11 miles from the northern frontier of the province; Tounghoo, in the Sitang valley, east of the Irawadi and the Pegu Yoma; Rangoon, in the eastern corner of the Irawadi delta, and 25 miles from the embouchure of the eastern branch of the river; Moulmein, on the Salwin, 24 miles above where it enters the Gulf of Martaban; and Mergui, on the sea coast of Tenasserim, near the southern extremity of the province.

The annual mean temperature of Akyab is 79°, or 2° higher than that of Chittagong, on the same coast but 160 miles farther north. That of January, which month, owing to the fall of the night temperature is 3° cooler than December, is 69°, and that of April and May 84°. In July and August the mean temperature sinks to 81°, the maximum of the day falling from 92° to 84° or 86°. The mean minimum temperature in January is 59°, and the lowest in the year varies from 47° to 55°. The highest yet recorded is below 100°, and has ranged between 93° and 99° in the past 13 years. In February the diurnal range averages 25°, but at the height of the rainy season sinks to 7° or 8°. The mean humidity of the station, which is situated on a narrow point of land between the sea and the broad estuary of the Koladyne river, is 80 per cent of saturation, falling to 70 in February and rising to 89 in July and August. The mean annual rainfall is 196 inches, but from December to March inclusive the fall is insignificant, and that of April under 2 inches. In May, however, rain is frequent, amounting to between 12 and 13 inches on an average on 14 rainy days, and the average fall in each of the months June and July is over 50 inches. In August the fall is lighter but not less frequent, and even in September it rains on two days out of three, and in October on more than one in three. The average number of rainy days in the year is 140.

For Mandalay we have complete registers for only 11 months, and relating to two different years, and rainfall registers for 17 months, including two monsoons. From the former, a mean temperature of 81° has been computed, but this is probably rather higher than the true normal average. That of January 1879 was 71°, and that of April 91°. The mean minimum of January was 59°, and 52·9° was the lowest reading recorded. In April and May the mean of the maxima was over 100°, and the highest reading registered 106·1°. The winter temperature was considerably higher than that of places in the same latitude in India, and only in January a little lower than that of Cuttack, 1½° farther south. After April the mean temperatures were about the same as those of Cuttack, but the rainfall being lighter the afternoon temperatures were higher at Mandalay in July and August. The mean atmospheric humidity of the year was 64 per cent of saturation, and that of March 43. September was the dampest month, when it rose to 77 per cent, and September and November (October being wanting) were the dampest months in the year. The rainfall, as far as can be deduced from the 17 months' registers, is only 27½ inches, and in no month apparently does it rain more frequently than about one day in three. The first three months of the year were rainless.

Thyet Myo has a mean temperature of 79°, being therefore *apparently* cooler than Mandalay, although it lies more than 2° farther south; but this must be considered as doubtful. The mean January temperature is 68°, and that of April, the warmest month, 87°. In January the mean of the morning minima is 54°, and the lowest readings of each year have varied between 40° and 49°. In April the mean of the afternoon maxima is 103°, and 110·2° is the highest reading yet recorded. In the first three months of the year the average daily range is from 31° to 36°, which is as great as in the driest months in the driest part of the Punjab, and must be the more trying to health that the atmosphere is by no means so dry as that of the latter province. The mean humidity of the year is 72, and that of the driest months, March and April, 56 per cent. In the rainy season (July and August) it is as high as 84. The rainfall averages about 45½ inches in the year, and the four months, December to March, are almost rainless. As in other parts of Burmah, the rainfall of May is considerable, viz. 5 inches; but during the summer monsoon it is only 7 or 8 inches in the month, and as much as 5 inches in October. It rains on an average on 107 days in the year.

Tounghoo is slightly cooler than Thyet Myo, the mean temperature being 78°. The temperature of January is, however, relatively higher, viz. 70°, and that of April lower, viz. 85°. The mean minimum of the mornings in the former months is 57°, and the lowest of the year between 47° and 54°. In April the afternoon maxima average 100°, the highest varying from 101° to 107° in different years. The range during the day in the first three months is not quite so great as at Thyet Myo, but still very considerable, viz. from 29° to 32° on an average, and in the rainy season it amounts to 12° or 15°. The mean humidity is 76 per cent of saturation, and in March not below 61 per cent. In July and August it averages 87 and 88 per cent. The mean annual rainfall is 78 inches, scarcely any of which falls in the first three months of the year. But the rainfall of May amounts to over 6 inches, and the rains continue till quite the end of October. The average number of rainy days in the year is 135.

Rangoon has a damper and more equable climate. The mean temperature is the same as that of Akyab and Thyet Myo, viz. 79°; but the temperature of the winter months is higher and that of the rainy season lower, so that there is a difference of only 3° between the mean temperature of January and those of July, August, and September. Indeed the afternoon maxima are lower from June to September than in any other months of the year. In January the mean temperature is 75°; that of the morning minima 64°, and that of the afternoon maxima 88°. Even here, on the margin of the Irawadi delta, within a few miles of the sea, and surrounded by broad rivers, the diurnal range is between 20° and 30° in the first four months of the year. In April, the mean temperature is 84°, the afternoon maxima average 98° and they often exceed 100°. But heavy rain begins in May, rapidly bringing down the heat of the daytime, and from June to September the mean temperature is only 78° or 79°, the afternoon maxima 85° or 86° on an average, and the diurnal range only 9°. The mean humidity of the station is 78, that of February 62, and that of July and August 90 and 91 per cent. The mean annual rainfall is but little below 100 inches, and of this less than one inch falls from December to March inclusive. From June to September, on the other hand, a rainless day is rare, and in October there are about 14 rainy days. The average number in the year is 153.

Moulmein has a very similar climate in point of temperature and general humidity, but a much higher rainfall. The winter temperatures are about the same as at Rangoon, but from April to September they are about 1° lower, and on the hottest days in the year the thermometer rarely reaches 100°. The diurnal range of temperature is either the same or rather less at the different seasons, and the humidity of the atmosphere less in the winter months, and about the same in the rains. From December to March the weather is equally rainless, but in the remaining eight months the rainfall averages not less than 188 inches. In May it is as much as 19 inches, and in July and August there is an average of nearly an inch and a half a day. In these months the atmosphere is always near saturation.

Mergui, the most southerly station in Burmah, resembles places on the south-west of Ceylon in the equability of its temperature throughout the year, but the mean is 3° lower, viz. 78°. July and December are the coolest months, and have the same mean temperature, viz. 76°, but while the average maximum of the day is 88° in December, in July it is only 84°. April, the warmest month, has a mean temperature of 81°, and an afternoon maximum of 92°. In most years the thermometer never sinks to 60° on the coolest morning, and it has never risen to 100° on the hottest days during the last 9 years. The mean humidity is high, viz. 82 per cent, and in no month does it fall below 73; while in August and September, which are here the dampest months, it rises to 90 per cent of saturation. The dry season is shorter and less dry than anywhere to the north. December and January are the only months in which the rainfall averages less than an inch, but in the summer monsoon the fall is less heavy than at Moulmein, and the mean rainfall of the year is about 160 inches. It rains on an average on 168 days.

Andaman and Nicobar Settlements.—The climate of these islands, which have been occupied as convict settlements since 1858, is almost equatorial in its uniformity and in many respects similar to that of Tenasserim. The Andamans are hilly, the hills being for the most part only a few hundreds of feet in height, and covered with forest, except where clearings have been made around the settlement. Mangrove swamps extend in places along the sea margin, and outside is a fringe of coral reef. The Nicobars, as far as they are known, are similar; but owing apparently to the magnesian character of certain of the rocks, there are large tracts, the dense clay soil of which is unfavourable to vegetation, and these constitute grassy plains, bare of forest. The Nicobar settlement of Nancowry has a much worse reputation than that of the Andamans for health, and this appears to be due to the different characters of the sites selected. On the latter islands clearings have been made in the once unbroken forest. The soil is porous, and admits of good drainage; but in the former, the grass land has been occupied, perhaps on account of its openness, but the stiff clay, which does not support vegetation, is impervious to drainage, and an obstinate form of fever is very prevalent among the settlers.

Both settlements are warmer than the Tenasserim mainland in the same latitude, the mean temperature of Port Blair and Nancowry being 80°, while that of Moulmein and Mergui is only 78°. This excess may possibly be due to both the former having an eastern aspect, and to their being therefore on the lee-side of the islands in the south-west monsoon. There is but little variation during the year; March and April are the warmest months, with mean temperatures of 82° and 83° respectively, and a mean daily maximum of 92° at Port Blair and 89° at Nancowry. A reading of 99·5° has been once recorded at Nancowry, but at Port Blair 96·4° is the highest, and 94° or 95° on an average, the highest in the year. The lowest yet recorded at Port Blair is 66°, and at Nancowry 70·2°. The average extreme range of temperature in the course of the year is only 26° at the former, and 22° at the latter. The diurnal range of temperature at Nancowry never averages more than 10° or 11° in any month, and varies but little at different seasons. But at Port Blair it is as much as 14° or 15° in the driest months, February, March, and April. Port Blair is the damper station of the two; the mean humidity being 83, while that of Nancowry is only 79 per cent of saturation; and this difference appears to hold good in all months. The rainfall of Port Blair is rather heavier, especially during the south-west monsoon; but Nancowry has more rain from November to April inclusive. At both stations the monsoon rains set in in May, and the rainfall of that month is little less than that of June. The number of rainy days is about the same at both stations, and amounts to nearly half the days in the year; but during the summer monsoon, while at Port Blair there are only 5 or 6 rainless days in the month, there are about twice as many at Nancowry.

WEATHER AND WEATHER REPORTS.

Climate and Weather.—The distinction of climate and weather in the language of everyday life is too familiar to need more than a cursory reminder of the sense in which the terms are here employed. By the former are meant the average conditions of heat and cold, damp and dryness, and the like, characteristic of a place or country, in so far as they vary regularly with the succession of the seasons. The latter has reference to the apparently irregular, though not always minor, atmospheric changes that take place from day to day. Climate depends on the latitude, on the height above sea-level, on the distance from the sea, on the form and slope of the land surface, on the nature of the soil and its vegetation, and a number of other circumstances, for the most part permanent, and generally easy to ascertain. The vicissitudes of the weather stand in very different case; and although of late years much has been done to determine fixed relations between weather changes and those great movements of the atmosphere that immediately give rise to them, the ulterior causes that produce these movements can be indicated only in very general terms. In great part they would appear to be beyond the range of our field of observation, and can be arrived at, if ever, only by a slow process of inference.

Anticyclones and Cyclonic Depressions.—The modern practice of weather forecasting is based on the known mutual relations of those tracts of relatively high and low barometric pressure that have been described at page 24 of this work; and which are respectively the source and goal of the surface winds. The former are generally termed anticyclones, the latter, barometric depressions, or cyclonic systems, or simply "cyclones"; though this extension of the latter term, from the hurricanes of the tropics (to which alone it was originally applied), to all barometric depressions, is attended with some inconvenience. As has been explained at page 46, the terms "cyclone" and "anticyclone" have reference to the course of the winds around the centres of low and high pressure respectively; which courses are different in the northern and southern hemispheres, though constant in the same hemisphere. In the northern hemisphere, with which we are now concerned, the winds blow spirally outwards from an anticyclone, the curve of the spiral being with the movement of the clock hands, or "clockwise"; and they blow spirally inwards to a cyclonic centre against the clock movement, or counter-clockwise. Each of these systems has its characteristic type of weather. In an anticyclone, as a rule, in all parts of the world and always in India, the weather is fine, and either the skies are cloudless, or the cloud, if any, is high; except that in damp regions, such as Assam, a high barometer is sometimes accompanied with a ground fog. In and around a cyclonic system the weather is more or less unsettled and cloudy, stormy or rainy, according to the season and the intensity of the depression. A partial exception to this rule occurs indeed in the Indus valley region throughout the summer monsoon, in so far that, where the barometer is lowest, the atmosphere is hot and oppressive, but the weather remains fine, and rain is restricted, as a rule, to the outskirts of the depression, where the barometer is somewhat higher. Owing to other causes, a clear sky and a calm atmosphere are also sometimes met with in the very centre of a cyclone.

The general prevalence of fine dry weather in the Indian cold season and of rainy and stormy weather during the summer monsoon, is primarily due to the fact that, in the former, Northern India is generally, for weeks together, a region of high barometer, and in the latter of low barometer; and it is in virtue of the comparative stability of these opposite conditions that the opposite characters of the two seasons are so marked. The anticyclone over Northern India in the cold season is, however, not quite stable. Now and then a barometric depression partly or entirely displaces it; and it is on such occasions that we have the cold weather rains of Upper India. How these depressions originate is at present by no means clear. In some cases they seem to be formed in North-western India, or even farther south. In others they probably reach India from the west; though, in the absence of weather reports from any of the countries that lie in that direction (beyond Quetta) it is impossible at present to verify this. It is, however, certain that when these cold weather depressions travel from the place of their first appearance, they always move in some easterly direction, and thus it is that a spell of rain in the early months of the year begins, as a rule, in North-western India, and on subsequent days extends to the North-west Provinces, and sometimes, though more rarely, to Bengal. In the rainy season they are formed either in some part of the permanent depression then existing in Northern India, or more frequently, on its outskirts, in the prolongation of its axis over the north of the Bay of Bengal; and their subsequent course is between west and north-west. In the intervals between the monsoons they originate over the seas around India, and more often on the Bay of Bengal than on the Arabian Sea. The direction in which they travel is then more variable, but still most frequently to the north-west or west-north-west.

It is to such travelling depressions as these, which, in their more intense form constitute cyclones (in the original sense of the term), and in their less intense form what I have termed cyclonic storms, that the more striking changes of weather are due. They are always accompanied with heavy rain; on the mountains in the winter and spring months with snow; and frequently with stormy winds.

At sea they are more violent than on land, and their characteristic features on Indian seas will be the subject of a special section of this work.

When such a storm has appeared in any part of the Indian weather report system, the general experience already gained of the storm tracks of the different seasons renders a certain amount of weather prevision possible; but such prevision is better grounded when the storm has already begun to travel, and the direction to which it tends is shown by the reports of two or more successive days.

How the weather is affected by the passage of these storms at different seasons will be described presently. But before doing so, it is necessary to notice another class

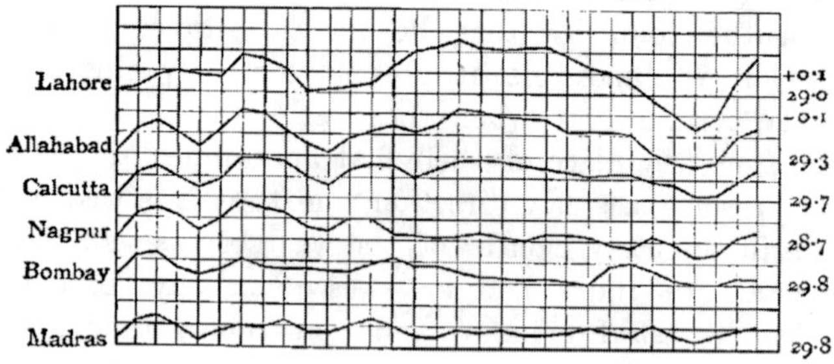

FIG. 10.—BAROMETRIC VARIATIONS IN MARCH 1887.

of barometric oscillations, the cause and origin of which are even more obscure than are those of cyclonic storms.

Barometric Surges.—These oscillations have already been shortly noticed at page 22. If the readings of the barometer at 8 A.M., as now published in the Simla weather reports, be plotted on ruled paper for say a dozen stations in different parts of India, and if this be repeated day by day for a month, a connecting line being run through all the readings of the same station, as in Fig. 10, it is found that these lines show certain oscillations common to all or nearly all the stations, and apparently simultaneous, but generally greater in amount at the northern than at the more southern stations. These oscillations are evidently independent of those produced by the passage of cyclonic storms, which are much more local and not simultaneous at distant places. They are irregular in period, each oscillation lasting from 3 or 4 days to a fortnight. As was remarked in a former place, they do not necessarily accompany or betoken any important change of weather, but they are apparently not without influence on the weather, since there is more tendency to cloud and the formation of summer squalls in the hot season and of cyclonic storms in the rainy season, when the barometer is below the mean, than when it is above it. Mr. Eliot has remarked, with regard to them, that their coincidence over so large an area "points to the probability of their being produced by the alternate expansion and contraction of the atmosphere."[1] In the absence, however, of any fuller explanation of this view, it seems hardly reconcilable with the fact that they are as great or probably greater in higher than in lower latitudes, that is to say, where the alternate expansion and contraction of the atmosphere are of less magnitude than in India. They occur in the temperate zone as well as in the tropics, though perhaps their occurrence is there more masked by the greater barometric changes arising from other causes. Mr. Abercromby has recognised them in his valuable work on *Weather*, and has termed them "surges," the name here adopted, remarking that they have a decided tendency to travel from west to east, or from south-west to north-east; and further, that while the passage of a surge exercises a moderate influence on the characteristic weather of any isobaric shape (cyclonic depressions, anticyclones, etc.), it exercises a very powerful influence on the formation of new systems. He states also that in the South Indian Ocean, during the period of the north-west monsoon (December to March), "hurricanes almost invariably form during the downward period of the surge, and practical forecasters, like Meldrum at the Mauritius, are always specially on the lookout for signs of serious bad weather, whenever there is the slightest symptom of a non-diurnal diminution of pressure." To a certain extent this is in accordance with Indian experience.

[1] *Report on the Meteorology of India in* 1877, p. 48.

In the Simla weather reports, the passage of these surges is shown by the column that gives the change of the barometer in the preceding 24 hours. If, as sometimes happens, the whole or the large majority of the stations in all parts of India show a rise or fall, this is due to the passage of a surge wave. As a rule, however, the surge is associated with changes of a more local character, and these may add to or detract from the general change, or even locally neutralise it. It is, of course, impossible so to analyse the total change as to assign to each element its exact value.

Weather of the Cold Season.— The most settled weather of the year in Northern India is from the cessation of the rains in September or October to the latter part of December. It is but seldom that the serenity of the climate of the greater part of extra-tropical India is disturbed by storms, either from the Bay of Bengal or of more northerly origin, during these two or three months. The barometer is high in the north-west; and light north-westerly winds or calms prevail down the Gangetic plain, becoming cooler as the year draws towards its close. Storms are, indeed, still formed in the south of the bay and often cross the peninsula, their tracks becoming more southerly as the season advances, until, by the end of December or a few days later, they cease to appear in Indian seas. The first cyclonic disturbances producing the winter rains of Northern India generally make their appearance in the latter part of December or the beginning of January, and they are repeated at intervals in this and the following months. One or two examples will show, better than any general description, the changes that precede and accompany this cold weather rainfall. The first instance is one in which the disturbance apparently originated in India.

On reference to Figure 3 on page 25, which shows the average distribution of pressure in the month of January, it may be observed that on the west coast of India the isobars all make a loop northward, indicating a trough of low pressure. In January 1883 this trough was more than usually developed, and on several occasions in the early part of the month extended northward, encroaching on the region of high pressure that, at this season, usually occupies North-western

India. This state of things is shown in Fig. 11, which is the weather chart of 10 A.M. on the 4th January. The wind arrows show that, in conformity with this anomalous barometric feature, light southerly winds prevailed at several stations in the west of India, and easterly winds in Kathiawar and Rajputana, blowing cyclonically around the

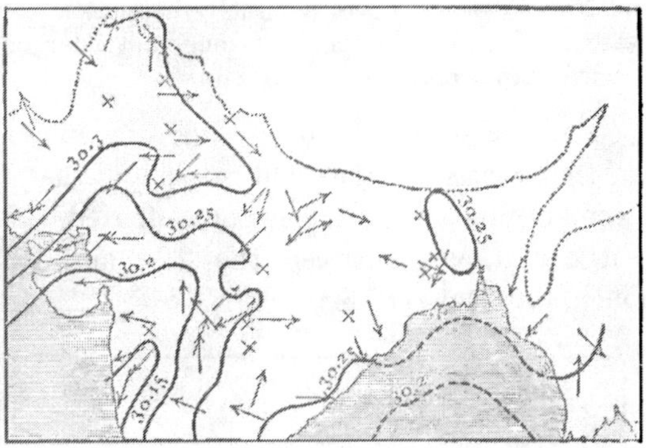

FIG. 11.—WEATHER CHART FOR 10 A.M., 4th January 1883.

loop of the isobars, and bringing damp oppressive weather to Western India. Up to the 9th, only a few light showers fell on the most northerly hills of the Punjab and at their foot. But on this day the depression extended so as to occupy the whole of Central and Western India, as shown on the weather chart for 10 A.M. of the 9th (Fig. 12).

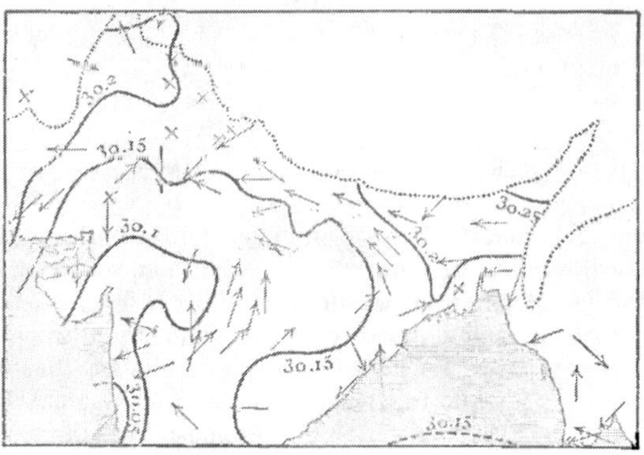

FIG. 12.—WEATHER CHART FOR 10 A.M., 9th January 1883.

Damp and warm southerly and easterly winds now prevailed over the whole of Central and Northern India, and on this and the following day rain fell over the greater part of the North-west Provinces, the Eastern Punjab, Central India, and Chutia Nagpur, up to the confines of Bengal, and snow on the North-west Himalaya. This rainfall tract is indicated by the shaded portion of the chart.

When the rain ceased, on the 10th, a band of high pressure was established from the Indus valley across Rajputana and Central India, and cool west and north-west winds prevailed for a day or two in the Punjab and North-west Provinces, as usually happens after falls of rain at this season. But this did not last long, and later on the trough again extended with a repetition of the rain.

The second instance is of a different kind and more typical of the storms of the cold season, although in some respects remarkable and, in one, unprecedented. It occurred towards the end, viz. between the 22d and 28th of the same month, January 1883.

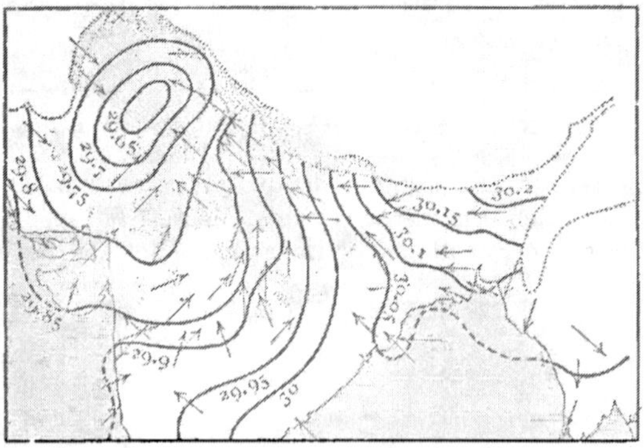

Fig. 13.—Weather Chart for 10 a.m., 25th January 1883.

On the 22d and 23d the barometer fell rapidly over the Indus valley to a minimum in Upper Sind, Rajputana, and the Punjab, and on the 23d and 24th rain set in over the whole of these provinces. On the 25th a well-defined cyclonic depression appeared in the Punjab, as represented in Fig. 13, which is the weather chart for 10 a.m. of that day; while in Bengal, Assam, and Burmah the pressure rose from 0·1 to 0·15 inch. A high barometric gradient was therefore established from east to west over Northern India, and steady southerly winds blew over the whole of India. Up to the forenoon of the 25th, the rain, which had ceased in Sind and Western Rajputana, extended eastwards over Guzerat, Eastern Rajputana, Central India, the whole of the Punjab, and the Gangetic plain as far as Cawnpore and the east of Oudh. Heavy snow fell on the Himalaya, and continued without intermission till the 27th, and on the latter day extended to the high plain at the foot of the Himalaya above the Salt Range, so that at Rawalpindi, on the night of the 27th, the ground was covered to a depth of 4 inches. No such occurrence has been recorded either previously or since at so low a level. On the 26th, 27th, and up to the 28th, a barometric minimum continued to occupy the Punjab, the pressure rising, however, from a minimum of 29·65 to 30 inches, while the rain continued extending eastward, the weather clearing up in the west; so that, on the afternoon of the 27th and forenoon of the 28th, rain was restricted to Bengal, Assam, the Himalaya, and the Upper Punjab. Probably, therefore, the main disturbance had moved eastwards over Tibet.

On the 28th a spell of cold weather with north-west winds set in over Northern India, the temperature falling for several days, and the lowest temperatures of the year were registered in the first week of February: at Simla and Darjiling the lowest on record. In all probability this cold was the effect of the enormous snowfall which, as mentioned above, descended to lower levels than has been recorded on

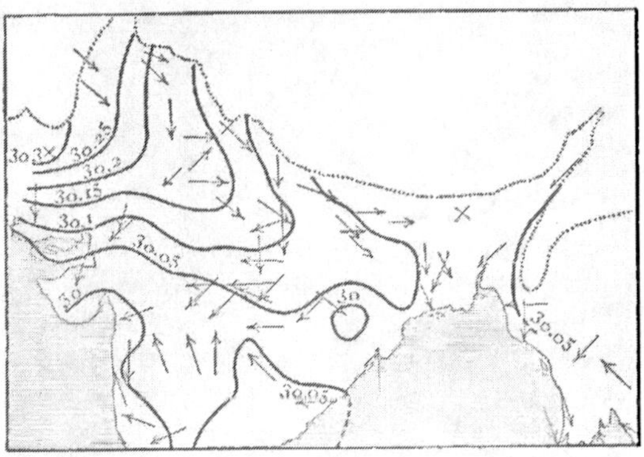

FIG. 14.—WEATHER CHART FOR 10 A.M., 26th February 1881.

any other occasion. A similar but less intense spell of cold weather invariably follows a storm of this character, and the distribution of pressure and the winds which accompany it are so characteristic, that the weather chart on one such occasion is almost exactly reproduced on any other. Fig. 14 represents an occurrence of this kind on the 26th February 1881, and in their main features that of the day which followed the storm of January 1883 and that published in the Simla weather report of the 22d February 1888 are almost exactly similar.

These instances will suffice to illustrate the chief vicissitudes of the weather of the cold season in Northern India. Warm close weather with light southerly winds precedes the storm. Then follows rain; and if the disturbance is one of those that first appear in the Indus valley and thence travel eastward, the rain, which always falls to the east and north of the depression, also travels eastward. Unless the course of the storm is much more to the south than is usually the case, more or less heavy snow falls on the Himalaya; and when the storm has passed away, the barometer rises rapidly in the north-west and cold weather

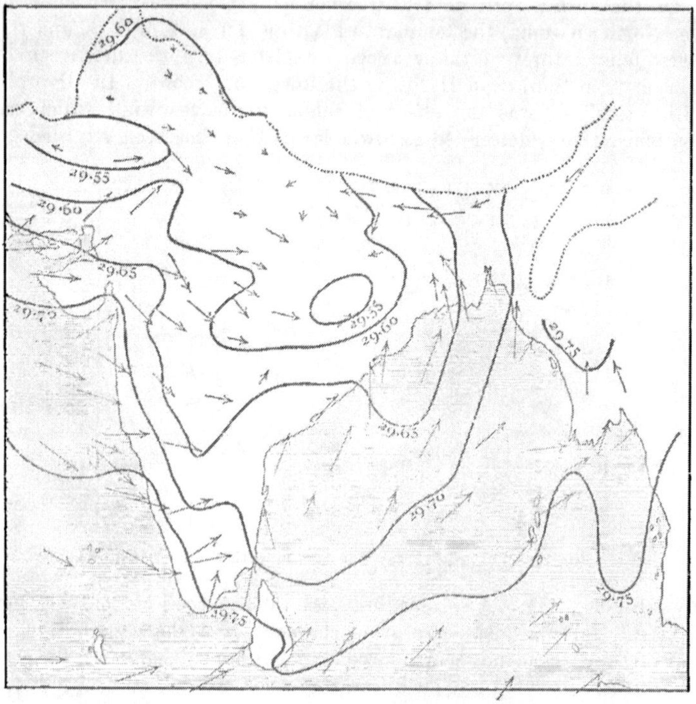

FIG. 15.—AVERAGE BAROMETRIC AND WIND CHART FOR MAY.

follows, with north-west winds in the upper provinces and easterly in the peninsula; the winds blowing clockwise around the anticyclone which is then formed over North-western India. If, however, the depression be formed over Western or Central India it is sometimes more persistent, and produces little or no snow on the Himalaya; and in such cases it is often not followed by cold weather.

Weather of the Hot Season.—The prevalent weather

of the hot season is determined by the heating of the land surface and the consequent reduction of atmospheric pressure in the interior of the country below that of the surrounding seas. The distribution of pressure thus resulting is illustrated in Fig. 15, which represents the normal barometric and wind chart for the month of May. The general arrangement of the isobars in this chart resembles that in the July chart (Fig. 4, p. 26), a broad trough of low pressure running obliquely across Northern India, with the isobars of higher value surrounding it concentrically on the east and south. But the greater intervals between these and the higher value of that encircling the region of minimum pressure indicate that the gradients are less steep, and therefore, as has been explained at page 25, the winds, shown in the Figure by arrows of varying length in proportion to their steadiness of direction, are less strong and steady. In the peninsula, too, there is a marked difference. In the May chart they form a southerly loop, conforming generally to the form of the land, which is not the case in July. It indicates that the pressure in the interior of the peninsula is lower than on either coast: the winds, in accordance with the cyclonic law, are therefore from south on the east coast and from between west and north-west to the west of the loop. These latter are dry winds, and accordingly, throughout Western India, from Sind as far south as Belgaum, it is very seldom that there is even a slight shower all through the spring months. Farther south, however, they have more the character of sea winds, and on the east coast, where they are from the south, they are decidedly sea winds. But at this season the winds are light over the sea, and the air drawn from the Bay of Bengal, though damper than that of Western India, is by no means so damp as it afterwards becomes in the stronger current of the summer monsoon. In the south and east of the peninsula, in Bengal, Assam, and Burmah, some rain falls in the spring months, but chiefly in the form of short-lived afternoon thunder-storms, and not continuously for a day or two, except on the comparatively infrequent occasions of cyclones being formed over the bay and travelling thence to the land.

In North-western India, as in the west of the peninsula, the chief winds of the hot season are land winds from between west and north-west, but less steadily in May than in March and April. They become less steady as the barometer continues falling in the Punjab, Sind, and Western Rajputana, but they occur at intervals up to the setting in of the rains and sometimes even during the rainy season itself. Meanwhile the sea winds, which are predominant in Bengal, gradually penetrate farther inland along the face of the Himalaya and the northern margin of the barometric depression of the Ganges valley. Accordingly, as westerly or easterly winds have for the time the mastery, so the hot season of Northern India alternates between dry heat, scorching but not intolerable, and the less intense but far more oppressive damp heat, which is only occasionally relieved for a few hours by a thunder-shower or simply a dust-storm.

These little storms do not originate with any great local disturbance of the barometer. They occur when the barometer is low, and are more frequent on the east and north of the tract of lowest pressure, along the foot of the Himalaya, in Bengal and especially in Assam, than elsewhere; that is to say, where southerly and easterly winds predominate. Everywhere they are most frequent in the neighbourhood of hills. The characteristic features of these storms, and the relation of the nor'-westers of Bengal to the dust-storms of North-western India, have already been described in the section on storms in the first part of this work.[1]

Cyclones scarcely ever occur in March, and seldom in April, on Indian seas north of the Equator. In the former month there are but two recorded instances in the Bay of Bengal, and in April I find but nine in considerably more than a century. All these were formed in the south of the bay, and but two reached the shores of Bengal; both in the last

[1] Page 81.

week of the month. The remainder either passed westward across the coast of Madras, or from the Andaman sea northward over Pegu. In May, storms of this class are more common. Including those of all degrees of severity, six have been recorded by Mr. Eliot in the 10 years, 1877-1886. Of these, two passed westward across the Madras coast; one on a west by north course from Balasore Bay to Central India, the usual track of monsoon storms; one northward across the Sunderbuns and over Jessore; one, originating to the west of the Andamans, reached the Arakan coast north of Akyab; and one from the south-west of the bay travelled north-westward to the mouths of the Kistna and Godavery, and then turning to north-east, skirted the Eastern Ghats of the Northern Circars, and finally reached Behar.

Of the May storms sufficiently severe to have attracted notice in years anterior to 1875, and which have passed over the land of some part of India and its dependencies, the following are extracted from the list published in the 1877 volume of the *Journal of the Asiatic Society of Bengal*:—

1787. Coringa. Very destructive storm wave.
1811. Madras. Great loss of shipping.
1820. Madras, across peninsula.
1823. Balasore.
1827. Madras.
1830. Felt slightly at Calcutta.
1832. Gangetic delta. Destructive storm-wave.
1833. Mouth of Hooghly. Storm wave.
1834. Kyaukpyu.
1840. Orissa coast.
1841. Madras.
1843. Ongole.
1844. Noakhally and Chittagong.
1849. Chittagong.
1851. Madras.
1852. Sunderbuns, east of Calcutta.
1858. Across peninsula. Also Chittagong and Dacca.
1869. Across Bengal.
1872. Madras.
1874. Madras.

Hence it appears that, in May, cyclones may be experienced on any part of the coasts of the Bay of Bengal, but that the coast of Madras is especially subject to them at this season. See also Fig. 26 (p. 231,) and the list of May cyclones in Appendix II.

On the west coast of India, cyclones are altogether less common; but according to a list drawn up by Mr. F. Chambers, they would seem to be relatively more so in the spring months than at the close of the monsoon. He enumerates two in March, nine in April, and thirteen in May, out of a total of 74 storms. But exception being made of those that have crossed the peninsula from the Bay of Bengal, and which became greatly weakened in the course of their transit, they have been felt only on the coast. I do not find any instance of a storm travelling inland from the west coast of India at any time of year.

Weather of the Summer Monsoon.—The transition from the hot season to the rains is gradual only in Assam, and to a less extent in Bengal and Arakan. In Western, North-western, and Central India, where land winds prevail, more or less, all through the hot season, the change is rapid; a few days only of light damp winds and calms being the forerunner of the monsoon. In some seasons, a day or two of rainy weather occurs in these provinces about a fortnight before the monsoon sets in permanently, and is called the "chhoti barsàt" or "little rains." It is, however, by no means a regular phenomenon. When it occurs, it is generally the result of an early cyclonic storm, similar to those which are frequent during the monsoon, and is due to an early and short-lived invasion of the monsoon. It is followed by a re-establishment of the land winds; a sequence that sometimes happens also in the middle of the rainy season, but this latter is then called "a break in the rains."

In Bengal, the average date for the setting in of the rains is the second week in June. The barometer falls steadily for three or four days beforehand, till it approaches its annual minimum; and the rainfall often sets in with a small cyclonic storm, giving squally weather at the head of the bay. This sometimes carries the rain at once up to Behar, and even farther to the north-west, but as a rule it takes from a week to a fortnight to extend to the North-west Provinces and the Eastern Punjab.

On the Bombay coast, the rains usually set in in the first week of June, a few days earlier than in Bengal, and quickly extend to the Central Provinces, and afterwards to Rajputana and Central India. On the Malabar coast they begin some days earlier than in Bombay, but do not extend very copiously beyond the Western Ghats; and in Mysore, the ceded districts of Madras, the Deccan and Hyderabad, more rain falls when the strength of the monsoon to Northern India relaxes, than when the interior plateau of the peninsula is swept by a strong current from the west coast. In Pegu and Arakan the rainy season sets in a week or so earlier than in Bengal, on an average, but not invariably.

The change, that takes place in the general distribution of the atmospheric pressure and winds on the setting in of the monsoon, will be best understood by comparing the May chart on a preceding page (Fig. 15) with that for July (Fig. 4, p. 26). From May to June or July, the barometer falls 0·1 or 0·15 inch lower in Northern India, and rises 0·05 inch on the Travancore coast; and the southern loop of the isobars, in the peninsula in May, disappears in July, these lines running obliquely across the peninsula from coast to coast. The western branch of the monsoon blows steadily and strongly across the peninsula as far north as the Satpuras, as is indicated by the lengths of the wind arrows on Fig. 4; less strongly over the Central Indian plateau up to the Ganges valley. In the latter and the Punjab, east winds predominate: but the whole region enclosed by the isobar 29·5" is one of barometric depression and generally light winds. Over the eastern half of this depression, and in Bengal and Assam, much rain falls without much local barometric disturbance, especially during the low stage of a surge; but heavier falls occur at intervals along definite belts of country, which are traversed by cyclonic storms, formed either over the head of the bay, or locally in some part of the permanent depression. The tracks of these storms are, with rare exceptions, between west and north-west, and they succeed one another at longer or shorter intervals all through the monsoon; reaching in some cases only as far as Central India or the North-west Provinces, but also occasionally to the Punjab, Sind, or Kathiawar; whence, in some rare instances, they pass out to the north of the Arabian Sea, with increasing intensity, and give rise to stormy weather on the Gwadar coast.

These cyclonic storms are so important a feature of the rainy season, and influence so greatly the distribution of the

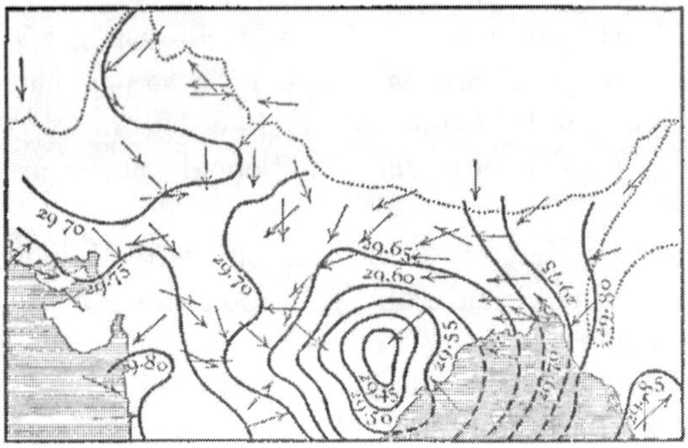

FIG. 16.—WEATHER CHART FOR 10 A.M., 8th September 1882.

rainfall, that a more detailed notice of one or two instances will much facilitate a due comprehension of the characteristic weather of this season.

Figs. 16 to 19 represent four stages of a storm that travelled from the Bay of Bengal to the Punjab and Sind at the close of the rainy

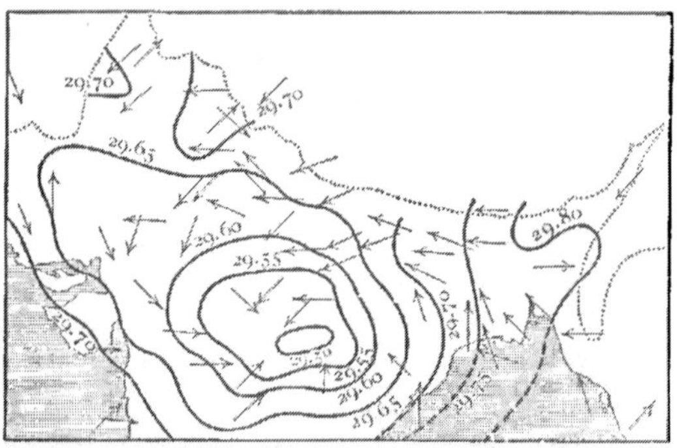

FIG. 17.—WEATHER CHART FOR 10 A.M., 10th September 1882.

season of 1882. It was formed over the head of the bay, off the Orissa coast, on the 6th and 7th September, reached the Punjab on the 14th, and finally disappeared in Upper Sind on the 18th, having thus lasted at least 12 days. It brought heavy rain to Orissa on the 7th and 8th, while in Eastern and Northern Bengal but little fell; as is usually the case when the centre of the storm lies so far to the south. But in the Central Provinces there was heavy general rain from the 8th to the 11th ; in Central India on the 11th and 12th ; in Northern Rajputana up to the 13th ; and in the Punjab between the 10th and 15th. On the other hand, the North-west Provinces, Oudh, and Behar, which lay at a distance to the north of the track, had little or no

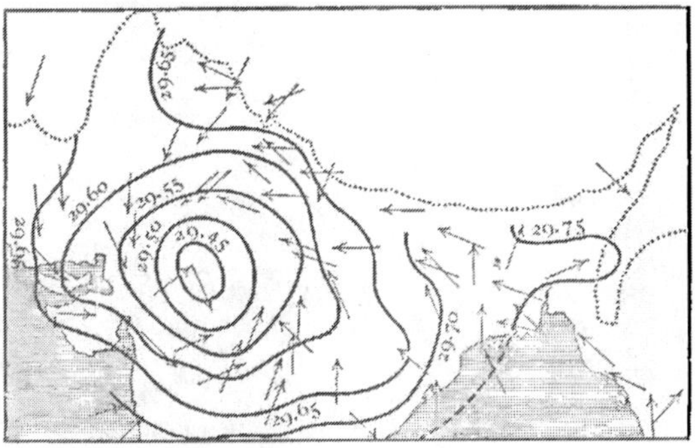

FIG. 18.—WEATHER CHART FOR 10 A.M., 11th September 1882.

rain ; and in Bombay there was a break of about a week from the 13th to the 20th, and the same in Hyderabad. The rainfall of this period or the greater part of it was therefore concentrated in the storm. The barometric depression in this storm, at least in its earlier

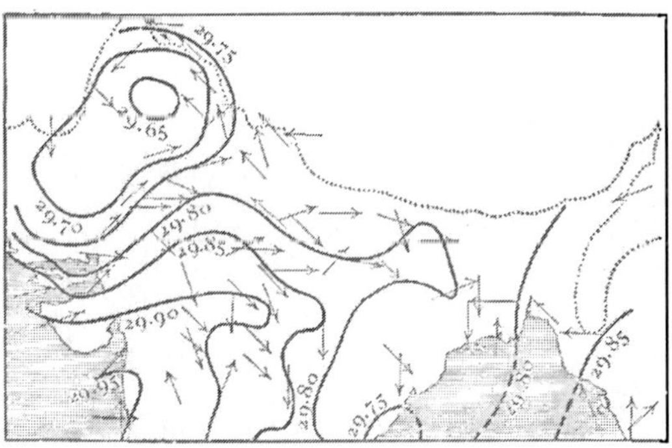

FIG. 19.—WEATHER CHART FOR 10 A.M., 14th September 1882.

stages, was somewhat greater than usual, and its duration was more prolonged.

Figs. 20 to 23 represent four stages of a storm that followed an almost due westward track, at the beginning of the monsoon of the following year, 1883. A full description of it, illustrated by charts of the weather for each day, was published by Mr. Eliot in the 53d volume of the *Journal of the Asiatic Society of Bengal*. It originated on the 27th June, between 20° and 21° north lat., and 89° and 90° east long., and during the next two days moved very slowly (at between 2 and 5 miles an hour) on a north-west course to Balasore Bay, crossing the coast on the evening of the 29th. While still over the bay, the south-west and south winds of the south and east quadrants had the force of a strong gale (10 to 11 on Beaufort's scale); while to the

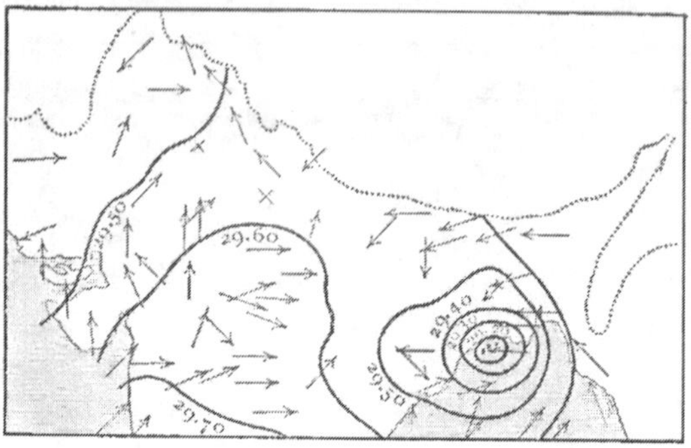

FIG. 20.—WEATHER CHART FOR 10 A.M., 29th June 1883.

north and west, the north-easterly and westerly winds were only moderate to strong (3 to 5); the barometric depression at the centre amounted to somewhat less than half an inch.

On reaching the land its course became almost due west, and this

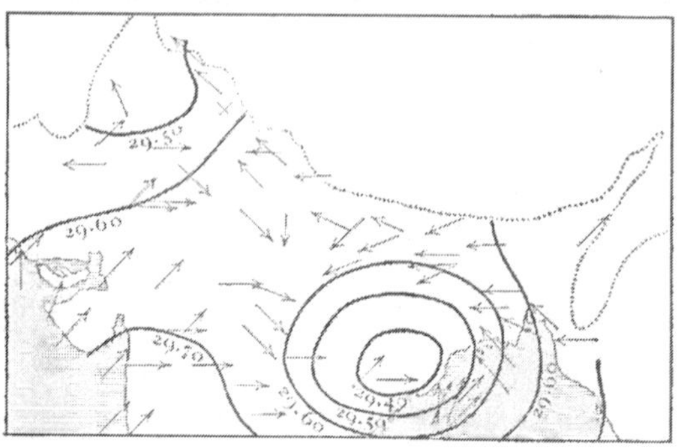

FIG. 21.—WEATHER CHART FOR 10 A.M., 30th June 1883.

direction was maintained across the whole width of the peninsula, at a rate of about 15 miles an hour during the greater part of the distance, but reduced to not more than 8 miles an hour while traversing the hilly country that intervenes between Orissa and Sambalpur. Its intensity also, as indicated by the depression of the barometer at its centre, was reduced by nearly two-thirds in this part of its course; but it recruited its strength, on being fed by the monsoon current from the Bombay coast, and after passing out to the Arabian Sea, it regained its original cyclonic force. The centre passed north of Cuttack and Sambalpur, close to Seoni, Indore and Bhuj; after which, on the morning of the 3d July, it left the land, travelling out to the south of Kurrachee. During the transit, the winds were strongest and the rainfall heaviest to the south of the central track. To the north

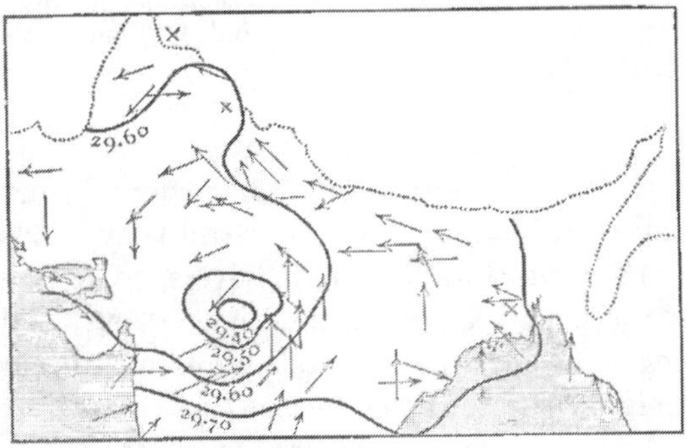

FIG. 22.—WEATHER CHART FOR 10 A.M., 2d July 1883.

of it, in equal proximity to the centre, the rainfall was not more than one-fourth as heavy. The strength of the wind in different parts of the storm was very variable; where strongest it did not exceed 43 miles an hour on the average of 24 hours, and at very few places amounted

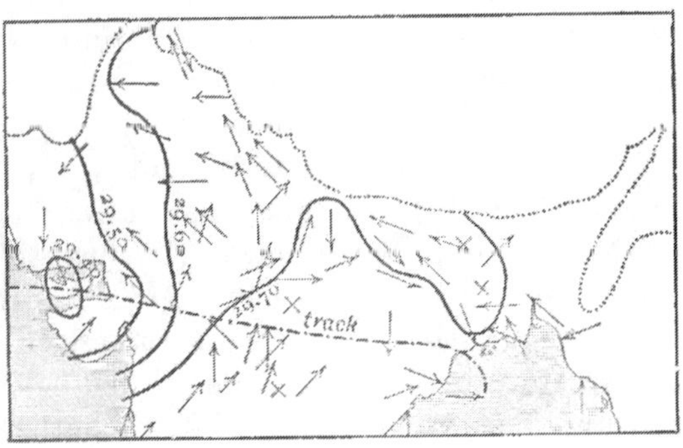

FIG. 23.—WEATHER CHART FOR 10 A.M., 3d July 1883.

to 25 miles. At one station it appears to have been as high as 64 miles an hour on the average of six hours, but this is the only instance of an inland station where it had the force of a strong gale until the storm again approached the sea.

There do not appear to have been, in this storm, any of those exceptionally high rainfalls of 20 or 30 inches in the 24 hours that have been registered on certain other similar occasions. The highest noticed by Mr. Eliot are 10·05 inches on the 3d July at Rajkote, and 9·44 inches on the 29th June at Pooree. But there was an average of 6·15 inches over the whole Pooree subdivision of the Cuttack district, 4·5 inches over that of Cuttack, and 2·66 inches over that of Balasore on the latter of these days; and of 10·56 inches, 9·6 inches, and 7·68 inches respectively on the same subdivisions in the three days 28th to 30th June. In the Central Provinces, an average of 6·13 inches fell on the Nagpur subdivision, 5·7 inches on that of Bhandara, 5·19 on that of Chhindwara, and 6·26 on that of Nimar in the space of two days.

Falls of equal and even greater amount are not infrequent on the Satpura range in cyclonic storms at this season, and at intervals of 2 or 3 years they produce heavy floods in the Tapti and Nerbudda rivers, sometimes causing great destruction. Arrangements have been made by the Meteorological Department to telegraph such occurrences to the districts on the lower course of these rivers, and thus to prepare them in some measure to meet the threatened catastrophe.

A storm, which followed a track almost identical with the above, occurred in September 1878. All these were comparatively long-lived storms; the majority do not travel beyond Central India or the North-west Provinces; and, after giving those provinces 1 or 2 days of heavy rain, they disappear.

At the beginning and end of the monsoon, and also during seasons of drought in Upper India, when there is a prolonged suspension of the rains, and westerly land winds hold sway in Rajputana, Central India, and the western half of the Gangetic plain, storms formed over the head of the bay sometimes take a more northerly course, across Bengal. Such was the case with four storms in succession in August 1877, the year of the last great dearth in the North-west Provinces; and one in August 1880, a month which was also one of prolonged drought and, both in the anomalous distribution of pressure and the unseasonable prevalence of land winds in Upper India, closely resembled the same month of 1877. In May and the first half of June, in September and October, the usual course taken by cyclones from the head of the bay is northerly.

Temporary interruptions of the rains, during which the monsoon falls weak, and ceases to penetrate Upper India, occur in all years, but, as a rule, do not last many days. But in the years just mentioned, and in 1883, they lasted several weeks, and west and north-west winds, similar to the characteristic winds of the hot season, blew all over the Upper Provinces, in Rajputana, and Central India and, even, for a time, into Berar and the Deccan. On all such occasions a very characteristic feature appears in the isobars of the weather chart, which is often referred to in the reports as "a shoulder of high pressure" in Western India. It consists in the greater northern protrusion of the northward bend of the isobars in Rajputana and Central India, and the somewhat greater obliquity of their slope from west to east. It is not always so marked as to be readily seized by an unaccustomed eye, unless by comparison with a normal chart, or with one of a day of well-distributed and normal rainfall; but though small, it is always significant. It means that the barometric pressure in Western India is higher than usual, relatively to that in the east of the country, in Bengal and Chutia Nagpur, and that the winds of this part of India are more northerly than is usual at the time of year. Now, it is this slight northing of the westerly winds that distinguishes the dry current coming from Baluchistan and Afghanistan from the rain-bearing monsoon of the Arabian Sea, and therefore indicates the replacement of the latter by the former. Generally this is associated with another feature of equally unfavourable import, viz. the northward displacement of the axis of the low pressure trough of Northern India to near the foot of the hills. This means that the easterly winds from the bay of Bengal, which prevail only to the north of that axis, fail to penetrate as usual up the Ganges valley, and, therefore, that the eastern branch of the monsoon is withheld from the Upper Provinces. The accompanying chart of the 9th August 1882 (Fig. 24) is a good illustration of these abnormal features. On this occasion there was a prolonged break in the rains from the 5th to the 20th of the month, during the whole of which time the charts exhibited these peculiarities of the isobars.

It was observed on a previous page that the serenity of the weather of a great part of Northern India is rarely disturbed during the 2 or 3 months that follow on the cessation of the rains in September or October. This is strictly true of Upper India, but somewhat less so of Bengal. In this province and Assam, the rains last well into October, and both in this and the early part of the following month severe cyclones from the bay sometimes reach the province. Indeed some of the most destructive storms on record have occurred in October and November, although several years

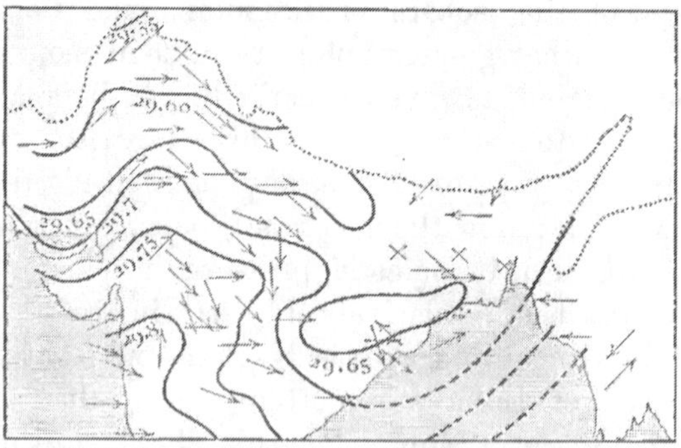

FIG. 24.—WEATHER CHART FOR 10 A.M., 9th August 1882.

may pass without such a visitation. But October is often a stormy month in the bay, and even though the hurricane core of the storm may not reach its northern shores, no great disturbance occurs in any part of the bay that does not make itself felt in Lower Bengal by close damp weather and cloudy skies. The signs and prognostics of these storms will, however, be more fitly noticed in the next section of this work.

When the monsoon has ceased to blow to Upper India, that branch which comes from the bay is still directed for a few weeks towards the north-east of the peninsula; but feebly and with frequent intermission. Mr. Eliot records nine cyclonic storms in the months of September and October, and two in November, in the 10 years 1877-1886, that crossed the coast between the Chilka lake and the mouths of the Godavery. Of these, one only traversed the whole width of the peninsula. The remainder either broke up against the Eastern Ghats or shortly beyond, or were diverted northward towards the Central Provinces or Chutia Nagpur. Hyderabad and the Deccan districts of Bombay enjoy, therefore, greater immunity from storms of this character than either Northern India or the south of the peninsula. In the Northern Circars, at the foot of the Eastern Ghats, the September or October rainfall is the heaviest of the year; but in Bustar, Hyderabad, and the Deccan, the rainfall of this season is small, and so soon as the westerly branch of the monsoon ceases to predominate, it is replaced by light north-easterly winds, with little or no rain.

About the middle of October or somewhat later, the heavy rainfall of the year sets in on the Madras coast. As in Northern India in its rainy season, so also here, much of this rain is attended with but little local barometric disturbance. But little cyclones and cyclonic storms are frequently formed over the south-west of the bay, and travel to the Madras coast, and though small, are sometimes locally destructive. The more severe cyclones of this month and November usually originate at a greater distance from land, either half-way across the bay, or still farther east, in the neighbourhood of the Andamans and Nicobars; and although the majority of them take a course about west by north towards the Madras coast, they occasionally travel northward up the bay, and reach the shores of Bengal. The disastrous Calcutta cyclone of the 5th October 1864, that of the 1st November 1867, and the Backergunj cyclone of the 1st November 1876, are well-remembered instances of such storms. The process of formation, and the subsequent movements of those that originate in the south-west of the bay, can generally be watched and traced from day to day by the indications of the barometers and the wind changes at Trincomalee, Negapatam, and Madras. These stations being at no great distance from the storm cradle, the barometer falls steadily during their incubation: and the wind directions, taken in conjunction with the relative rates of fall of the barometers at the different stations, afford a fair indication of the storm's position. Warnings are now sent to the coast ports from the Calcutta office, the meteorological reporter for Bengal having special

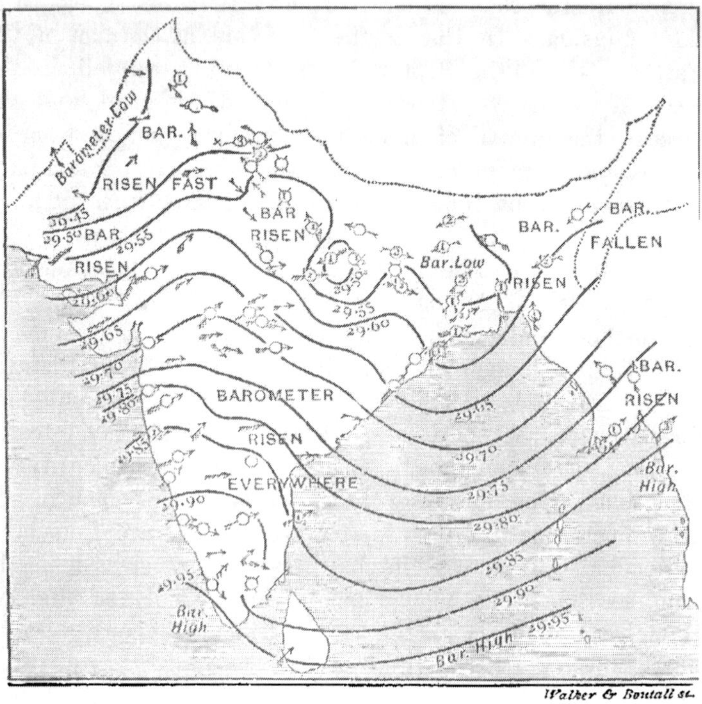

FIG. 25.—SIMLA WEATHER CHART OF THE 26TH JULY 1888, 8 A.M.

charge of the storm warning system for all the coasts of the bay.

Weather Reports.—The Indian daily weather reports having been frequently referred to in the foregoing pages, and many of the daily charts given in illustration of certain characteristic states of weather, the signification and utility of these reports and charts will be fairly well understood by any one who has studied the preceding descriptions of the weather, and the explanation of its constituent elements as set forth in the first part of the work. It may, therefore, suffice in this place to recapitulate in due order the tabular statistics of one of these reports, with references to those pages of this work in which each of them is explained; and, in conclusion, to give one more example of a chart, casually selected from the recent issues, drawing attention to the chief points worthy of notice. I must suppose that the reader has one of these reports before him, the sheet being too large for reproduction on these pages. The chart which illustrates it is given on a reduced scale in Fig. 25.

The report for the 26th July 1888 gives telegraphic reports from 96 stations, despatched shortly after 8 A.M. Those of three other stations which are deficient, together with all other deficiencies in the reports of the month, are published in a supplementary issue at the end of the month.

Col. 1 shows the readings of the barometer at 8 A.M., corrected to the Calcutta standard (p. 21) and to the temperature of freezing (p. 21), and except in the case of hill stations, to its equivalent value at the sea-level (p. 23). These readings are the data from which the isobars, showing the distribution of atmospheric pressure at 8 A.M. of the 26th July, are laid down on the chart, Fig. 25 (see p. 24). Col. 2 gives the changes of the barometer since the same hour on the preceding day, when, therefore, as regards the regular diurnal variation, it was in the same phase. In this report, nearly all the readings show a rise, greatest at the most northerly stations, Sialkot, Rawalpindi, and Peshawar, where it exceeds 0·2 inch; and least on the west coast of the Konkan, where it amounts to only 0·02 or 0·03 inch. This is an instance of a surge (pp. 22, 200). Its effect has been to diminish the barometric gradient (p. 24) from the west coast to the Punjab and Sind, which determines the strength of the western branch of the monsoon towards Upper India; since the barometric difference is reduced by nearly 0·2 inch. But the tract around Allahabad has not shared in the general rise of pressure; in this direction, therefore, the gradient is slightly increased, and therewith the strength of the winds towards it.

Cols. 3 and 4 give the wind direction at 8 A.M., and its mean rate of movement per hour during the preceding 24 hours. These are indicated on the chart by arrows, and the number of the feathers (from 1 to 7) show the strength or velocity of the wind according to a scale printed below the chart. A calm is represented by a cross.

Cols. 5 to 9 give the temperature of the air at 8 A.M., the highest and lowest temperatures of the previous 24 hours, as recorded by self-registering thermometers; the mean of the two, as representing approximately the mean temperature of the 24 hours; and the difference of this mean temperature and that given in the report of the previous day. These changes are not very great in the rainy season, and they depend very much on rain, the air being cooled by rain, and becoming warmer again on its cessation.

Cols. 10 and 11 show the relative humidity of the air (p. 46) at 8 A.M., and its increase or decrease since the same hour of the preceding day. As the air becomes drier, so does the prospect of rain diminish, and *vice versâ*. As a general rule, the humidity increases when the temperature falls and decreases when the latter rises.

Col. 12 gives the estimated cloud proportion (p. 61) at 8 A.M., on the scale 0 to 10. 0 indicates a cloudless sky, and 10 one entirely overcast.

Cols. 13 to 16 are the rainfall data—viz. 1*st*, the rain measured at 8 A.M., that had collected in the gauge during the previous 24 hours; 2*d*, the total quantity registered at the station since the 1st June, (*i.e.* approximately that of the summer monsoon up to date); 3*d*, the average of the same period deduced from the registers of past years; and 4*th*, the excess or deficiency of the actual fall as compared with this average. The rainfall of the 24 hours at any station is shown on the chart by a small circle over the position of the station, with an inscribed figure if the fall has exceeded half an inch.

Col. 17 contains a single word or short phrase descriptive of the general character of the weather during the previous 24 hours.

The chart gives a general view of the state of the weather as regards the three important elements—barometric pressure, winds, and rainfall. Temperature, humidity, and cloud proportion may also be represented on a chart, but they are omitted to avoid overcrowding and confusion, and any features of importance are noticed in the remarks on the first page of the report.

If we compare the chart (Fig. 25) with the normal chart for July (Fig. 4, p. 26) an important difference appears in Northern India. The trough of low pressure which, in the normal chart, extends obliquely across Northern India, occupying nearly the whole of the Ganges basin as far east as Behar and the confines of Lower Bengal, is divided, in Fig. 25, by a shoulder of higher pressure occupying the western half of the North-west Provinces and Oudh; and its eastern half is extended so as to include nearly the whole of Behar and Bengal. Within this are two small cyclonic systems, indicated by the wind arrows, the one with its centre between Allahabad and Benares, tolerably definite, and further indicated by a small oval isobar; the other less distinct, but apparently centring between Berhampore and Purneah, and indicated only by the wind arrows. In both, the barometric depression is very slight, amounting to only a few hundredths of an inch, as is frequently the case in these local depressions in the rains; but their influence is shown, not only by the winds, but also by the concentration of the rainfall of Northern India around them. The more easterly of the two first appeared on the chart of the 23d, two days previously, and was short-lived, since it had entirely disappeared from that of the 27th, having probably amalgamated with the western depression. This latter was much more durable. It originated about the 18th, off the coast of the Sunderbuns, and in the course of eight days had travelled very slowly to the position, between Allahabad and Benares, shown in Fig. 25. It lasted some ten days longer, moving slowly westward to between Agra and Jhansi, still concentrating around it the greater part of the rainfall of Northern India. Simultaneously with its advance from the coast, was formed the shoulder of high pressure noticed above, viz. on the 23d; and, on this day, a break of the rains set in in the Punjab and lasted until the 28th, when the former began to yield to the encroachment of the advancing storm. During this time, north-west winds prevailed at Delhi, Agra, and Jeypore, and a strong south-west wind blew over Lower Sind and Guzerat, as shown by the six feathered arrows, which indicate a mean rate of over 26 miles an hour. The greater part of this current appears to have been diverted eastward across Rajputana, to feed the rainfall around the depression. On the 28th, another very slight depression, similar to that of the 18th, appeared over Bengal, and like the former, moved westward and amalgamated with its forerunner on the 1st August, with the effect of recruiting its strength. It finally disappeared between the 3d and 4th.

During the fortnight or 16 days that these changes were in progress, the strength of the western branch of the monsoon varied considerably, accordingly as the depression in the North-west Provinces became enfeebled or recruited its strength. The Bengal branch was weak throughout after the 22d, and was chiefly directed towards Arakan.

THE STORMS OF INDIAN SEAS

Piddington's Work.—At a time when the science of storms was in its infancy, and long before, even in the western homes of science, there existed any official organisation for collecting information on the meteorology of the ocean, the spontaneous and patient labours of Henry Piddington laid the foundation of our knowledge of the storms of Indian seas. For this work, in so far as regards the collection of marine observations and the interpretation of their teachings in language that appealed to the minds of seamen, his own early seafaring experience gave him an advantage which he turned to the best account; and his name is deservedly held in honour as that of a pioneer of science, and a benefactor whose claims to the gratitude of at least two generations of sailors in Indian seas stand unquestioned and universally recognised. The practical rules laid down by him for the guidance of seamen have undoubtedly been the means of saving many a ship from destruction, and if these rules are in some respects defective, and his conception of the nature of cyclonic storms erroneous, it must be borne in mind that, in his day, the whole science of meteorology was so little advanced, that even had he lived less remote from the centres of scientific activity, it would not have been possible for him to bring the results of his work into harmony with the physical laws of the atmosphere as we now know them. That the winds blow in circles around a storm centre, as taught by Reid and Piddington, is, as we now know, only a rough approximation to the actual facts, even in high latitudes, and in low latitudes is considerably wide of the truth; and practical rules based on that assumption may, therefore, seriously mislead in tropical seas; still, the work done by Piddington in ascertaining the directions in which storms usually travel, and in describing their premonitory indications, is of very great value, even admitting, as we must do, that his preconceived idea that the winds blow in circles sometimes led him astray in his conclusions.[1]

[1] In the charts that illustrate Piddington's later Memoirs, the observed wind directions are not shown; but as the logs that furnish the data are given in full in the text, it is easy for any one to lay them down for himself, and he will then find, not only that they afford ample evidence that in these storms, as in all those since observed, the wind really blew in spirals, but that, this being admitted, many of the difficulties which Mr. Piddington met with, and which obliged him sometimes to suppose that two or more storms existed simultaneously in near proximity to each other, disappear. His fourteenth and fifteenth Memoirs give good instances of this. And in his first Memoir, in which he deduces a west-south-west track for the storm of the 3d to 5th June, 1839 (a track of which I find no other example in all the records before me), it is obvious that, when once it is admitted that the winds blow in spirals, the course of the storm was in a northerly direction, indeed almost due north, not an uncommon course at that time of the year.

It must be evident to any one who goes carefully through Piddington's

The general characters of cyclones have already been described briefly in the first part of this work. In the present section, which is written for the practical guidance of seamen more especially, it will be necessary to notice them more in detail, with reference to their local characters and their relative frequency and behaviour at different seasons; and in the first place, I shall deal with the storms of the Bay of Bengal.

1. The Bay of Bengal

Storm Seasons in the Bay of Bengal.—The following table of cyclones of the Bay of Bengal was drawn up from the records available up to the end of 1876. At that time little was known of the less severe cyclonic storms of the summer monsoon, and, with a few exceptions, they may be considered as omitted from the list, having failed to attract prominent notice. As will presently be seen, when they are included, the summer monsoon, instead of being comparatively free from storms, is found to be the stormiest time of year; but except that the bay is swept by the strong south-west gale that feeds them, they are restricted to the north of the bay and to the coasts of Orissa and Bengal.

Number of Recorded Cyclones on the Bay of Bengal up to 1876

Month	No.	Month	No.	Month	No.
January	2	May	21	September	6
February	0	June	10	October	31
March	2	July	3	November	18
April	9	August	4	December	9

Total, 115 storms.

February is the only month in the year in which no storm

Memoirs on the Law of Storms that he never sought to evade or ignore the difficulties and apparent inconsistencies that presented themselves in his evidence. But his method was faulty, and of this he was unaware. Having assumed the truth of the view that the winds blow in circles, he determined the position of the storm or storms on that assumption, often on quite insufficient evidence; and if one storm centre would not fit in with the wind observations, he assumed two, and regarded the result as proving the truth of the original assumption. He seems never to have had the alternative of spiral winds present in his mind, and although he admits the occasional incurvature of the winds, he regards it as exceptional and irregular.

has been recorded in any part of the bay. The two January storms occurred in the first week of the month, in the extreme south of the bay, one passing over Pondicherry, the other over Trincomalee and the southernmost districts of the peninsula. Both of the March storms occurred in the last week of the month on the Madras coast. It may therefore be said that, during the two and a half months from the 7th January to the 26th March, storms were, so far, unknown in the bay; and as appears from Mr. Meldrum's investigations, this is precisely the season when they are most frequent in the South Indian Ocean.

All the April storms of the above list, with two exceptions, both in the last week of the month, all the December storms and those of the last half of November, were entirely restricted to the southern half of the bay. But later experience has shown that the north of the bay does not enjoy complete immunity from stormy weather up to the early days of December; since, in 1883, an ill-defined cyclonic disturbance of no great severity, and producing but a slight depression of the barometer, was formed at the end of November in the south-east of the bay, and traversed its entire length to the Eastern Sunderbuns and Cachar. And again, in 1885, a storm formed off the coast of Ceylon on the 14th and 15th November, after moving towards the Madras coast, turned away to the north-east, and reached the Arakan coast, near Akyab, on the 23d.

The cyclonic storms of the winter and early spring, that first appear in North-western India and travel eastward (p. 87), very rarely reach to the Bay of Bengal. There has been, however, one recent instance of such a storm, the only instance of the kind on record. At the end of January 1888, a storm crossed India, on an east-south-east course, from Baluchistan to Gopalpur, and thence passed across the bay to Akyab.

According to our present experience, then, with the single exception just noticed, the *northern half* of the bay is always quite free from storms from the end of the first week in December to nearly the end of April, an interval of about four and a half months, and they are very rare after the middle of November. In March and April strong south-west winds, sometimes amounting to a moderate gale, blow at the Sand-heads and in the north-east of the bay, but they are not cyclonic at sea.

In May and the first half of June, and in October and November, cyclones may occur in any part of the bay (see Figs. 26 to 28); but in the first half of May the large majority are restricted to the southern half; in the second half of the month, though formed in the middle or south of the bay, they more frequently travel to the coasts of Orissa and Bengal. In the first half of October, they are about twice as frequent on the coasts of Bengal and Orissa as on those of Madras; in the second half these relative proportions are reversed; and in November storms are comparatively rare in the north of the bay. In June, one cyclone has been recorded at Madras and one at Akyab. All others have affected Bengal and Orissa and their coasts. And the cyclonic storms of the summer monsoon have all been formed either in the north of the bay (generally between False Point and Chittagong) or over the adjacent land.

Hence, for two months at the beginning of the storm season, viz. in March, April, and the first half of May, and again for two and a half months at its close, viz. from the middle of October to the beginning of January, the southern half of the bay is chiefly or exclusively the stormy region. The intervening months, including the whole of the summer monsoon, are the stormy season of the north of the bay.

The relative frequency of storms in these latter months will be better shown by the following table, drawn up from the lists prepared by Mr. J. Eliot, of the storms recorded on the Indian weather charts from 1877 to 1886. It includes all the severe cyclones and the milder or smaller cyclonic storms of those years, excepting, perhaps, some whirls of small magnitude that have been promptly dissipated, and those formed over the land which have not been felt as cyclonic in any part of the bay, though, for the time being, they have increased the strength of the monsoon.

CYCLONES AND CYCLONIC STORMS ON THE BAY OF BENGAL IN THE TEN YEARS 1877 TO 1886, FROM MAY TO DECEMBER INCLUSIVE

Month	Storms	Month	Storms
May	6	September	17
June	11	October	10
July	18	November	16
August	17	December	4

Total, 99 storms.

Place of Formation, Tracks, and Rate of Travelling.—So much with regard to the seasons at which the seaman may be prepared to encounter stormy weather in the north and south of the bay respectively. I shall now notice, in more detail, the parts of the bay in which storms originate, the directions in which they move, and their rate of progression. We shall see that, while courses between west-north-west and north-west are much more common than others, exceptions sometimes occur in which they deviate very considerably from these most usual directions; but even the exceptions do not surpass certain limits, and with the single exception of the January 1888 storm, noticed on page 226, all storms yet recorded have travelled towards some point of the compass comprised between west and north-east.

The following results are mainly deduced from the work of Mr. John Eliot, who has made a special study of the storms of the Bay of Bengal, under far more advantageous circumstances than were ever at Mr. Piddington's command. The data are more extensive than any that have yet been brought to bear on the questions at issue. Not to mention his elaborate and detailed Memoirs, and those of Messrs. Willson, Pedler, and others, describing the life history of several of the more striking and important cyclones of late years, Mr. Eliot has given lists of all the recorded storms, of all degrees of severity, that have occurred during the last ten years, together with charts of their tracks. All these, in addition to Mr. Piddington's Memoirs, have been consulted in drawing up the following summary. Lists of all the storms referred to and their principal data, summarised in the following tables, are given in Appendix II. at the end of this work, and Figs. 26, 27, and 28 are charts of the tracks in the three months May, October, and November.

In order to exhibit in the most convenient form those results that are of most importance to seamen, I have summarised them in four tables, showing for each month from May to December the following data:—

1. The number of storms that *have originated* in each square of 4° of latitude by 4° of longitude.
2. The number that *have been experienced* in each square of 4°×4°. This includes, in addition to the cases of the first table, all those in which the vortex of a storm has traversed any portion of a square.
3. The directions of the storm tracks. In this table storms that, in the course of their passage across the bay, have changed their direction through two or more points, appear in more than one column.
4. The rates at which storms have travelled across the bay. Here again the same storm sometimes appears under different headings. Probably in most, if not in all cases, a storm moves forward from its place of origin at first slowly, and advances more rapidly as it approaches the land; but it is only in a small number of instances that the available data show such an increase as to double or treble the initial rate thus determined.

TABULAR SUMMARY OF RECORDED STORMS ON THE BAY OF BENGAL, ACCORDING TO THEIR PLACE OF ORIGIN, AND THE PARTS OF THE BAY TRAVERSED BY THE CENTRAL VORTEX.

Months.	Between Longitudes.	Storms Originated between Latitudes					Storms Experienced between Latitudes				
		6 to 8.	8 to 12.	12 to 16.	16 to 20.	20 to 22.	6 to 8.	8 to 12.	12 to 16.	16 to 20.	20 to 22.
April and May	80 to 84	1	2	0	0	...	1	4	4	1	...
	84 to 88	0	1	1	0	0	1	1	1	2	3
	88 to 92	0	1	0	2	2	0	1	1	3	3
	92 to 96	0	0	0	0	0	0	0	0	1	1
June	84 to 88	0	0	0	0	2	0	0	0	2	7
	88 to 92	0	0	0	4	7	0	0	0	5	10
July	84 to 88	0	0	0	0	0	0	0	0	4	16
	88 to 92	0	0	0	5	15	0	0	0	5	16
August	80 to 84	0	0	0	0	...	0	0	0	1	...
	84 to 88	0	0	0	2	0	0	0	0	3	11
	88 to 92	0	0	0	1	12	0	0	0	2	12
September	80 to 84	0	0	0	0	...	0	0	0	0	...
	84 to 88	0	0	0	2	1	0	0	0	13	11
	88 to 92	0	0	0	11	5	0	0	2	13	8
	92 to 96	0	0	2	1	0	0	0	2	1	0
October	80 to 84	0	1	2	0	...	0	2	2	6	...
	84 to 88	0	0	2	3	0	0	2	7	10	7
	88 to 92	1	3	2	3	0	1	4	5	9	8
	92 to 96	0	0	1	2	0	0	0	0	2	1
November	80 to 84	0	4	0	0	..	0	7	8	1	...
	84 to 88	1	3	2	0	0	1	7	4	3	0
	88 to 92	0	2	0	1	0	0	3	1	3	2
	92 to 96	0	4	0	0	0	0	4	3	1	0
	96 to 100	0	0	1	0	...	0	0	1	1	...
December	80 to 84	1	0	0	0	...	2	2	1	1	...
	84 to 88	0	1	0	0	0	0	1	1	2	0
	88 to 92	0	0	0	1	0	0	0	0	1	1

DIRECTION OF TRACK AND RATE OF PROGRESS OF STORMS ON THE BAY OF BENGAL

Months.	Direction of Movement.								Miles per Hour.			
	W.	W. by N.	W.N.W.	N.W.	N.N.W.	N.	N.N.E.	N.E.	Below 4.	4 to 8.	8 to 12.	Above 12.
April and May	0	1	3	4	2	2	1	1	2	6	1	0
June . .	3	0	2	6	2	4	0	0	4	4	1	0
July . .	3	6	8	5	1	1	0	0	7	12	1	0
August . .	1	3	4	1	2	2	0	1	6	3	0	2
September .	2	10	4	5	3	1	0	0	8	9	4	2
October . .	2	3	3	4	5	8	3	1	0	14	8	1
November .	1	2	8	5	4	5	4	1	5	13	7	1
December .	0	0	1	3	1	1	1	0	1	3	1	1
Total .	12	25	33	33	20	24	9	4	33	64	23	7

The first and second of the above tables entirely confirm the conclusions already drawn from the 1876 list of storms (p. 225), as to the parts of the bay in which storms are liable to occur at different seasons; and these need not be repeated. But it must be observed that these tables show the positions of the storm vortex only, where the winds are of hurricane violence (force 10 to 12), and stormy weather, with winds of force 6 to 8, prevails far to the east and south of the storm centre, but only to distances of 150 to

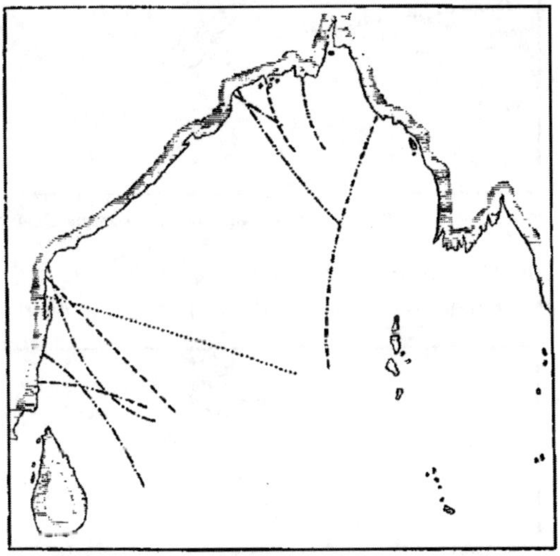

FIG. 26.—STORM TRACKS OF MAY IN THE BAY OF BENGAL.

200 miles to the north and west. On this point some further information will be given presently.

The third table shows that the most usual direction, in which storms travel, is to west-north-west and north-west. In the months of the south-west monsoon (from July to September), nearly all move on some path between west and north-north-west; and only in May and the last three months of the year, especially in October and November, are those to north and north-east at all common; the latter, with one exception in November, only in the northern half of the bay. The accompanying Figures 26, 27, and 28, show the storm tracks of these three months.

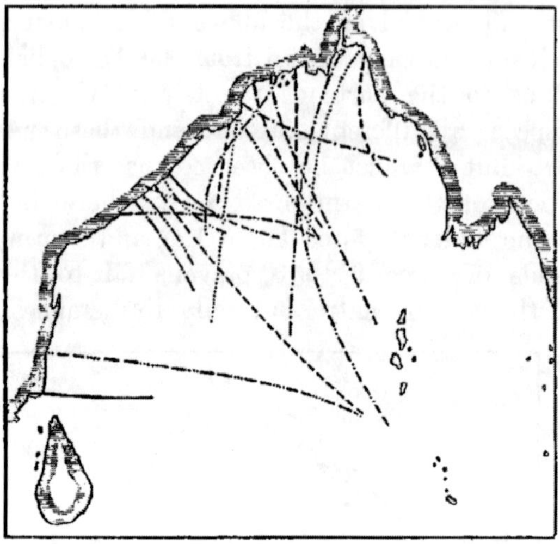

FIG. 27.—STORM TRACKS OF OCTOBER IN THE BAY OF BENGAL.

The fourth table shows that the rate at which storms travel across the bay does not, as a rule, exceed from 4 to

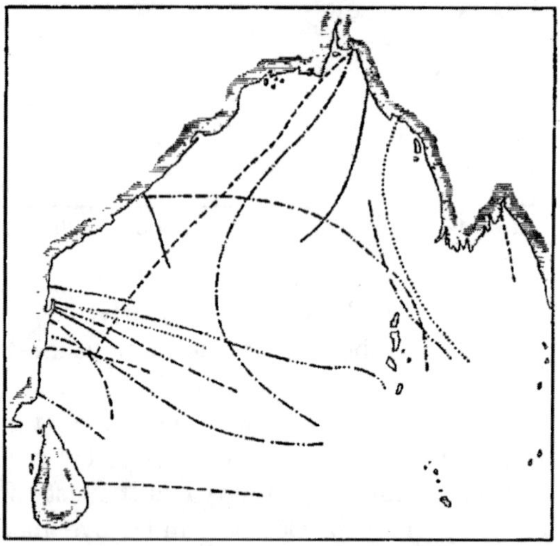

FIG. 28.—STORM TRACKS OF NOVEMBER IN THE BAY OF BENGAL.

8 miles an hour. The chief exceptions, exceeding these rates, occur in September, October, and November. The

highest rate recorded in my list of storms (with one doubtful exception) is 14 or 15 miles an hour, which was satisfactorily determined by Mr. Pedler as that of the False Point cyclone of the 22d September 1885, when traversing the north-west corner of the bay.

Incurvature of Winds and Bearing of the Storm Centre.—The next point I shall notice is one of extreme importance to the seaman, viz. the bearing of the storm centre with reference to the direction of the wind, since he must depend on this, even more than on the fall of his barometer, for judging of his position in the storm, and what course to pursue in order to escape from the violent winds that blow immediately around the central calm. The most convenient way of expressing this relation is to suppose a line drawn from the ship's position to the centre of the storm, and another line intersecting it, which shows the direction of the wind, and to determine the angle between them. Thus let *s* in the accompanying Figure 29 be the position of the ship, *c* the storm centre, and *ws* the direction of the wind at *s*; then *sc* is the bearing of the storm centre, and *wsc* the angle to be determined. According to the view held by Reid and Piddington, this angle is a right angle, and the practical rules for avoiding the centre, laid down by these writers, are based on that assumption. But, as already remarked, Mr. Piddington's own data, whenever they are sufficient for determining the position of the centre, cannot be reconciled in many cases with that view, and, as his Memoirs show, he was sometimes driven to suppose that two storms existed, side by side, in order that the observed wind directions might conform to those required by his theory. As a fact, abundant experience shows that the winds blow in spirals, in towards the centre of the storm; and, in low latitudes, in tolerably sharp spirals, as shown in Fig. 8, p. 85; and, on this point, observation accords with the theory of wind movements as developed by Mohn and Guldberg.

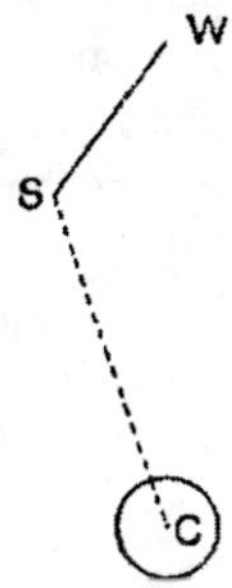

FIG. 29.—BEARING OF STORM CENTRE.

In order to determine as nearly as possible the law of the wind's incurvature for the Bay of Bengal, I have measured with a protractor the angle *wse* (Fig. 25) on the numerous storm charts published at different times by the Meteorological Offices of India and Bengal, rejecting none on account of any apparent discrepancy with theory, excepting one or two cases, in which it would appear that the wind, as reported, has been reversed in direction. I have restricted the measurements to wind observations made at sea, or at good observatories on the coast, since there is reason to believe that the incurvature is greater on the land than at sea; and I have taken the means of all the measurements with the following results:—

132 observations between N. lat. 15° and N. lat. 22°, within 500 miles from the centre, give for angle *wsc* the mean value 122°.
12 observations between N. lat. 15° and N. lat. 22°, within 50 miles from the centre, give for angle *wsc* the mean value 123°.
68 observations between N. lat. 8° and N. lat. 15°, within 500 miles from the centre, give for angle *wsc* the mean value 129°.

The observations within 50 miles of the storm centre in the south of the bay are too few to afford any trustworthy result.

These results show (1) that, in the north of the bay, running before the wind, and with the wind due aft, the bearing of the centre is, *on an average, about 3 points before the port beam;* and (2) that *this relation holds good at all distances from the centre,* as far, at least, as the winds are stormy;[1] (3) that, in the south of the bay, the bearing to port is from

[1] Such, as stated in the text, is the result shown by the storm charts for the Bay of Bengal, and a similar apparent constancy (with greater incurvature) was found by Professor Loomis for the typhoon of the 20th October 1882, which passed over Manilla. But it must be mentioned that as the result of a very extensive discussion of storms both in the tropics and the temperate zone, Professor Loomis finds that the incurvature of the winds decreases as the centre is approached, and he remarks, "This change in the angle of inclination is not simply an occasional occurrence, but is an invariable characteristic of great and violent storms." Theory leads to the same conclusion. It is probable, therefore, that such is also true of the storms of the bay, but the difference is probably small, and it would require a more accurate determination of wind direction and the position of the storm centre than is usually available to determine its amount.

half a point to a point less than in the north of the bay, or *on an average* between $3\frac{1}{2}$ and 4 points before the port beam. Fig. 8 represents the wind directions in a cyclone in the north of the bay as expressed by this law.[1]

These, then, are the average results of actual experience. As far as they go, they are conformable to theory, for theory requires that the incurvature should be the greater the lower the latitude, and also probably greater on the land than on the sea, owing to the greater friction of the wind on the irregularities of the land surface. But individual observations sometimes deviate widely from this average This may, in some instances, be due to an error in the position of the ship; but in any case the law would only be strictly true if all storms were circular vortices, and this we know not to be the case. As a general rule, it appears that their form is rather oval than circular; and, moreover, there are frequently local irregularities and changes in the barometric pressure in these storms, all of which must influence the local wind directions. It becomes necessary, therefore, to ascertain from the observations what amount of deviation from the law must be reckoned on as probable, and for this due allowance must be made. The deviation may be such that the real position of the centre may be either less or more than 3 points before the port beam of a ship running before the wind, and both are equally likely.

Now I find from the 132 observations recorded in many different storms in the northern half of the bay, that the position of the centre, as inferred from a single observation, is subject to a mean error of 21° or about 2 points before or aft the average direction. So that the utmost that can be inferred from such an observation is, that *the centre probably*

[1] These measurements show a smaller amount of incurvature than that determined by Professor Loomis, on the evidence of the Indian storm charts. His method of procedure, however, is different, since he determines the angle between the wind and the nearest isobar, and the majority of his measurements are made on land observations of the winds. To this last circumstance the difference is no doubt chiefly due. The method I have adopted seems more suitable for practical seamen, and the result is, I think, more useful.

lies somewhere between 1 *and* 5 *points before the port beam,* under the conditions above specified.

This amount of uncertainty, however, holds good only of observations made generally at any distance within 500 miles of the storm centre, and is partly due to errors of the ship's position, which would not affect any judgment of the bearing of the storm centre formed on the spot. Within 50 miles of the centre, when the winds are of hurricane violence (force 12), the observations show a much closer agreement, and that the true position of the storm centre, in the north of the bay, may be judged to lie between 2 and 4 points before the port beam of a ship running before the wind.

In the south of the bay the storm centre bears more to leeward. In storms off the coast of Ceylon it is probably between 2 and 6 points before the port beam. And since the amount of deviation depends on and varies with the latitude, it may be expected that the same rules hold good for similar latitudes in other tropical seas north of the equator, and, substituting *starboard* for *port,* in tropical seas of the southern hemisphere; but not for storms in any higher latitudes.

The Barometer in Storms.—It has been, and perhaps is still, a common belief that the barometer falls steadily for many hours, if not days, before the approach of a cyclone, and hence that a careful watch on the behaviour of the barometer should put any one in a position to anticipate and prepare for the coming storm. As a fact, this holds good only when a cyclone is forming in the neighbourhood; if the cyclone is travelling up from a distant part of the bay, the local barometer is often quite unaffected until the appearance of the skies and the characters of the wind and weather alone are such as to leave little room for doubt that a cyclone is approaching, and at no great distance; and even when, as sometimes happens, a slight fall for a day or two precedes the advent of the storm, it is no greater than such as constantly occurs at all times of year, when it has no special significance. Before the great Calcutta cyclone of the 5th October 1864, there was no appreciable fall of the barometer until 8 P.M. of the 4th, $16\frac{1}{2}$ hours before it reached its lowest point; and before the False Point cyclone of the 22d September 1885, in which the barometer fell lower than has ever before been recorded at the sea-level, although there had been a slight fall from the 18th, the whole amount of that fall up to the afternoon of the 21st was scarcely more than one-twentieth of an inch, and the mercury column rose and fell with the daily tides till $10\frac{1}{2}$ hours before the central calm passed over the station. It was not till 8 P.M. of the 21st that a steady fall set in, but the weather had begun to assume a threatening appearance between 4 and 5 P.M., the wind being from north-east and blowing fresh in squalls, with a heavy overcast sky and heavy banks of clouds rolling up from north-east.

The observations of ships in the north of the bay, when a cyclone is advancing from the south, afford similar experience. Thus, on the approach of the Midnapore and Burdwan cyclone of October 1874, the *Coleroon* pilot-vessel at the Sand-heads had the barometer, at 4 P.M. of the 14th, somewhat higher than on the previous day at the same hour; it then began to fall slowly, but no rapid fall set in till 10 A.M. of the following morning, when it had been blowing a gale for 9 hours; and at 1 P.M. the central calm passed over the ship. And in the case of the May cyclone of 1887, in which the pilgrim steamer *Sir John Lawrence* foundered, the barometer on board the P.V. *Cassandra*, at the same station, remained quite unaffected when the advancing storm centre was but 200 miles to the south-east, and began to fall slowly only at 2 A.M. of the 25th, when the centre had approached to within 120 miles. From this time to noon, when its distance had decreased to about 50 or 60 miles, and the force of the wind had increased from 6 to 10 on Beaufort's scale, the total fall was only about one-tenth of an inch. Three hours later the ship was in the central calm of the storm, the barometer having fallen in the interval nearly 1 inch.

When we compare the curves, plotted from barometric readings recorded at short intervals during the passage of different cyclones, such as that of the Midnapore (1874) storm, represented in Fig. 30, we find that they have this character in common. The barometer begins to fall steadily as soon as the winds become strong and squally (force 6 to 8), which may be when the storm centre is yet between 100 and 200 miles distant. For some hours, the rate of fall increases but slowly; but at length, when the centre has approached to within from 20 to 50 miles (according to the extent of the vortex), and when the winds have increased to hurricane force (12), the fall suddenly becomes very rapid,

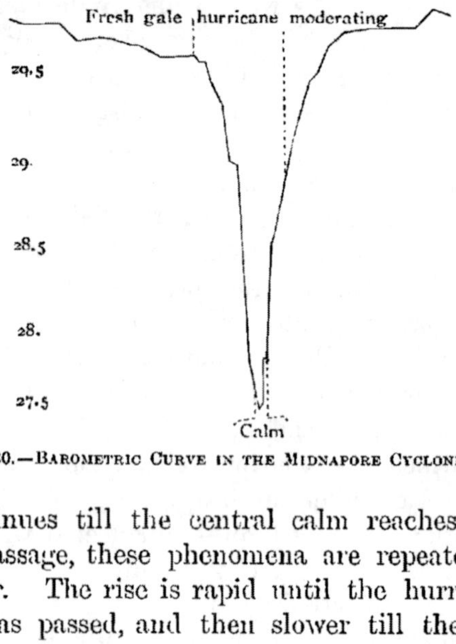

FIG. 30.—BAROMETRIC CURVE IN THE MIDNAPORE CYCLONE, 1874.

and so continues till the central calm reaches the station. After its passage, these phenomena are repeated in the reversed order. The rise is rapid until the hurricane core of the storm has passed, and then slower till the winds have fallen to the strength of an ordinary fresh breeze, and the barometric pressure has become nearly normal. Frequently the rise is somewhat slower than the fall, but not invariably, as this was not the case either in the Calcutta cyclone of 1864 or in the False Point cyclone of 1885. It probably depends on whether the intensity of the storm is increasing or diminishing during its passage.

These characters are distinctive of those fierce and destructive storms to which I have restricted the term "cyclone." In those which I have termed "cyclonic storms," the hurricane core and its central calm are not developed, and neither do the winds at sea exceed the strength of an ordinary gale (6 to 9), except sometimes to the east and south of the storm centre, nor does the barometer at the centre fall more than three or four-tenths below its reading outside the vortex. Mr. Eliot, who first drew particular attention to these distinctive features, describes the more violent class of storms in the following words:—

"They may be described as consisting of three parts; an outer storm area, an inner storm area, and a central calm. In the outer storm area the weather is very similar to that in the smaller storms of the rains. The winds circulate in the usual spiral manner, and their direction and shift enable the sailor to determine the bearing of the centre with approximate accuracy, especially when he is near the inner edge or is approaching the inner storm area. In the outer storm area winds are more or less violent, frequent squalls occur, the sea is high, and there are strong currents, such as always obtain in any cyclonic storm at sea. The barometer falls very slowly, and stands very little lower than its ordinary height at the season. Within this outer storm area is the inner storm area, usually of much smaller extent, surrounding the central calm area. In it the barometer falls with great rapidity from the inner edge of the outer storm area to the calm central area. . . . The transition from the inner storm area to the calm centre is always very sudden and sharply marked. The transition from the outer to the inner storm area appears also to be more or less sharply marked, but to a much less degree than [the former]. Within the inner storm area the winds are of hurricane force, the squalls and rainfall of excessive violence, and the sea tremendous. The weather in the calm centre is too well known to require description. In many of the largest cyclonic storms in the bay, the sky is frequently almost clear over a portion of the calm centre, the atmosphere hazy, and the sun or [stars] visible."

Weather around the Storm Area.—There is a marked distinction in the character of the weather to north and south of a cyclone in the Bay of Bengal. Cyclones are generated and fed by the damp stormy south-west wind that blows from the equatorial sea and, when it prevails right up the bay, is known as the south-west monsoon. Hence, whenever a cyclone or cyclonic storm is in process of formation, or is travelling across the bay, this wind prevails everywhere to the south of the storm, and as a south or south-east gale far to eastward of the vortex. For a distance of 400 or 500 miles in these directions the weather is squally and rainy with strong winds, and these conditions continue until the storm is exhausted or has reached the land and travelled far inland. But beyond a couple of hundred miles or less to the north and west, the weather remains fine, often calm and sultry, during the formation of the cyclonic vortex, and up to the time when, in its advance, its outer margin reaches the place of observation; which is, as a rule, when the centre of the storm is not more than 200 miles distant, and sometimes less. Thus, before the storm of the 15th October 1874, a strong north-east breeze, the first effect of the approaching vortex, was experienced at False Point when the storm centre was but 180 miles distant; and at Saugor Island, although other premonitory signs were not wanting, there was no bad weather until the centre had approached to within 150 miles of the mouth of the Hooghly. Also, before the great Calcutta cyclone of the 5th October 1864, the weather was fine at the Sandheads and the wind light from the east up to the night of the 3d, when the storm centre was 270 miles to the south. In the night the wind increased to a fresh breeze, but fell light in the early morning, and it was not until some hours before noon that it again freshened, and in the afternoon increased to a gale, with a falling barometer. At this time the centre had advanced to within 200 miles of the pilot station. The experience afforded by other storms is of a similar character.

But although to the inexperienced eye there may be little either in the behaviour of the barometer or the appearance of the weather in the north of the bay to give warning of an approaching storm, such approach is always heralded by significant indications, which may be duly noted by the observant seaman. A still oppressive atmosphere or light variable winds from between north and east, followed after a day or two by light easterly winds, a strong westerly set of the current across the head of the bay, and by a long swell from the south-east; light wisps of cirrus cloud moving from south-west and gradually thickening to a sheet of cirro-stratus, through which the moon shines surrounded by a luminous corona; a ruddy glow in the atmosphere, especially at sunrise, and a low dark bank of clouds on the south-eastern horizon, lit at night by distant lightning, are among the commonest and most ominous signs of the brewing storm. Their validity has been abundantly verified and confirmed, not only by frequent remarks in the logs that, at different times, have been sent in to the Meteorological Office, after cyclones in the bay, but also by the long observation of Mr. S. R. Elson, an experienced branch pilot on the Hooghly establishment, who for many years past has given close attention to the weather signs of the north of the bay, and, in his little work *The Sailor's Sky Interpreter*, has embodied them in a metrical description which is well worthy of attention. Waterspouts are sometimes formed in the still vapour-charged atmosphere that broods over the surface of the bay, before it is stirred by any in-draught towards the advancing vortex.

Practical Conclusions and Rules for Guidance.—It now remains to show how the foregoing information may be turned to useful account by seamen navigating the Bay of Bengal in the stormy season. Obviously, the main point is to avoid being involved in the hurricane core of the storm, which, perhaps, never exceeds a couple of hundred miles in diameter, and is generally much less. This, at least, may often be accomplished, even though it may be impracticable to escape the gale of the outer storm, which blows for a distance of 100 to 150 miles beyond it to the north and west, and to a greater distance to the east and south of the hurricane vortex. And, first, let us consider the case of ships running up from the south with a strong south-west wind, occasional squalls and rain, and a slowly-falling barometer, indicating that bad weather prevails somewhere to northward. If this happens any time between the middle or even the beginning of June and the middle of September, it is almost certain that the position of the storm vortex is somewhere to the north of lat. 16°; if in July or August, probably not farther south than lat. 19°, though there have been one or two exceptions to this rule. Since storms, at this season, move on tracks between north and west, and indeed, with rare exceptions, between west-north-west and north-west, the best course for a sailing vessel will be to keep away to the east, with a view to rounding the eastern quadrant and getting the advantage of the south and south-east winds to run up the bay. But care must be taken not to run up too rapidly and to get involved in the storm before it has moved far on its westerly course. On reaching the neighbourhood of the vortex, and, in any case, if the weather is getting rapidly worse, it is best to lie to, or, if the vessel be a steamer, to run to eastward till the storm has travelled to a distance. This will be indicated by the rising barometer, and by a general improvement in the weather.

In May, October, and November, the cyclone may be in any part of the bay, and the best guides to its position are—the rate of fall of the barometer as the ship holds on its course, and the direction of the wind; bearing in mind the rule that, running with the wind, the centre lies not at right angles or abeam, but probably between 1 and 5 points before the port beam, if in the north of the bay; between 2 and 6 points in the latitudes of Madras, Ceylon, and the Nicobars. In these months, storms still most frequently travel to north-west, but the possibility of their course being to north or even north-east must be taken into account, in any attempt to sail round the eastern quadrant of the storm (see Figs. 26 to 28, pp. 231, 232).[1] In any case, to run directly before the wind is to be carried infallibly into the heart of the storm, for the rate of the storm's progress is, as a rule, much less than that of ship driving before the storm wind.

Now, to turn to the north of the bay. A ship leaving the Hooghly, in the months from June to September, will be amply warned by the easterly winds and falling barometer,

[1] On this passage, Captain Henry Toynbee remarks, "I would advise, whenever running in a Northern Hemisphere cyclone, to keep the wind well on the starboard quarter, when force, etc., will permit it."

in any case by the storm signals at the telegraph stations, of the state of the weather in the north of the bay. In July and August it is unlikely that the winds in any part of the storm will be of hurricane strength, but there is always a possibility that a cyclonic storm, even in these months, may temporarily develop into a small cyclone with its hurricane core and central calm, and it is better to avoid the risk of encountering it. The storms of this season almost always travel towards Balasore Bay and False Point (rarely, and only during prolonged breaks in the rains of the upper provinces, northwards to the Sunderbuns); so that, when such a storm is forming at the head of the bay, a ship running down to south will probably get into the thick of it. In June and September, some of the most violent cyclones have visited the shores of Bengal and Orissa.

In May, October, and November, the birthplace of the storm is generally far down in the bay, and it has time to develop into a furious cyclone before reaching the coast. A shipmaster, proceeding to sea in these months, should give careful heed to the premonitory signs of the weather noticed in a preceding paragraph (p. 240), and detailed more at length in Mr. Elson's little work. Storms are less common in May than in the subsequent months, but such as do occur are, as a rule, very severe and destructive. It may be noticed that, in the list of notable cyclones on p. 225, the number recorded in May is much greater, proportionally to those of other months, than is the total number of storms as exhibited in the table of storms of all degrees of severity on p. 228. An east or north-east wind at the head of the bay in May, with a strong westerly set of the current at the Sand-heads, is very ominous.

If, in any of these months, the weather experienced at the mouth of the river is already squally, the wind rising in gusts, and low long-drawn masses of cloud are driving before an easterly wind, it is certain that the storm is not far distant, and a prudent seaman will remain at anchor under shelter of the land until the worst is over. If, however, he have proceeded to sea with apparently fine weather, but such as warns him that a cyclone is brewing to southward, all that he can do is to watch the changes of the wind and his barometer, and—remembering that, in these months, the course taken by a cyclone is somewhat uncertain, and though most likely to be north-west, may be in any direction between west and north-east, and that its rate of progress may be anything between 4 miles and 12 miles an hour—to estimate the bearing of the storm centre and its probable course, according to the directions and information given in the preceding paragraphs, and to take his measures accordingly. His position is then a critical one, requiring close observation and the greatest judgment. If the course of the storm should happen to be to west or north-west, by running to south-west he incurs the risk of crossing its path and finding himself involved in the inner vortex, possibly the centre; whereas, if he lies to, and the storm should be moving northward, there is the chance of its overtaking him, with the like result. Probably the best alternative, on nearing the storm, is to heave to, until the wind, which we may suppose to be at north-east, shows a decided tendency to veer to east-north-east (indicating that the storm is travelling on a north-west or west-north-west course, to the south of him), or to back to north (which will be the case if the storm is moving northward, and to the east of his position). In the former case, he should remain hove to on the starboard tack. In the latter, as soon as the wind is decidedly north, he may safely run to south-west, keeping the wind as much as possible on his starboard quarter. In doubtful cases it may be better to remain hove to, since, as Figs. 26 to 28 and also the track table on p. 230 show, even in these months more storms move to north-west and west-north-west than to north and east of north. He must bear in mind that, with the wind at north, the probable bearing of the centre is between east by south, and south-east by south of his position; with the wind at east-north-east it bears probably between south by east and south-west by south of him in the north of the bay; or a point more to the west in the south of the bay.

Ships lying in the Madras roads or off any of the Carnatic ports, when a cyclone is near at hand, approaching from the bay, usually try to run to south round the western quadrant. Sailing ships, under such circumstances, have probably not much choice, since it might be impracticable for them to make sufficient way against the strong north winds and currents of the north-western quadrant to escape the vortex of the storm; especially if it happens to be travelling towards some point to the north of the port. But in the case of steamers, leaving Madras as soon as the barometer begins to fall steadily, and when the storm centre is yet at a distance of 100 or 150 miles or more, it may be a question whether the safer course is not to steam full power to north-east, and thus avoid the risk of encountering the vortex, since with the wind at Madras from north or north-north-west the storm may be moving towards Pondicherry. In judging of the direction and course of the storm, it is most necessary in this case to bear in mind that, in this latitude, running with the wind, the centre is probably as much as 4 points before the port beam, and that its course is generally between west by north and north-west. Thus, *e.g.* in the storm of the 5th to 9th November 1886, the wind at Madras was north-north-west, when the storm centre bore east by south at a distance of 250 miles; and at Negapatam due west, with the centre bearing east-north-east at a rather greater distance; and 24 hours later, when the centre bore about 80 miles north east of Madras, the wind at that station was due west. Most of the cyclones that reach the Madras coast in May, October, and November are formed off the coast of Ceylon; but occasionally, as in the instance just mentioned, a storm, generated to the west of the Andamans and Nicobars, crosses the bay on a course nearly due west.

Ships leaving Rangoon or Moulmein, and encountering stormy north-east winds, with a falling barometer, indicating the existence of a cyclone in the Andaman Sea, will do well to delay their departure until the storm has passed over the land, should it take a northerly course, or until it has crossed the islands into the Bay of Bengal on the more usual track to the north-west. This will be indicated by the rise of the barometer and the wind changing to east and south of east.

2. The Arabian Sea

Our knowledge of the storms of the Arabian Sea is still very imperfect. Some few notices of them are to be found in Mr. Piddington's Memoirs; one has lately been investigated by Mr. Eliot, and made the subject of a memoir in the official publications of the Indian Meteorological Office; and a remarkable storm that traversed the whole length of the Gulf of Aden in June 1885 has been noticed in several European publications, and is now being fully investigated by the London Meteorological Office. A list of storms known to have affected the west coast of India, and to have been met with in the Arabian Sea, has also been published by Mr. F. Chambers, but this, being a mere list, furnishes little information beyond affording some idea of the seasons at which stormy weather is most prevalent. It is therefore impracticable to give any comprehensive estimate of the relative frequency of storms in different parts of the Arabian Sea like that already given for the Bay of Bengal, or more than some general indications of their tracks.

Storm Seasons on the West Coast of India, etc.—The following tabular summary of Mr. Chambers's list shows the relative frequency of storms in the different months of the year, as far as can be inferred from the existing records.

Number of Recorded Storms on the West Coast of India and in the Arabian Sea up to 1881

January	4	May	13	September	3
February	3	June	20	October	4
March	2	July	2	November	10
April	9	August	2	December	2

From this it would seem, at first sight, that in the Arabian Sea the stormy season is less definitely restricted than in the Bay of Bengal. Thus, three storms are recorded in February and four in January out of a total of 74; whereas, in the Bay of Bengal, none are known to have occurred in the former, and only two in the latter month, out of a total of 115. This difference is, however, probably rather apparent than real. The notices of most of the storms enumerated in the above table are very fragmentary, but they afford grounds for believing that at least two of the February storms and two of those in January were not experienced at sea, and are hardly to be reckoned as Arabian Sea storms; they were felt as gales only on the coast of Sind, the seat of the storm being over the land. They were, probably, instances of the cyclonic storms of the cold season of Northern India (already described in the preceding section), which travel from west to east. It is yet doubtful whether such storms reach India from the west across the continent, or are formed over the land of India. Most probably some are derived and others of local origin.

One other respect in which the above table contrasts with that previously given (p. 225) for the Bay of Bengal, consists in the much greater proportional frequency of storms in May and June, and the relative paucity of those in October and November, especially the former month. But the list contains few references to any storms of the Arabian Sea that have not been felt on the west coast of India; indeed, only seven such are included in Mr. Chambers's list, and one other is known to have occurred since the list was drawn up. And it is improbable that storms of this class are relatively so rare in the Arabian Sea as this record would seem to show. It seems likely, therefore, that when the storm records of the Arabian Sea shall receive the same attention that Mr. Piddington formerly, and Mr. Eliot and others of late years, have devoted to those of the Bay of Bengal, some of the leading features of the above table may be found to require modification.

Bearing in mind, then, that the information as yet available to us on this subject is very imperfect and fragmentary, it remains to see what positive facts can be deduced from such data as we have before us.

Storm Tracks of the Arabian Sea.—Of the whole number of storms enumerated in Mr. Chambers's list, eleven were originally formed over the Bay of Bengal, and travelled westward across the Indian peninsula; and one further instance of the same kind, that has since occurred, has been worked out by Mr. Eliot. The distribution of these twelve storms according to season is as follows:—

January	1	October	1
April	1	November	4
May	4	December	1

These have all crossed the southern half of the peninsula on a west by north or west-north-west course, and most of them have been traced for a considerable distance beyond, across the Arabian Sea; one, viz. that of October 1842, as far as the coast of Arabia, where it caused a great destruction of native craft. This west-north-west course seems therefore to be that most usual in the Arabian Sea, as it is also in the Bay of Bengal. But a very important variation of the latter part of its course was presented by the November cyclone of 1886, the history of which has been worked out by Mr. Eliot; and it is the more important to notice this, because it took place midway in the steamer track between Aden and Bombay, and not anticipating it, the Peninsular and Oriental mail steamer *Peshawar* became involved in the hurricane, while endeavouring to escape from it by steaming to the north-west. Having reached the latitude of Bombay and the meridian of E. long. 64°, the storm changed its course rapidly from west-north-west through north-west and north to north-north-east, and travelling faster than the steamer, overtook it, the centre passing within 20 miles of its position. It is not improbable that others of these storms may undergo a similar change of direction when reaching the northern part of the sea. At all events, the contingency is one that must always be regarded as possible.

The cyclonic storms generated in the north of the Bay of Bengal during the summer monsoon, as a general rule, break up and disappear after a more or less protracted course over the land; but exceptional instances occur, in which they traverse the whole width of Central India on a nearly due west course to Kathiawar and Cutch, and thence pass out to the extreme north of the Arabian Sea. Such storms are rare. One only occurred during the 10 years, 1877-1886, viz. that of the 25th June to the 4th July 1883. Its history has been fully worked out by Mr. Eliot, in a memoir published in the 53d volume of the *Journal of the Asiatic Society of Bengal.* Like all storms that cross India, its intensity was greatly reduced while traversing the land, but it recruited its strength as it neared the Western Sea, and a British India steamer, that became involved in it in N. lat. 22°, E. long. 63½°, experienced winds of hurricane strength (force 12 on Beaufort's scale).

In 1886 a storm, originating over the middle of the Bay of Bengal at the end of September, also crossed the peninsula in lat. 17°, on a due west track ; passing out to the Arabian Sea about Rutnagiri, on the morning of the 3d October. After leaving India there is no further record of its progress, but it probably continued its course across the North Indian Ocean.

Another direction, sometimes taken by cyclones in the east of the Arabian Sea, is north by west, parallel with the Indian coast. An instance of the kind, in April 1847, is discussed at length in Mr. Piddington's fifteenth Memoir. This storm was apparently formed to the west of Cape Comorin, and its progress was very distinctly traced, first to north-west and then to north by west as far as the latitude of Goa. Shortly afterwards it *may have* broken up, since, on the following day, the weather moderated and cleared up. But its influence was felt as far north as the Gulf of Cutch,[1]

[1] Mr. Piddington makes two storms of this. That which originated to the west of Cape Comorin he supposes to have travelled to the north-east and to have broken up before reaching land, since it was not felt at the Travancore observatory. The other, which travelled northwards to the latitude of Goa, he supposes to have originated close to the Maldives. But it is clear that in this, as in many other instances in which he infers two independent cyclones to have existed simultaneously in near proximity to each other, he has been misled by the error of his theory that the winds blow in circles. The north-east track is a figment required by his theory, and is inconsistent with the evidence.

and three days after the *Buckinghamshire* had been involved in the centre of this cyclone off Goa, a severe cyclone swept over Muscat. If, as seems possible, this was the same storm, then, instead of breaking up, it must have suddenly changed its course from north by west to north-west, and crossed the North Arabian Sea at an average rate of about 14 miles an hour. Its rate of progress up to Goa had been 9·2 miles an hour according to Mr. Piddington.

In the same Memoir, Mr. Piddington mentions another storm in June 1811, which he supposes also to have travelled in a north by west direction in E. long. 69°-71° and N. lat. 14°-18°. On general grounds there is nothing improbable in this track, but the evidence is, I think, insufficient to establish it or to admit of any trustworthy inference.

Of other June storms recorded in Mr. Chambers's list little can be said, except that the majority of them appear to have been formed on the setting in of the monsoon rains on the Bombay coast; and, judging from the analogy of the Bengal storms of this month, they probably originated not far south of the latitude of Bombay. It is at least doubtful whether these or indeed any other storms have ever moved eastwards across the land. In Mr. Chambers's list, only one storm of hurricane strength is recorded in June at Bombay, and none at any other coast station. The track of this one is not indicated, but there are notices of several severe southerly gales at Bombay in June, and the information quoted, respecting these, seems to indicate that, in many instances, a cyclone was travelling northwards in the immediate neighbourhood of the coast.

Of the storms that have been recorded in the western half of the Arabian Sea, but which, so far as is known, have been unfelt in their earlier stages on the Indian coast, the existing notices are for the most part too fragmentary to throw any light on their place of origin and the direction of their tracks. The sole exception is that of the cyclone already noticed, which, on the first days of June 1885, traversed the whole length of the Gulf of Aden on an almost due west course. A very good though not complete history of this storm has been published by the German Seewarte of Hamburg.[1]

This storm appears to have originated somewhere to the east of Socotra on the last days of May. It was fully formed before the 31st, on the early morning of which day the *Mergui* was involved in it in N. lat. 13°, E. long. 59°. Thence it travelled on a due west course, passing a little to the north of Socotra; and on the 2d and 3d June centrally up the Gulf of Aden to Obrok and over the African mainland. Its rate of progress averaged between 9 and 10 miles an hour throughout till it neared the land, when its rate of progression seems to have increased. It was small, but very severe and destructive, four steamers, including a German corvette and a French despatch boat, having foundered in it with a total loss of 426 souls. Admiral Cloué computed its diameter as 150 miles on the 30th May, decreasing as it passed up the Gulf to no more than 50 miles when it reached Obrok.

It is noteworthy that the appearance of the skies and the character of the weather in front, *i.e.* to the west of this storm, in its advance up the Gulf of Aden, were in all respects similar to those already described as prevailing to the north and in front of cyclones that advance up the Bay of Bengal. The *Inchulva*, passing the islands of the Red Sea on the 1st June, when the storm had not yet reached Cape Guardafui, experienced light variable airs with hot sultry weather and a smooth sea. At sunrise of the 2d June, the day preceding that on which she became involved in the cyclone, the sky was partially overcast "with a very fiery appearance," and at sunset it is described as "looking very bad, green, red, and yellow, all mixed up together, and the atmosphere most oppressive." At night the lightning was incessant, the sky having a red glare all over, and light hot airs came from all points of the compass,

[1] *Revidirte Auszug aus Ann. d Hydr.*, etc., XIV Jahrgang (1886), Hefte V. and VI. It is now being more fully investigated in the London Meteorological Office.

with showers of rain occasionally. And these signs were followed by a swell making up from the eastward and gradually getting heavier. Other ships describe similar appearances and a halo round the moon.

The barometer on the *Inchulva* began to fall slightly on the forenoon of the day preceding that on which she encountered the cyclone, and when the centre was yet some 450 miles distant; but no rapid fall set in until 8.30 A.M. on the morning of the 3d, only three hours before the centre passed close to the ship, and when it could hardly have been more than 30 miles distant.

The course of this storm seems to have been exceptional, inasmuch as no similar instance has been experienced at Aden since it was first occupied as a British possession in 1839; and it is probable that most storms that originate off Socotra take a somewhat more northerly course towards the coast of Hadramant or the Khooria Mooria islands. In April 1856, the *Haddington* and s.s. *Queen* were involved in a cyclone on this coast to the south of the above-named islands, in which the steamer narrowly escaped shipwreck. On the whole, however, there seems reason to believe that cyclones are less frequent in the west of the Arabian Sea than in the neighbourhood of the Indian coast, and much less so than in the Bay of Bengal.

Storm Warnings and Signals. Bay of Bengal.—The Calcutta Meteorological Office, under the control of the Meteorological Reporter to the Government of Bengal, is now charged with the duty of issuing storm warnings to all the principal ports on the coasts of the bay. A weather report for the Bay of Bengal, illustrated with a weather chart, is issued daily at Calcutta, giving the telegraphic reports for 8 A.M. from the following stations:—

Diamond Island.	Jessore.	Cuttack.	Masulipatam.
Akyab.	Calcutta.	False Point.	Madras.
Chittagong.	Saugor Island.	Gopalpur.	Negapatam.
Burrisal.	Midnapore.	Vizagapatam.	Trincomalee.
Dacca.	Balasore.	Cocanada.	Galle.

A table shows the readings of the barometer at 8 A.M.,

corrected and reduced to sea-level values; its change since the same hour of the preceding day; the wind direction and velocity at 8 A.M., and its mean rate of movement in miles per hour for the previous 24 hours; the maximum, minimum, and mean temperature of the 24 hours, and the change of the latter since the previous day; the thermometer reading, humidity, and cloud proportion at 8 A.M.; and the total rainfall of the 24 hours, together with a word descriptive of the state of the sea at these ports where the observations are situated on the shore. These tabulated data are followed by a short description of the general results, and a weather chart shows the isobars, with notes of the barometric changes, the direction of the winds and the state of the sea and general weather. This report is published daily about 2 P.M., and sent to all Government offices and subscribers.

In the event of these reports affording evidence of stormy weather in any part of the bay, a warning is telegraphed to the principal ports of the coast most likely to be affected, with instructions to display the appropriate storm signals. Except at the port of Calcutta and its approaches, the signals are of two kinds only, one being the *warning signal*, to intimate that a storm exists in the bay; the other the *danger signal*, indicating that the storm appears to be approaching the port at which the signal is displayed. The system includes day and night signals of both classes. They are—

Storm Signals, Bay of Bengal

	Day.	Night.
Warning Signal .	A ball.	Three lamps suspended one above the other.
Danger Signal .	A drum.	Two lamps suspended one above the other.

The official directions and instructions for the use of these signals are given in Appendix IV.

The ports at which these signals are displayed are the following:—

In Bengal—

Chittagong, Balasore, Chandbally, False Point, and Pooree.

In Madras—

Gopalpur, Vizagapatam, Bimlipatam, Cocanada, Masulipatam, Madras, Negapatam, Tuticorin.

In Burmah—

Akyab, Bassein, Rangoon, Moulmein.

For the port of Calcutta and its approaches a more elaborate system is in use; the day signals include distinctive signals for large and small storms (cyclones and cyclonic storms of this work), and also indicate the probable course of the storm, whether towards the Sunderbuns east of the Hooghly, direct to the Hooghly estuary, or towards the Orissa coast to the west of it. The night signals are the same as at other ports.

The Calcutta day signals are as follow:—

CALCUTTA STORM SIGNALS (DAYTIME)

	DESIGNATION.	SIGNAL.		INDICATION.
No. 1.	Bad weather signal.	Ball,	●	Cyclone of unknown intensity in bay.
,, 2.	Danger signal.	Cone, point upwards,	▲	Large cyclone moving northward, east of Saugor, west of Chittagong.
,, 3.	Do.	Cone, point downwards,	▼	Large cyclone moving northward, west of Saugor and north of False Point.
,, 4.	Great danger signal.	Drum,	■	Large cyclone approaching Hooghly estuary and Calcutta.
,, 5.	Bad weather signal.	Two cones, upper inverted,	▼ ▲	Small cyclonic storm in bay. Probable course to south of Chittagong and False Point.
,, 6.	Warning signal.	Two cones, lower inverted,	▲ ▼	Small cyclonic storm approaching coast north of Chittagong and False Point.
,, 7.	Warning signal.	Ball below cone,	▲ ●	Small storm moving northward to Sunderbuns, between Saugor and Chittagong.

Designation.	Signal.		Indication.
No. 8. Warning signal.	Ball below inverted cone,		Small storm moving northward, to west of Saugor and north of False Point.
,, 9. Danger signal.	Ball below drum,		Small storm approaching Saugor roads.

These and the night signals are displayed at Saugor Lighthouse, Mud Point, and Diamond Harbour in the Hooghly estuary, at Budge Budge below the port, at Shalimar Point and the Port Commissioners' office, Calcutta. The notification of the Bengal Government in Appendix V gives further details with respect to the signification of the several signals, and the precautions to be taken by seagoing vessels, in the event of their being hoisted.

Storm Signals. Bombay Coast.—The meteorological office at Bombay is charged with the duty of issuing storm warnings to the Bombay coasts, for which purpose it receives daily telegraphic weather reports from the following places :—

Kurrachee.	Calicut.
Surat.	Cochin.
Rutnagiri.	Negapatam.
Karwar.	Madras.
Mangalore.	Masulipatam.

And in the case of storms which travel westwards from the Bay of Bengal and appear likely to cross the peninsula, intimation is telegraphed from the Bengal office and the Imperial office at Simla. No weather report is published at Bombay,[1] but in the event of threatening weather, warnings are issued to the following ports :—

Kurrachee.	Vingorla.
Bombay.	Karwar.
Rutnagiri.	Kumta.

And a signal, consisting of a cone for the daytime and three lamps in a triangle at night, is displayed until the disturbance has passed away.

[1] Since the above was written, arrangements have been made for issuing a daily weather report at Bombay similar to that published by the Bengal Meteorological Office.

Rainfall in Relation to Water Supply and Drainage —Evaporation and Wind Pressure

The subjects to be discussed in the present section are those of most importance to engineers; the chief of which is the quantity and distribution of the rainfall, as affecting the water supply available for the maintenance of the flow of rivers, for tank storage and irrigation, and also as a guide to the requisite provision of culverts and bridges to carry off the drainage. Some information on this head has already been given in the section on rainfall in the first part of this work. But for many purposes further details are necessary, and more particularly those of the heaviness of the rainfall within short intervals of time. Into this subject I shall enter as fully as the available data and the scope of this work admit of, and I shall also give some statistics of the mean rainfall on the catchment areas of some of the Indian rivers, which may hereafter be useful for comparison with the discharge volumes of those rivers.

Nearly connected with these subjects is the question of the amount of evaporation, which, strictly speaking, is also a meteorological datum, and one of great importance. It is, however, one on which there has been but little direct observation in India, and that little, as far as it has any practical value, has been contributed chiefly by engineers, and is to be found in their professional publications. Some of these will be noticed in the sequel. The few observations that have been made at meteorological observatories have been so much affected by purely artificial conditions, that little or nothing of scientific or practical value can be safely concluded from them; and for this reason, and not owing to any under-estimate of the importance of the subject, evapometer readings have not been included in the regular scheme of work at the observatories established by the Meteorological Department.

The first point to be dealt with is the average and

maximum rainfall within short periods of time, and the frequency of falls of stated amount. On this head, Mr. Lowis Jackson, at page 13 of his *Hydraulic Manual*, specifies the following items of rainfall statistics as those required by the engineer, to furnish him with a correct basis for works which demand the greatest amount of accurate information :—

1. The mean, maximum, and minimum monthly rainfall.
2. The mean and maximum daily falls in 24 hours for each month.
3. Special occurrences, hourly falls, longest continuous falls and droughts.

For the first of these series the published rainfall statistics of India afford very ample data, relating to a large number of stations in all parts of the empire, and these can be easily taken out and tabulated. For the second, and in part for the third, the requisite materials exist in the Indian meteorological offices, and a considerable portion have been published periodically in the daily weather reports and in the *Gazettes* of certain of the local Governments. But the labour of examining them for the extraction of the full data in question would be very great; and it must suffice for the present work to give a few examples, relating to different parts of India, which may serve as a general guide in the absence of strictly local data. If required, in special cases, these latter may be obtained from the Indian or provincial meteorological offices, for any place the daily rainfall of which is published in the Government *Gazettes.* But it will be found in the sequel that the relative frequency of falls of different amounts, and the probability of exceptionally heavy falls, are so uniform over large areas, and indeed, with some rare local exceptions, throughout India, that the instances given, when taken together with the local statistics of the mean monthly rainfall and its variability, will generally suffice to enable the engineer to arrive at pretty trustworthy conclusions on any problem involving the consideration of the more detailed data.

Mean, Maximum, and Minimum Monthly Rainfall.— In a recently-published memoir on the rainfall of India, forming vol. iii. of the *Indian Meteorological Memoirs*, are given the full existing records of the monthly rainfall of 451 stations in India and its dependencies, including Ceylon. Of these, the large majority extend over periods of from 20 to 30 years and upwards; and they suffice, therefore, to represent, very fairly, the variability of the rainfall. They include, on the one hand, the remarkably dry years, 1876 and 1877; and, on the other, the very wet years, 1867, 1874 and 1878. From these I have selected 110 stations, most of them on account of their importance as cities or military cantonments, or as representing important agricultural districts, and have tabulated the mean monthly rainfall and the highest and lowest as yet recorded in each month of the year at each station. These data are given in Appendix III.

These tables show that the extreme variations of the rainfall, as well as their mean or average monthly amounts, are very uniform at stations in the same part of the country. The general features of the rainfall distribution, according to locality and season, have already been briefly described in Part I. of this work, and need not here be repeated. The rainfall of the winter and spring months, in most parts of the country, is very variable and uncertain, and, as a general rule, affords no supply to local tanks, and but little affects the streams and rivers. That of the rainy season is also variable, but scarcely ever fails completely in the months of July and August; nor indeed in June and September, except in the drier provinces. At this season, the maximum of any single month appears to be about double, and in some of the drier provinces as much as three or four times its average amount. It is scarcely practicable to generalise to any useful purpose on these irregularities; very probably they will eventually be found to conform to empirical laws very similar to those which regulate the variations of the daily rainfall, the subject next claiming our attention.

Mean and Maximum Daily Rainfalls.—It has already been shown, as resulting from the tabular statistics given at pages 72 to 75, that on the plains of Bengal and the North-west Provinces, and in parts of Central India and the Central Provinces, the average fall of each rainy day is between five and seven-tenths of an inch.[1] On the hills of the outer Himalaya and Assam, on the coasts of Arakan and Tenasserim, and on the Western Ghats and some parts of the Konkan and Malabar, it is higher; at Matheran, Mahableshwar, Akyab, and Moulmein, between 1 and 2 inches, and at Cherrapunji as much as 2·6 inches per rainy day. But in the Deccan and Mysore, and in the Indus valley, only between three and four-tenths. These figures represent the general average of the year; but, as might be expected, the rain of the summer is heavier than that of the winter months, and more lasting than that of the thunder-storms of the spring; and hence the average fall of each rainy day is greatest in the former season. Thus, at Calcutta in the last 13 years, the average fall of a rainy day, in the months November to April, has been four-tenths of an inch; in the months May to October, 0·54 inch; and at Lahore, the average of the former is three-tenths of an inch, that of the latter 0·73 inch.

The comparative heaviness of the rainfall at different seasons and in different parts of the country is, however, better shown by the following tables, which have been drawn up from the daily rainfall returns of the last 13 years (in the case of Lahore, 12 years), and exhibit the total number of days on which falls of given amounts have been registered in corresponding months during the whole period, at the six stations, Calcutta, Lucknow, Lahore, Nagpur, Bombay, and Madras; and also, in the right-hand column, the heaviest falls recorded on any one day, in each month of the year, during the same period (1875 to 1887).

[1] This is more accurate than "between six and seven-tenths," as stated on page 75.

DAILY RAINFALLS OF GIVEN AMOUNT IN EACH MONTH OF THE YEAR DURING 13 YEARS

CALCUTTA

	Inches.							Highest.
	Below ¼.	¼ to ½.	½ to 1.	1 to 2.	2 to 3.	3 to 5.	5 to 7½.	
January .	14	2	3	2	...	...	...	1·9 ins.
February .	20	4	4	6	1	...	...	2·1 ,,
March .	31	3	9	7	...	...	...	1·6 ,,
April . .	28	7	10	4	1	...	...	2·5 ,,
May . .	82	25	26	20	2	1	...	4·5 ,,
June . .	109	41	39	34	8	2	...	4·4 ,,
July . .	138	83	51	45	8	3	2	6·2 ,,
August .	152	70	56	39	9	5	1	6·2 ,,
September .	135	48	31	29	6	3	1	6·2 ,,
October .	51	24	15	11	1	...	1	6·8 ,,
November .	14	1	...	1	...	...	...	1·4 ,,
December .	11	2	1	...	1	...	...	2·1 ,,
13 years .	785	310	245	198	37	14	5	6·8 ins.

LUCKNOW

	Inches.								Highest.
	Below ¼.	¼ to ½.	½ to 1.	1 to 2.	2 to 3.	3 to 5.	5 to 7½.	7½ to 10.	
January . . .	17	5	4	2	...	1	...	...	3·7 ins.
February . .	15	3	1	...	...	...	...	...	0·5 ,,
March . . .	20	2	1	...	...	...	...	...	0·7 ,,
April . . .	8	1	1	...	...	...	...	...	0·7 ,,
May . . .	21	6	5	2	...	...	...	...	1·2 ,,
June . . .	34	20	17	14	1	...	...	1	9·0 ,,
July . . .	101	38	31	30	7	1	1	...	5·9 ,,
August . . .	78	28	44	21	12	5	2	...	7·0 ,,
September . .	66	20	24	8	6	2	1	...	5·3 ,,
October . . .	12	5	6	3	...	...	...	...	1·8 ,,
November . .	1	...	...	...	...	...	...	...	0·2 ,,
December . .	12	1	1	3	...	...	...	...	2·0 ,,
13 years . .	385	129	135	83	26	9	4	1	9·0 ins.

LAHORE (12 years)

	Inches.							Highest.
	Below ¼.	¼ to ½.	½ to 1.	1 to 2.	2 to 3.	3 to 5.	5 to 7½.	
January .	24	4	7	1	...	...	...	1·5 ins.
February .	21	7	4	3	1	...	...	2·2 ,,
March .	21	6	5	1	...	...	...	1·1 ,,
April . .	28	7	4	1	...	...	...	1·2 ,,
May . .	18	12	2	...	2	...	...	3·0 ,,
June . .	23	7	8	6	...	1	...	3·5 ,,
July . .	22	20	19	15	5	4	2	7·2 ,,
August .	35	8	14	10	5	2	1	5·1 ,,
September .	18	6	6	4	...	1	...	5·7 ,,
October .	9	1	2	...	1	...	...	2·0 ,,
November .	6	1	2	...	...	...	...	0·8 ,,
December .	7	4	4	...	...	...	...	0·8 ,,
12 years .	232	83	77	41	14	8	3	7·2 ins.

NAGPUR

	Inches.							Highest.
	Below ¼.	¼ to ½.	½ to 1.	1 to 2.	2 to 3.	3 to 5.	5 to 7½.	
January .	11	3	5	1	...	...	...	1·0 ins.
February .	9	5	1	1	...	..	...	1·5 ,,
March .	17	2	4	1	...	...	...	1·8 ,,
April . .	22	2	3	1	...	...	...	1·1 ,,
May . .	34	9	4	1	...	1	...	3·9 ,,
June . .	98	44	24	27	7	5	1	7·2 ,,
July . .	122	38	54	37	18	6	5	6·2 ,,
August .	120	35	38	33	10	2	1	6·2 ,,
September .	94	34	28	22	9	2	2	7·2 ,,
October .	33	9	10	12	1	2	...	4·9 ,,
November .	9	2	1	2	1	...	...	2·7 ,,
December .	10	...	6	2	...	...	...	1·8 ,,
13 years .	579	183	178	140	46	18	9	7·2 ins.

Bombay

	Inches.									Highest.
	Below ¼.	¼ to ½.	½ to 1.	1 to 2.	2 to 3.	3 to 5.	5 to 7½.	7½ to 10.	Above 10.	
January .	7	1	...	...	...	...	...	...	...	0·5 ins.
February	5	1	...	...	...	...	...	...	...	0·3 ,,
March .	6	...	...	...	...	...	...	...	...	0·1 ,,
April .	4	...	...	...	...	...	...	...	...	0·1 ,,
May .	27	1	2	2	...	...	...	...	...	1·9 ,,
June .	123	29	35	38	15	17	5	...	3	16·1 ,,
July .	175	52	58	31	24	21	9	2	...	8·2 ,,
August .	181	61	43	32	7	3	2	...	2	11·3 ,,
September	141	49	38	21	14	8	...	2	...	8·1 ,,
October .	51	8	14	6	5	4	1	...	...	5·8 ,,
November	6	2	2	3	...	...	...	...	...	1·5 ,,
December	4	...	1	...	...	...	...	...	...	1·0 ,,
13 years .	730	204	193	133	65	53	17	4	5	16·1 ins.

Madras

	Inches.									Highest.
	Below ¼.	¼ to ½.	½ to 1.	1 to 2.	2 to 3.	3 to 5.	5 to 7½.	7½ to 10.	Above 10.	
January .	21	3	4	2	...	...	...	...	...	1·3 ins.
February	1	...	...	...	1	...	...	...	...	2·5 ,,
March .	7	2	3	1	...	...	...	...	...	1·2 ,,
April .	2	1	2	1	...	...	...	...	...	1·7 ,,
May .	21	4	3	6	...	1	1	...	1	13·0 ,,
June .	101	20	9	8	...	...	...	...	...	2·0 ,,
July .	129	20	21	10	...	...	...	...	...	1·9 ,,
August .	123	30	19	11	4	...	...	...	...	2·8 ,,
September	104	16	17	7	6	2	...	...	...	4·0 ,,
October .	98	29	21	24	12	9	1	...	...	5·8 ,,
November	73	34	18	34	17	12	6	1	...	8·2 ,,
December	67	13	16	13	6	4	1	...	...	5·1 ,,
13 years .	747	172	133	117	46	28	9	1	1	13·0 ins.

The relative frequency of light and heavy falls, as exhibited in these tables, may be represented graphically by curves as

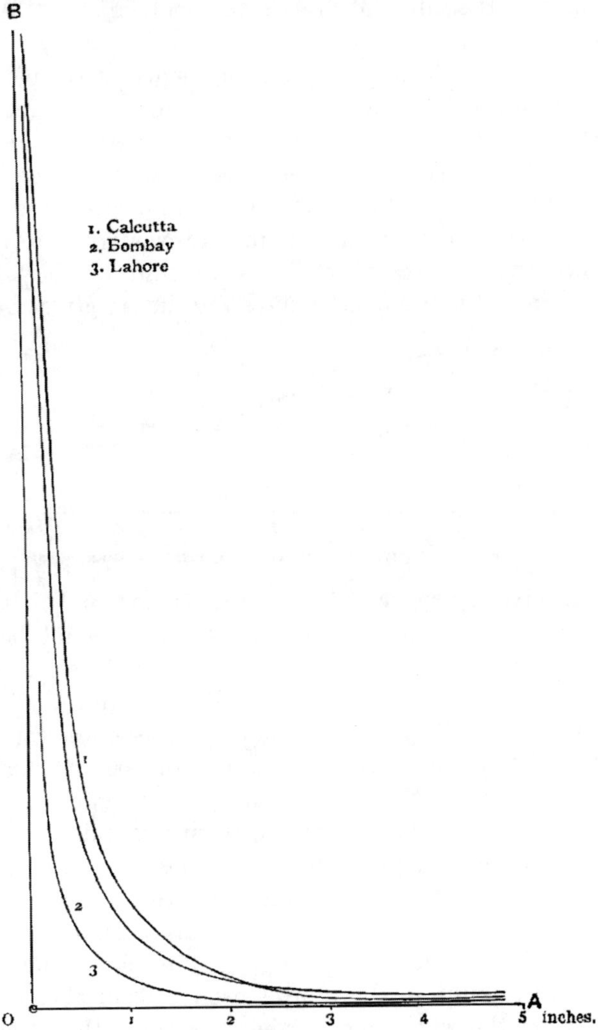

FIG. 31.—CURVES OF RELATIVE RAINFALL FREQUENCY AND QUANTITY.

in the accompanying figure (Fig. 31),[1] which gives the varia-

[1] The right hand portion of the curves beyond 3 or $3\frac{1}{2}$ inches is incorrectly drawn. The curves should be assymptotic, not parallel to the line of abscissæ.

tion curves of the three stations, Calcutta, Bombay, and Lahore. If on the horizontal line OA (the line of abscissæ), we set off from its origin O, a length proportionate to the amount of the fall, and draw a perpendicular to OA at the point so determined, the comparative frequency of falls of that amount will be shown by the length of the perpendicular (the ordinate) between the base line and its intersection with the curve. The curves bear a general resemblance to hyperbolas. Were they true hyberbolas, the lines OA, OB, being their assymptotes, this would indicate that the frequency of a fall of given amount is inversely as its quantity, or, otherwise expressed, that the amount of the fall multiplied into the number expressing its relative frequency,

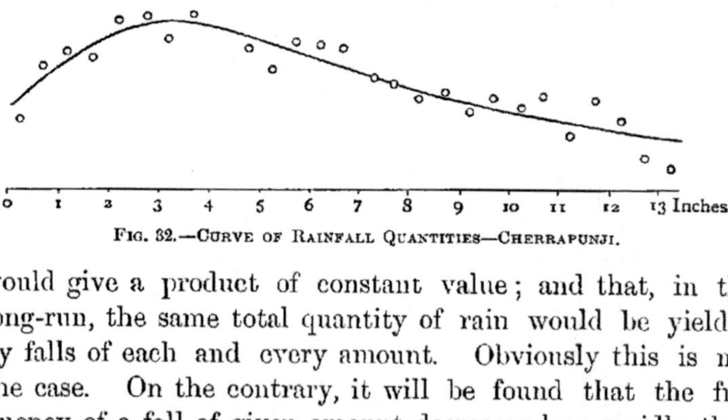

FIG. 32.—CURVE OF RAINFALL QUANTITIES—CHERRAPUNJI.

would give a product of constant value; and that, in the long-run, the same total quantity of rain would be yielded by falls of each and every amount. Obviously this is not the case. On the contrary, it will be found that the frequency of a fall of given amount decreases less rapidly than the quantity increases, up to a certain point, which differs for different stations, and then at a more rapid rate. Moreover, that the greatest quantity of rain is yielded by falls not differing very much from those of average heaviness. Thus, at Calcutta, the greatest quantity of rain appears to be contributed by falls of rather less than half an inch in the 24 hours, the average of each rainy day being 0·52 inch. At Cherrapunji, in the rainy season, falls of about $3\frac{1}{2}$ inches yield, on the whole, more rain than any other, the average of each rainy day being $2\frac{2}{3}$ inches.

The curve in Fig. 32 represents this variation of total quantity, according to the heaviness of the fall, at Cherrapunji. It has been obtained in the following manner. The whole number of rainy days, in fifteen rainy seasons, were tabulated according to increments of half an inch. The total quantity of each term of the series was then computed, by multiplying the number of days into the mean value of the increment, and the results plotted on ruled paper as the ordinates of the curve. Finally, the curve was drawn with a free hand between the points thus determined, and which are shown by the encircled dots on the figure.

On comparing together the foregoing tables, it may be noticed that light falls of rain are proportionally more frequent at Calcutta than at any other station, and that heavy falls are not only relatively less frequent, but of less absolute amount. Thus, falls of more than 2 inches in the 24 hours occur on only 3 per cent of the total number of rainy days at Calcutta, whereas they form 5 per cent of the total at Lucknow and Lahore, 6 per cent at Nagpur and Madras, and 10 per cent at Bombay. Again, the highest fall at Calcutta during the 13 years was 6·8 inches; at Lahore and Nagpur it was 7·2 inches; at Lucknow, 9 inches; at Madras, 13 inches; and 16 inches at Bombay.

Special Occurrences.—A few instances of exceptionally heavy rainfalls, varying from $10\frac{1}{2}$ to nearly 41 inches in the 24 hours, have already been given at page 77 of this work, and in the foregoing analytical tables are noted falls of 13 inches in one day at Madras and 16 inches at Bombay within the last 13 years. This last, which occurred on the 18th June 1886, is probably the highest ever recorded at Bombay, since from 1847 to 1874 the highest, as quoted by Mr. C. Chambers,[1] was 15·3 inches, on the 27th June 1869.

The highest fall yet recorded on one day in India is 40·8 inches at Cherrapunji; and the highest on the plains of India, 35 inches, viz. at Purneah in the Bhagulpore division of Bengal. Nearly as much, viz. 28·5, 30·4, and 32·4 inches, was registered on the 18th September 1880 at three stations in Rohilkhand in the North-west Provinces.

[1] *Meteorology of the Bombay Presidency*, p. 86.

At Calcutta, the heaviest falls on record are one of 12 inches on the 1st May 1835, and another of 12 inches, as measured, and an unknown excess lost by the overflow of the gauge, on the 21st August 1844.

These falls at Bombay and Calcutta are much lower than those of Purneah and Rohilkhand and even in the Punjab, as noticed at page 77; but it must not be concluded that such as these latter are restricted to the interior of the country, since the same list contains an instance of 20·3 inches in one day at Chandbally, a small station on the coast of Orissa, and a fall of 20·58 inches was registered at Madras on the 21st October 1846. The fact, however, remains, that rainfalls of excessive amount occur occasionally in the drier, and even the driest provinces of India—regions where the average annual rainfall is below 30 and even below 10 inches. The average rainfall of the year in the province of Sind varies between 3 and 11 inches; but even this province is not exempt from occasional falls of excessive amount. In the years 1862 and 1866 it was visited by disastrous floods, due to heavy local rainfall, some notice of which is given in the following extract from a letter of Mr. G. P. Bidder, published in the *Proceedings of the Institution of Civil Engineers*:—[1]

"Referring for a moment to the Kurrachee table, in which the total annual rainfall is so small, it appears that in 1853, 3·08 inches fell on the 31st July, out of a total of 6·29; in 1855, 3½ inches on the 7th January, out of 12·28; in 1857, 3·15 inches on the 12th August, out of 11·33; in 1863, 4·88 inches on the 10th July, out of 13·52; and in 1866, 7·40 inches on the 5th August, out of 12·91, the total rainfall up to 30th September of that year; but this is far surpassed at other parts of the line. Thus, at Hyderabad, in 1865, out of the annual rainfall of 15·88 inches[2] 13·41 inches fell on the 5th, 6th, and 7th August, of which 10·24 inches fell on the 6th.

"In 1866 the results were still more exceptional. Thus, at Dorbajee, out of a total fall of 44·67 inches, 36 inches fell on the 4th, 5th, and 6th of August, of which 20 inches fell on the 4th, and 14 inches on the 5th; and at Joongshaie, out of 44·27 inches, 32·22

[1] *Op. cit.*, vol. xxvii., session 1867-68.

[2] This is the total of that year. The average rainfall of Hyderabad is only 8 inches.

inches fell on the 5th and 6th August, of which 19·1 inches fell on the 5th, and 13·12 inches on the 6th.

"I regret that no returns exist relative to the year 1862, except of the rainfall at Kurrachee, because considerable damage was done to the line in that year, and it was represented as one quite exceptional. There is no doubt, however, that the rains of 1866 far exceeded those of 1862."

The following list, of falls of 10 inches and upwards in the 24 hours, has been compiled chiefly from the annual reports on the meteorology of India, with a few additions from other sources. It comprises only such as are specially mentioned in the works referred to, and a search through the registers of the past 12 years alone would probably greatly add to their number as experienced on the plains of India. In the case of the wetter stations of the Assam hills and the Ghats, the additions would be very numerous.

RAINFALLS OF 10 INCHES AND UPWARDS IN 24 HOURS

PROVINCE.	STATION.	INCHES.	DATE.
Punjab	Delhi	19·5	9th September 1875.
	Gurgaon	10·5	,,
	Hoshiarpur	14·0	19th September 1878.
	Ludhiana	12·2	4th July 1880.
	Delhi	12·0	18th September 1878.
N.W. Provinces	Nagina	32·4	18th September 1880.
	Danipur	30·4	,,
	Najibabad	28·5	,,
	Hurdwar	19·5	,,
	Mawana	15·5	,,
	Meerut	14·6	,,
	Jauli Jansath	14·6	,,
	Bijnor	12·4	,,
	Champawat	13·0	,,
	Naini Tal	12·4	,,
	Ranikhet	12·0	,,
	Meerut	16·0	17th September 1880.
	Jauli Jansath	16·0	,,
	Mozuffernagar	12·9	,,
	Bijnor	12·4	,,
	Roorkee	10·7	,,
	Allahabad	15·5	29th July 1875.
	Nawabganj	13·4	2d August 1879.
	Bareilly	11·8	,,
	Paranpur	11·4	,,
	Pilibit	13·6	,,
	Dehra Dun	10·7	2d August 1885.
	Gorakhpur	10·0	10th September 1879.

RAINFALL OF 10 INCHES AND UPWARDS IN 24 HOURS—Continued.

PROVINCE.	STATION.	INCHES.	DATE.
Bengal	Purneah	35·0	13th September 1879.
	,,	11·5	4th September 1879.
	Chandbally	20·3	16th September 1879.
	Cox's Bazar	15·7	9th July 1886.
	Furreedpore	14·3	5th July 1877.
	Chittagong	14·0	3d June 1886.
	,,	10·3	9th August 1877.
	Mozufferpore	12·5	17th June 1883.
	Rangamatia	12·2	8th July 1886.
	Balasore	12·0	12th July 1877.
	Nattore	11·1	6th July 1877.
	Kooshtea	11·1	20th August 1877.
	Mymensing	10·8	21st August 1877.
	Calcutta	12+?	21st August 1844.
	,,	12·0	1st May 1835.
Assam	Cherrapunji	40·8	14th June 1876.
	Jowai	40·1	11th September 1877.
Central India	Rewah	30·4	16th June 1882.
	Sutna	22·1	,,
Rajputana	Abu	13·0	2d September 1884.
Central Provinces	Jubbulpore	12·5	29th June 1877.
	Sambalpur	11·0	9th August 1879.
Berar	Chikalda	12·0	7th September 1884.
	Makhla	11·0	,,
Sind	Joongshaie	19·1	5th August 1866.
	,,	13·1	6th August 1866.
	Doorbaji	20·0	4th August 1866.
	,,	14·0	5th August 1866.
	Hyderabad	10·2	6th August 1866.
Cutch	Mandvi	14·8	12th July 1881.
Bombay	Bombay	16·1	18th June 1886.
	,,	15·3	27th June 1869.
	,,	14·6	19th June 1877.
	,,	11·3	3d August 1881.
	,,	10·3	15th August 1885.
	,,	10·1	3d June 1882.
	Rajkote	10·1	3d July 1883.
	Baura	15·5	17th June 1886.
	Mahableshwar	13·0	18th June 1886.
	,,	10·8	22d July 1886.
	Honawar	11·3	17th June 1886.
	Rutnagiri	10·6	22d June 1877.
Madras	Madras	20·6	21st October 1846.
	,,	18·0	24th October 1857.
	,,	13·0	18th May 1877.
	,,	12·1	9th May 1827.

In all these instances, the day which forms the time unit is the 24 hours intervening between the fixed hours at which the rainfall is regularly measured, and the quantities may, in some cases, have been greater in an interval of 24 hours beginning and ending otherwise. The following instances of falls exceeding 14 inches in two, and 18 inches in three days, are subject to the same remark, and are so far from being exhaustive that they can be regarded merely as a few illustrations of excessive falls.

RAINFALLS OF 14 INCHES AND UPWARDS IN 2 DAYS

PROVINCE.	STATIONS.	INCHES.	DATES.
	Delhi	22·4	8th-9th August 1875.
	Gurgaon	19·5	,,
	Rohtak	14·5	,,
Punjab	Jullundur	15·0	10th-11th August 1875.
	Ludhiana	15·5	4th-5th July 1880.
	Amritsar	14·5	9th-10th July 1881.
	Hoshiarpur	14·2	2d-3d September 1877.
	Nagina	41·0	17th-18th September 1880.
	Dhanipur	39·1	,,
	Najibabad	38·6	,,
	Hurdwar	31·5	,,
	Jauli Jansath	30·6	,,
N.W. Provinces	Bijnor	24·8	,,
	Naini Tal	21·4	,,
	Roorkee	18·7	,,
	Mozuffernagar	17·1	,,
	Ranikhet	15·7	,,
	Dehra Dun	18·5	8th-9th August 1885.
	Jhansi	15·0	23d-24th July 1869.
Bengal	Mymensing	16·9	21st-22d July 1877.
Assam	Cherrapunji	63·6	14th-15th June 1876.
	Jowai	52·2	11th-12th September 1877.
Central Provinces	Pachmarhi	14·0	23d-24th July 1884.
Sind	Doorbaji	34·0	4th-5th August 1866.
	Joongshaie	32·2	5th-6th August 1866.
Cutch	Mandvi	20·5	12th-13th July 1885.
	Bombay	22·6	18th-19th June 1886.
	,,	20·4	19th-20th June 1877.
Bombay	,,	14·4	14th-15th August 1885.
	,,	14·4	2d-3d June 1882.
	Mahableshwar	20·6	22d-23d July 1886.
	Rutnagiri	17·8	22d-23d June 1877.
Madras	Madras	19·2	17th-18th May 1877.

RAINFALLS OF 18 INCHES AND UPWARDS IN 3 DAYS

PROVINCE.	STATIONS.	INCHES.	DATES.
Punjab . .	Sialkot . .	18·8	9th-11th August 1875.
Assam . .	Cherrapunji .	79·0	12th-14th June 1876.
Sind . . .	Doorbaji . .	36·0	4th-6th August 1866.
Kathiawar . .	Drol . .	33·3	11th-14th July 1881.
	Nawanagar .	28·9	,,
	Morvi . .	25·5	,,
Bombay . .	Rutnagiri .	23·9	21st-23d June 1877.
	Bombay . .	28·6	17th-19th June 1886.
	,, . .	22·3	18th-20th June 1877.
	Karwar . .	20·3	15th-17th June 1886.
Madras . .	Negapatam .	25·0	14th-17th November 1885.

Imperfect as these lists undoubtedly are, the above instances of remarkably heavy falls suffice to show how large a quantity of rain, within an interval of two or three days, has sometimes to be reckoned with by engineers in providing for flood drainage. Applying to these cases the law, already found, of the relative frequency of different amounts, it may safely be concluded that the probability of minor but still excessive falls is greater than in the inverse proportion of the quantities. The instances of Bombay and Madras, the registers of which have been better examined than those of other stations, show that such falls may be expected at irregular intervals, averaging from 3 or 4 to 10 years; in the case of the more arid provinces, perhaps somewhat more. But there is probably no part of the country that can be regarded as absolutely exempt from such visitations.

It has been mentioned, in the earlier part of this work, that these heavy falls occur only in cyclones and cyclonic storms, most frequently the latter; and their frequency at any given place is likely, therefore, to vary with the liability of that place to come within the track of such storms. Now the charts of the storm tracks of the 10 years, 1877 to 1886, drawn up by Mr. Eliot, show that the cyclonic storms of the summer monsoon most frequently traverse India in a west-north-west direction from the Orissa coast; and accordingly, the greater number of the falls enumerated in the foregoing lists, including some of the heaviest, have occurred in this belt of maximum storm frequency. The Satpura range, the greater part of which lies within it, is notoriously subject to heavy falls in the monsoon, producing such floods in the Tapti and Nerbudda as were followed by disastrous consequences in 1864 and 1866, noticed by Mr. A. C. Howden in the *Proceedings of the Institution of Civil Engineers*,[1] and those of 1877 and 1883.

The central part of the peninsula, including the Deccan and Hyderabad, are less frequently visited by cyclonic storms than the Satpura region and extra-tropical India on the north, or the Carnatic on the south; but in the later months of the monsoon they are sometimes reached by storms from the Bay of Bengal, and the absence from the lists of any instances of rainfalls exceeding the adopted limits is probably due rather to the want of complete data than to the fact of their non-occurrence.

The extent of country simultaneously affected by these heavy rainfalls is sometimes very great, usually some thousands of square miles. In the case of the remarkable floods of the 17th and 18th September 1880, in Rohilkhand and the adjacent divisions of Meerut and Kumaon in the North-west Provinces, an average of more than 10 inches in the two days was measured over the greater part of the districts Saharunpore, Mozuffernagar, Meerut, Bijnor, Moradabad, Pilibhit, and Kumaon, a total area of not less than 10,000 square miles. This was doubtless quite exceptional, but rather in the heaviness of the fall than in the extent of its horizontal distribution.

On the 14th and 15th July 1882, an average fall of 5 inches took place over the greater part of the Central Provinces, south of the Nerbudda and Sone valleys, an area of some 30,000 square miles; and in his *Report on the Meteorology of India* in 1877, Mr. Eliot states that from the 6th to the 9th October, the average rainfall of the whole of the North-west Provinces and Oudh amounted to $3\frac{1}{4}$

[1] *Op. cit.*, vol. xxvii., session 1867-68.

inches in the four days, the greater part of which fell on the 7th and 8th. The area of these provinces is upwards of 111,000 square miles.

Special Hourly Rainfalls.—With rare exceptions, it is the practice at Indian observatories to measure the rainfall once only in the 24 hours, and hence the majority of the registers afford no information of its heaviness within shorter periods, and the available statistics on this head are very scanty. A few casually-recorded instances will, however, serve to afford at least some idea of what occasionally occurs in India.

The heaviest fall of short duration, within my own recollection, at Calcutta, was one of 4 inches in an hour and a quarter. But a heavier and more prolonged fall took place on the 11th May 1835, when 12 inches were registered within 3 hours, viz. from midnight to 3 A.M.

According to Mr. C. Chambers, the greatest amount registered at Bombay in one hour, during the 28 years, 1847 to 1874, is 4·22 inches, viz. between 3 and 4 A.M., on the 12th June 1847. And on the 12th September 1872, 7·2 inches fell in the 2 hours between 10 P.M. and midnight. Other falls of 3·07 inches, 3 inches, and five between 2 and 3 inches within an hour, are mentioned by Mr. Chambers as having occurred during the same 28 years.

At Madras, 17 inches fell in 12 hours on the 21st October 1846, 6·22 inches in 5 hours on the 20th November 1856, and 12·21 inches in 12 hours on the 24th October 1857. These instances are quoted from Mr. Lowis Jackson's *Hydraulic Manual.* Other instances mentioned in the same work are 1·2 inches in a quarter of an hour, 1 inch in 35 minutes, 1·6 inches in 40 minutes, and 2·49 inches in 2½ hours at Bangalore; these occurred respectively in September 1860, May 1859, and May and September 1861.

In Mr. Howden's notice of floods in the Nerbudda valley, in the *Proc. Inst. Civil Engineers,* is mentioned a fall of 10·5 inches in 18 hours at Hurda, on the G. I. P. Railway, on the 5th August 1864, and in a letter of Mr. G. P.

Bidder, published among the remarks on this paper, an instance is quoted of a fall of 3 inches per hour at a station in Sind in 1866.

Mr. Binnie also, in a paper on the Ambàjhàri tank at Nagpur, in the same serial, gives an instance of a fall of 3·55 inches in 45 minutes at Nagpur on the 6th October 1872, and another of 3·5 inches in $1\frac{3}{4}$ hours on the 10th August of the same year; also two other instances of falls lasting respectively 20 and 30 minutes, at rates of between 2 and 3 inches in the hour, all occurring in the same monsoon.

From these instances it appears that falls of from 3 to 4 inches in the hour are not rare in India, though but few are recorded at one and the same station; and, further, that such falls sometimes last for 2 or 3 hours in succession.

Proportion of Surface Drainage to Rainfall.—The proportion of the rainfall which flows off the ground surface, and either fills the rivers or is available for tank storage, is a subject which concerns the physical geographer and the engineer rather than the meteorologist; but a knowledge of this proportion is so indispensable to the engineer, in order that he may turn to practical account the information set forth in the foregoing pages, that it seems desirable to supplement the rainfall statistics with a brief notice of such observations as have been recorded in India, and in other countries which, like India, are subject to the annual alternation of dry and rainy seasons. The various formulæ used by engineers for expressing the relations of flood drainage to rainfall need not here be referred to. Some of the elementary facts by which their validity must be tested will suffice for the purpose of the present work.

In planning the Vehar reservoir at Tanna for the water supply of Bombay, Mr. Conybeare states [1] that it was assumed that six-tenths of the total rainfall on the catchment area of 3918 acres would be available for tank storage, and that the rainfall would be at least equal to that of the low-lying station Tanna, which he estimated at 124 inches. Nearly the whole of this falls during the four months of the monsoon.

[1] *Proc. Inst. Civil Engineers*, vol. xvii., session 1857-58.

The mean rainfall of Tanna was here overestimated; it is really less than 100 inches; nor does it appear to be higher, if indeed so high, on the hills that feed the Vehar tank. In other respects, the estimate has been better justified by experience. According to Mr. Russell Aitken,[1] the proportion run off and stored has been as follows. In 1865, when the recorded rainfall was 89 inches, it amounted to 70 per cent; in 1870, when the former was only 65 inches, to 75 per cent; and in 1871, a very dry year, when the rainfall was only 39 inches, to 50 per cent. The mean of these is a little over 60 per cent.

In the case of the Ashti tank, near Pandharpur, on the Bhima river, in the dry climate of the Deccan, it was estimated that the "run off" would be one-fourth of the average monsoon rainfall, taken at 23 inches. But actual measurements, during the monsoon of 1881, gave an average storage of only one-sixth of the total of 18 inches, and the proportion varied, in different months, from one-five-hundredth to one-third. In July, when on only one day the fall exceeded half an inch, the proportion was one-five-hundredth; whereas, in September, when there were 13 days on which the rainfall was from half an inch to over 2 inches, the proportion was one-fifth; and in October, on the first 5 days of which 2·75 inches fell, one-third.[2]

At Nagpur, observations were made by Mr. Binnie on the discharge of the Ambàjhàri drainage area (4224 acres) in the monsoons of 1869 and 1872. He found that, at the beginning of the monsoon, a fall so heavy as 2¼ inches within 1 hour and 20 minutes gave no appreciable surface drainage, the whole being absorbed by the desiccated soil; while, in September, 98 per cent of an almost identical fall (2·2 inches in the same interval of time) flowed into the tank in the space of 2 hours and 50 minutes. The total results of the several months were as follow:—

1869.		
June and July	12·76 inches	10 per cent.
August	9·61 "	35 "
September	7·41 "	44 "
Totals of season	29·78 "	26·5 "

1872.		
June	6·77 inches	4·7 per cent.
July	12·70 "	22·7 "
August	11·82 "	55·8 "
September	7·99 "	74·4 "
October	4·37 "	39·4 "
Totals of season	43·65 "	40 "

The rain of October 1872 was preceded by an interval of dry weather,

[1] *Proc. Inst. Civil Engineers*, "Discussion on Nagpur Waterworks," vol. xxxix., session 1874-75.

[2] *Idem*, "Burke on the Ashti Tank," vol. lxxvi., session 1883-84, Part ii.

allowing of the partial desiccation of the ground; and to this fact was doubtless due the decline of the drainage percentage from 74·4 to 39·4.

All the above observations have reference to places where the geological formation is a basaltic or other trappean rock, covered with a highly absorbent layer of *regur* or black soil, which becomes cracked and crumbling after long drought. Evidently, under these conditions, the drainage proportion varies from 0 to nearly the whole of the rainfall, according to the condition of the soil; and the proportion collected in any season depends chiefly on the continuity of the rain, and only in a subordinate degree on the total quantity that falls. The high average of the Vehar tank is probably due partly to the greater raininess of the site, and partly to the higher slopes of the catchment area, which is situated among hills, while those of the Ashti and Ambàjhàri tanks are on the plains of the Deccan and Nagpur.

The experience of engineers in New South Wales and Cape Colony may profitably be taken into consideration in connection with the above Indian statistics, inasmuch as, while there is much similarity in the climates, it has reference to larger areas. The *Proc. Inst. Civil Engineers* contain two papers bearing on the present subject, one by Mr. Coghlan on the relation of the rainfall to the drainage of the Nepean and Cataract rivers, in the neighbourhood of Sydney, and one by Mr. Tripp on that of the river Buffalo, at the Cape of Good Hope. From these the following data are extracted:—

The area of the Nepean basin is about 284 square miles, that of the Cataract about 70 square miles. They lie to the west of the Illawarra or Mittagong range, at an average elevation of 1200 feet above sea-level, and are, for the most part, uninhabitable. Numerous swamps, chiefly in the Cataract basin, retain the water, and serve to equalise the flow of the rivers, but do not occupy a very large proportion of the area. The remainder is a rugged plateau, intersected by steep and broken ridges. The whole country is barren, and the surface soil poor, resting directly on sandstone. Vegetation is stunted, though the scrub is in many places thick, and in the valleys the gum trees attain goodly proportions. The annual rainfall varies from 75 to 34 inches, the average being 54 inches at the Cataract, and 44½ inches at the Nepean. It has been found that the average proportion between rainfall and discharge in the case of the Cataract is 44·7 per cent, while in that of the Nepean it is 44·1 per cent for an extended period, including every condition of flood and drought. The average of freshets is 54 per cent of the rainfall, while, after the ground has been saturated with previous rain, the volume discharged is nearly equal to the volume of rain falling. In dry seasons, on the other hand, the average yield of the streams is not more than 10 or 12 per cent. The quantity of water carried off is, however, not directly proportional to the rainfall, but the greater the quantity of rain falling within a given time, the greater is the percentage of that rain carried off by the streams.

The area of that portion of the drainage basin of the Buffalo river, dealt with in Mr. Tripp's paper, is about 105 square miles, and the surface is covered with forest and bush. Owing to this circumstance, unlike many South African rivers, the Buffalo has a small perennial flow, sustained by more or less permanent springs, and varying from 87 cubic feet to 2880 cubic feet per minute, with an average of 933·8 cubic feet. The flood discharge, due to surface drainage, varies from 600 to 285,000 cubic feet per minute above the ordinary flow, and the total ordinary flow is to the flood discharge as about 1 to 2¼. The mean rainfall of the drainage basin, on the average of three years, was about 27 inches; but these seasons were dry and abnormal. Its annual distribution was very irregular; an almost complete drought in one winter being followed by heavy spring rains, while, in the following year, the chief fall occurred in the winter; and, in both the second and third years, the springs and summers were very dry. The highest rainfall in any one day was 2 inches, and there were two falls of over 3 inches in 48 hours. Three-eighths of the whole rainfall was contributed by falls under half an inch, and the remainder equally divided between falls below and above 1 inch. Under these circumstances, the proportion of the rainfall flowing off the ground from all sources, during the whole period, varied from 0·019 to 0·715 in different months; the total result being 0·225, or between one-fifth and one-fourth of the rainfall.

Evaporation from Water Surfaces.—The most trustworthy and indeed the only practically useful observations on the evaporation of water surfaces are such as are afforded by the same authorities whose works have been quoted in the preceding paragraphs, and to these therefore our attention may be restricted.[1]

[1] For ascertaining the rate of evaporation of an extensive sea surface, a datum much required by meteorologists and physical geographers, it may be hoped that some day the unique advantages afforded by the Suez Canal and Bitter Lakes will be turned to account. The evaporation of this great sea-

In the case of the Vehar tank, 2 square miles in area, Mr. Conybeare found that, during the first dry season, after the completion of the dam, the surface was not lowered more than 6 inches in the month, and he observes that this loss was principally due to leakage through a temporary plug, closing the orifice of a large pipe. In 1874, 16 years after the completion of the work, Mr. Russell Aitken stated, as the result of subsequent experience, that the leakage, together with evaporation from the lake, did not exceed 5 inches per month, and he calculated the leakage from the various dams at 1½ inches; so that the evaporation during the dry season would be about 3½ inches a month or an average of 0·116 inch per day.

The Ashti tank in the dry Deccan climate (how dry is shown by the figures for Sholapur given below) has an area of 1412 acres. During the six months, December 1880 to May 1881, when no water was drawn off from the tank, the loss by evaporation, absorption, and leakage was as follows in each month. I add for comparison the mean temperature, humidity, and wind movement of the same months, as registered at the Government observatory at Sholapur, less than 40 miles distant, and under similar climatal and topographical conditions.

	Reduction of Surface Level.	Temperature.	SHOLAPUR. Humidity.	Wind Daily.
December 1880	0·12 foot	71·4 Fahr.	46 per cent.	140 miles.
January 1881	0·40 ,,	70·7 ,,	41 ,,	169 ,,
February ,,	0·65 ,,	75·6 ,,	34 ,,	170 ,,
March ,,	0·65 ,,	82·2 ,,	27 ,,	154 ,,
April ,,	0·95 ,,	88·6 ,,	28 ,,	154 ,,
May ,,	1·02 ,,	90·4 ,,	30 ,,	250 ,,
Total	3·79 ,,			

Mr. Burke gives no estimate of how much of this total loss of 3·79 feet may be attributed to leakage, etc. But assuming that the whole was evaporated, it would give an average for this very dry, and in the later months, hot climate, of 0·37 inch per day as the maximum that could possibly have occurred. Careful experiments on a neighbouring tank with an area of 100 acres water surface, gave a total loss from

water surface is made good almost entirely by the influx from the Mediterranean and the Red Sea, through two canals of uniform section, the rainfall being very small; and it only needs a constant gauge of the flux and reflux in these two channels, say for a twelvemonth, to ascertain the net excess of the flow over the ebb. Arrangements might certainly be made to obtain this by self-registering current gauges, so placed as not to interfere with the canal traffic; and with one or two rain gauges at Lake Timseh and the Bitter Lakes to register the small rainfall, and a set of thermometers, hygrometer, and anemometer, to record the temperature, humidity, and wind movement, we should have all the data requisite provided at little expense. The subject is well worthy the attention of the British Association or the Académie des Sciences.

October to May, both months inclusive, of 4·41 feet, an average of 0·29 inch per day.

At the Ambājhāri tank, Nagpur, Mr. Binnie registered the fall of the water surface, foot by foot, from the 10th October 1872 to the 9th June 1873, and deducting 200,000 cubic feet daily, for the estimated "draw off," loss by leakage, etc., found the residual loss due to evaporation as follows. The mean temperature and humidity of each period are given in parallel columns.

	Evaporation.	Temperature.	Humidity.
10th Oct. to 14th Nov.	0·19 in. per day.	74·5 Fahr.	77 per cent.
14th Nov. to 24th Dec.	0·15 ,,	70·9 ,,	80 ,,
24th Dec. to 4th Feb.	0·13 ,,	69·3 ,,	82 ,,
4th Feb. to 16th March	0·14 ,,	77·2 ,,	63 ,,
16th March to 15th April	0·23 ,,	88·7 ,,	51 ,,
15th April to 7th May	0·37 ,,	91·7 ,,	38 ,,
7th May to 9th June	0·19 ,,	92·6 ,,	63 ,,

Adding to the total loss 3 inches of rainfall, registered during the period, the general average for the whole season was 0·198, or say 0·2 inch per day. The validity of this result depends, of course, on the accuracy of the estimated reduction of the water level from other causes, but it is to be observed that the allowance made for this reduction is only 46 per cent of the whole, the total fall of level averaging 0·0289 feet or 0·347 inch per day, so that it cannot be very greatly in excess. Moreover, some observations conducted by Mr. Culcheth, near Ajmere in Rajputana, with evaporation gauges, partly on land and partly floated on a tank, gave a result almost identical with that computed by Mr. Binnie.

Allowing for differences of climate, the results of these several observations are sufficiently consistent, and it may be concluded with some confidence that the loss by evaporation of large fresh-water tanks in India varies from rather more than one-ninth of an inch daily, on the average of the dry season, in the comparatively damp climate of Bombay, to about one-third of an inch as a possible maximum in the very dry hot climate of the Deccan; probably less.

Some measurements of the evaporation of a large reservoir (Van Wyk's Vley), in the driest part of Cape Colony, reported by Mr. J. G. Gamble, hydraulic engineer to the Cape Government, gave an annual average of 6 feet, and others near Port Elizabeth, in the south-eastern district, one of 3 feet. The former would correspond to an average of one-fifth of an inch, the latter one-tenth of an inch per day, agreeing therefore well with the Indian results.

Rainfall on Indian River Basins.—A knowledge of the average quantity of rain falling annually on the drainage basin of a river, although of less immediate importance to the engineer than is that of flood-producing rainfalls, is nevertheless a fundamental datum in questions of water supply; and not less so in many problems that concern the physical geographer. Hitherto, as far as I am aware, no one has attempted to frame such an estimate in the case of Indian rivers, and that now put forward can be regarded only as a first attempt, aiming at an approximative result, and subject to indefinite future correction, as the available registers of the rainfall become more numerous, and come to represent longer periods of time.

In the basins of the southern tributaries of the Ganges and Jumna, and of those rivers of the Bengal and Bay of Bengal drainage systems that take their rise in Central India and the Satpura highlands, or on the Deccan plateau at some distance from its western margin, the inequalities of the mean annual rainfall in different parts of the area are comparatively moderate, and provided the rain gauge stations are evenly distributed, and so situated as fairly to represent the country around, the error of the average of all their registers, as representing the average rainfall of the whole area, is probably not much larger than that of the mean annual rainfall of each individual station, which, as a general rule, is about 1 inch. But in the case of rivers, which, like the Kistna and its tributaries, rise on the crest of the Western Ghats, where the annual rainfall is 200 inches and upwards, and then traverse a broad belt of country where it barely exceeds one-tenth of that amount, there is very considerable difficulty in so selecting the stations that each portion of the basin may be represented in proportion to its area; and an approximation to within 2 or 3 inches, for the general average of the whole, is the utmost that can perhaps be hoped for.

I do not include in these statistics any of the Himalayan rivers. In their case, or at least that of all the larger streams which bring down the waste of the melting glaciers, to the difficulty arising from the great inequalities of the rainfall on the face of the Himalaya, and on the Punjab and Gangetic plains at its foot, is added that of assigning any probable value to the precipitation of the mountains beyond their outermost spurs; a difficulty which is all the greater, since the snowfall on the high peaks and glaciers may be very different from the rain and snowfall at the few valley stations for which we have registers. Instead, then, of attempting to give an estimate involving so much conjecture as to be almost valueless, I have deemed it better to omit these rivers.

I have taken for the fundamental data for the computation, the rainfall averages up to 1886, and have corrected those of the minor stations accordingly. I think that the adopted values of the general average of the river basins are rather under than over the true amounts. I begin with the rivers of the Central Indian tableland, that drain into the main arterial system of the Jumna and Ganges; then follow with those of Western Bengal and Orissa, and the chief river systems of the eastern slope of the peninsula; and conclude with the great rivers that debouch into the Gulf of Cambay and the Gulf of Cutch.

Rivers of the Central Indian Tableland.—The chief rivers of this plateau, omitting some minor but still considerable streams, are, in order from west to east, the Chumbul with its great tributary the Banàs, the Sind, the Betwa, the Cane, and the Sone. The first four drain Eastern Rajputana, the states of the Central India Agency, and portions of the North-west Provinces, Saugor, and Dumoh; and all but the Sind take their rise on the southern margin of the tableland. The Banàs, and eventually the Chumbul, receives most of the drainage of the eastern slope of the Aravalis as far south as Udaipur, the driest portion of the whole plateau. The rainfall of this region is little over 20 inches on its western margin, and it increases southwards and eastwards to over 40 and locally over 50 inches in the extreme south and east of the tableland. The Sone alone rises at one of the highest points of the Satpura system, where the annual rainfall exceeds 50 inches, and some of its southern feeders in hills of equal elevation, where it exceeds 60 inches, but its main course lies through a country where, as far as the somewhat scanty data show, it is between 40 and 50 inches, and its final portion traverses South Behar, where the fall scarcely exceeds 40 inches. The averages of these five basins I make respectively as follows:—

Chumbul and Banàs	28 inches.
Sind	34 ,,
Betwa	39 ,,
Cane	42 ,,
Sone	49 ,,

Rivers of Western Bengal.—These rivers drain the highlands of the Sonthal country and the northern portion of Chutia Nagpur. They all empty themselves into the Bhàgirathi arm of the Ganges and the Hooghly. Excepting two small tracts drained by the Damooda and the Adjie, the region has an annual rainfall above 50, and in part above 60 inches. In order from north to south they are—

Mor	54 inches.
Adjie	52 ,,
Damooda	50 ,,
Roopnarain	59 ,,
Cossye	56 ,,

Rivers of Orissa.—Of these, the Soobunrika and Brahminee drain the southern part of Chutia Nagpur, all of which is hilly country with a rainfall of 50 inches and upwards. The northern tributaries of the Mahanadi, a much more important stream, take their rise near the head waters of the Nerbudda and Sone, in hills where the annual fall is from 50 to over 60 inches, and the chief southern tributary, the Tel, in Bustar, where it also exceeds 60 inches, and in places perhaps reaches 70 inches; but the western feeders traverse the plain of Chhattisgarh, where the fall is, for the most part, between 40 and 50 inches. The general averages are—

Soobunrika	. . . . . .	57 inches.
Brahminee	. . . . . .	56 ,,
Mahanadi	. . . . . .	53 ,,

Rivers of the Deccan.—The two great rivers which discharge their waters through contiguous deltas, midway down the eastern coast of the peninsula, together bring down nearly the whole drainage of the Deccan tableland. The head waters of all their western feeders lie close under the crest of the Western Ghats, and the torrential rainfall of this hill range fills their channels in June and July, many weeks before rain has fallen in abundance on the rolling plains and flat-topped hills of the Mahratta country and Hyderabad. The northern tributaries of the Godavery, the Paingunga, Wardha and Waingunga, which, united in the Pranhita, form a junction with the Godavery at Sironcha, are of sufficient importance to be treated as independent streams. So also the Bhima, the great northern, and the Tungabhadra the chief southern tributary of the Kistna. Of the former, the Wardha and Waingunga rise in the Satpura range, where the rainfall generally varies from 45 to 60 inches, and locally, as at Pachmarhi, exceeds 70 inches. The subsequent course of the Waingunga is wholly through plains with a rainfall of 45 inches and upwards; but the Wardha, skirting the drier plain of Berar farther west, drains a considerable area, where the average is not more than 35 inches. The basin of the Paingunga, which river drains the southern slopes of the Indhyàdri range, except its lower part, lies wholly in a region where the annual rainfall is below 40 inches. The main stream of the Godavery rises near Nasik, under the lee of the Ghat range, but it receives the drainage of only a small extent of this range, and for upwards of 100 miles to the eastward, traverses a plain where the fall varies between 20 and 30 inches. Beyond this, all the northern portion of the Hyderabad state, which is drained partly by the main river and partly by its tributary the Manjira, has a rainfall between 30 and 40 inches. The Godavery joins the Pranhita at Sironcha. Below this point, the only important additions to its volume are from the hill country of Bustar to the north of the main stream; of the rainfall of this tract nothing is known, except for a few places about the head waters of the Indravati, where it amounts to between 60 and 70 inches. I do not therefore attempt any estimate of the drainage of the Godavery below Sironcha.

The Bhima drains about 100 miles of the eastern face of the Ghat range, and its subsequent course to its junction with the Kistna, a distance of more than 250 miles in a direct line, lies entirely through a country where the rainfall is below 30 inches, in places below 20. The Kistna, above this junction, receives the drainage of more than 250 miles of the Ghats, and then traverses nearly 200 miles of the same dry tract of the Deccan; and the Tungabhadra, its great southern feeder, drains about 150 miles of the Ghats, and then, also for about 200 miles, crosses the same plain and the equally dry and even drier districts of Northern Mysore and Bellary. The lower course of the Kistna, below the junction of the Tungabhadra near Kurnool, is partly through a country where it is below 30 inches, but in the Nullamalai hill tract below Kurnool, and in that traversed in the last 80 miles of its course, the rainfall exceeds this amount. The estimated averages of these several rivers are as follow :—

Godavery and tributaries .	Waingunga .	52 in.	Pranhita and tributaries	41 in.
	Wardha .	40 ,,		
	Paingunga .	39 ,,		
	Godavery and Manjira, above Sironcha . .			37 ,,
Kistna and tributaries .	Bhima and Sina .	49 in.	Whole Kistna basin . .	49 ,,
	Kistna above junction	59 ,,		
	Tungabhadra . .	43 ,,		
	Kistna below junction	33 ,,		

Rivers of the Carnatic.—The very dry tract in the north-east of Mysore and the rocky hills between Bellary and Cuddapah are drained by the Pennaur, which flows past Nellore to the Coromandel coast; the remainder of Mysore, to the south of this river basin and that of the Tungabhadra, by the Palar, which enters the sea near Chingleput; the Panar, which debouches at Cuddalore; and the Cauvery. Of these rivers, the last only has its sources under the crest of the Western Ghats. The others take their rise in the hilly but not very elevated country of Eastern Mysore, where the rainfall is about 30 inches or somewhat less, and then descending the Eastern Ghats, cross the Carnatic plain, the Palar and Panar receiving a considerable accession from the isolated hill groups that intervene between the Ghats and the coast plain, and which, to judge from the rainfall register of Yercaud on the Shevaroys, have a rainfall in some places exceeding 60 inches. A smaller river, the Vellaur, which enters the sea at Porto Novo, also brings down part of the drainage of these hills. The Cauvery alone receives a considerable flood drainage from the Western Ghat range during the summer monsoon. It then traverses the comparatively dry tableland of Southern Mysore, and descending to a lower level through the Baramahal, where the rainfall exceeds 30 inches, crosses another dry tract lying to the west of the hill groups above mentioned. Finally, passing Trichinopoly, it spreads out its delta in the fertile district of Tanjore, which it irrigates with its flood waters, and itself receives a rainfall of over 40 inches, chiefly at a later season.

To the south of the Cauvery the only stream of importance is the Vigay, which drains the eastern slopes of the Pulni hills, where the rainfall is probably about the same as on the Nilgiris, which they rival in height. In the lower hills around Madura, which also, in part, drain into the Vigay, the rainfall is over 30 inches; but the plain beyond, intervening between these hills and the coast, is for the most part very dry, and, like the Carnatic, receives its chief rainfall towards the end of the year. The annual average around the lower course of the river is between 20 and 30 inches, and very precarious. The mean rainfall on the drainage basins of these several rivers has been computed as follows:—

Pennaur	26 inches.
Palar	36 ,,
Panar	38 ,,

Vellaur	40 inches.
Cauvery	44 ,,
Vigay	32 ,,

The Tapti and Nerbudda.—These two rivers, together, receive the whole drainage of the Satpura range, except that of its eastern extremity, which is shared by the Sone, the Mahanadi, and the Godavery. The Nerbudda carries down little beyond this and the rainfall of the long narrow plain through which it flows, since the Central Indian tableland, which rises a mile or two to the north of the main stream, contributes nothing of importance to its volume. The rainfall of the Satpura range is locally very variable. In places it is upwards of 60 or 70 inches, but most of the stations at 2000 feet or less have a fall of about 50 inches or somewhat under. The eastern end of the Nerbudda valley has as much as 50 inches, but the rainfall diminishes to between 30 and 40 inches in the western half, below Khandwa. Still farther west, the Malwa Ghats, to the north of the river, have a rainfall between 40 and 50 inches, and some part of this drains into the Nerbudda; but the monsoon of the Arabian Sea, which brings its torrential rainfall to the Western Ghats south of the Tapti, reaches only occasionally so far north, and under ordinary circumstances is more or less diluted and intermingled with the drier currents of the Indus region.

The Tapti and its main tributary the Poorna have their sources in the Gawilgarh hills, where the rainfall is nearly as high as at Pachmarhi. But after leaving these hills they flow through the high plains of Berar and Khandesh, in a great part of which the annual average is below 30 inches, receiving from the south only the scanty drainage of a tract of the northern Deccan, where the fall is of even less amount. Only the last 80 miles of its course, where it skirts the northern extremity of the Sahyadri or Western Ghat range, and crosses the coast plain of Surat, is through a wetter country. The rainfall here amounts to between 40 and 50 inches, and is doubtless higher on the hills that hem in its channel.

Both the Nerbudda and Tapti are subject to occasional floods of great destructiveness, arising from cyclonic rainfalls on the Satpuras, such as have already been referred to in the earlier part of this section. The mean rainfalls of their basins are respectively as follow:—

Nerbudda	46 inches.
Tapti	39 „

The Looni.—This river carries off the uncertain and intermittent drainage of the west face of the Aravali range to the north of Mount Abu; and its main course lies through the arid plain of Western Rajputana, where rain at any season is very precarious, and the averages of many years vary between 10 and 15 inches. The soil of the country around the lower course of the Looni is strongly impregnated with salt, and in the dry season the water of the river, as described by Mr. W. T. Blanford, is not merely brackish, but decidedly salt; whence the name of the river.

Looni	18 inches.

Wind Pressure.—The pressure of the wind, as an element to be considered in providing for due stability in bridges and other structures, is usually so low in India as to be practically negligible. It has been mentioned in the former part of this work that, except in some few parts of the country during the height of the summer monsoon, the wind is too light to allow of the introduction of windmills as a source of motive power, with any prospect of success. But there are few parts of the country that are not liable to occasional storms, either the temporary squalls of the hot season, or the more lasting and destructive cyclones; and therefore the pressure of the wind on the vertical surfaces of structures cannot be left entirely out of account, though the occasions on which it may put their stability to the proof are rare and may not occur more than once in many years.

There exist very few continuous registers of wind pressures for India. The best and longest available to me is that of the Calcutta observatory, which extends over the 10 years from April 1867 to March 1877. The instrument used was a self-registering Osler's pressure gauge, exposing a pressure plate of 1 square foot of surface to the direct action of the wind. It was verified more than once, and the indications may therefore be accepted with confidence, in so far as such an instrument is capable of affording trustworthy data. It was set up at an elevation of 73 feet above the ground, on the old office of the Surveyor-General, in Park Street, Chowringhee. During these 10 years the following pressures were registered, omitting all below 10 lbs. to the square foot:—

10 to 19·9 lbs.	34 days.
20 to 29·9 ,,	7 ,,
30 to 39·9 ,,	6 ,,
40 to 49·9 ,,	3 ,,
50 lbs.	1 ,,

this last being the highest the instrument was capable of registering. This was recorded in the cyclone of the 9th June 1869, a by no means severe storm as experienced at Calcutta, and one that caused but little damage. The lowest barometer reading at Calcutta was 28·713 inches. In the great cyclone of 1864, which swept the port of its shipping and the city of its trees, and was one of the most destructive storms on record, the pressure gauge was blown away when it had recorded 33 lbs. to the square foot; and, at the time of that of the 1st and 2d November 1867, it was dismounted and under repair. There is therefore no register of the maximum pressure in a great destructive storm. But it must certainly considerably exceed 50 lbs. It cannot be computed directly from the mean recorded velocity of the wind, because this is by no means constant; the wind comes in violent intermittent gusts, the pressure oscillating from about 10 lbs., during the lulls, to its maximum.

Nearly all the remaining instances above tabulated occurred in nor'-westers, and rarely more than one gust of the maximum pressure, lasting a few seconds only, has been registered in the same storm.

At Bombay an Osler's pressure gauge was in operation

from 1847 to 1865. Mr. Chambers states that the maximum pressure of the wind recorded by it was "in the hurricane of the 1st and 2d November 1854, and exceeded 29·5 lbs., on attaining which the instrument became deranged." He estimates its maximum force at 50 lbs. A pressure of 35 lbs. in the same storm was recorded with a Lind's wind gauge.

These are the only instrumental records of wind pressure, known to me, that relate to Indian storms. In the Midnapore and Burdwan cyclone of the 16th October 1874, it is recorded in Mr. Willson's report, that a railway train of twenty-five carriages, standing on the line at Kanoo junction, was blown over. Mr. Willson obtained the dimensions and weights of the several carriages composing it, and considering them as empty carriages (which they were not), computed the minimum pressure requisite to overturn each vehicle separately. This he found to range from 34·1 to 55·6 lbs. per square foot. But he remarks that the rails at the place of the occurrence are about 6 feet above the level of the surrounding country, and the lifting force of the wind, under the carriages, which was not considered in the calculation, probably helped to upset them.

APPENDICES

APPENDIX I

The following tables give the mean and extreme daily and annual temperatures, and their daily, monthly, and annual range; the humidity, cloud proportion, rainfall (quantity and frequency), and the mean monthly barometric pressure and daily range, at 92 stations, selected from those of the Indian Government observatories. With one or two exceptions the mean values are taken from the tables in the Official Report for 1885, neglecting fractions of Fahrenheit degrees, hundredths of the rainfall inch, and thousandths of the barometric inch. The mean temperature-extremes and ranges, the mean barometric range (diurnal), and the mean number of rainy days have been computed specially for this work from the published reports of the Meteorological Offices of India and Bengal. The monthly rainfall data, those of the wettest and driest years, and the annual extreme temperatures have been corrected up to 1886. The elevations are those of the barometer cisterns above mean sea level.

1. Quetta—Elevation 5501 feet

	Temperature.					Hum.	Cloud.	Rainfall.		Barometer.	
				M. Range.							
	Mean.	Mean Max.	Mean Min.	Daily.	Month.	Mean.	Mean.	Ins.	Days.	Mean.	Daily Range.
January .	40	53	29	24	48	65	4·1	1·6	7	24·68	·08
February .	40	52	31	21	46	66	4·3	1·9	7	·64	·07
March . .	50	64	40	24	47	59	4·9	2·4	8	·63	·07
April . .	58	72	46	26	47	52	3·9	1·3	8	·59	·07
May . .	66	83	53	30	47	42	2·2	0·5	3	·55	·06
June . .	74	79	59	31	42	41	1·2	0·1	1	·44	·07
July . .	77	92	65	27	43	47	1·7	0·8	2	·40	·07
August. .	75	91	63	28	42	43	1·7	0·6	2	·45	·08
September .	67	86	52	34	50	37	0·7	0·2	1	·55	·08
October .	56	75	40	35	53	41	0·5	0·1	...	·69	·08
November .	46	64	30	34	52	49	0·9	0	...	·73	·08
December .	41	57	28	29	43	54	3·2	0·4	1	·71	·08
Year .	58					49	2·4	9·9	40	24·59	

Mean highest temperature of year 99°
,, lowest ,, ,, 15°
,, annual range of temperature 84°
Highest recorded reading (9 years) 103° (1878)
Lowest ,, ,, ,, 10·8° (1878)

Absolute range of temperature . 92·2°
Rainfall of wettest year (9 years) 21·6″ (1885)
,, driest ,, ,, 4·2″ (1879)
Rainy days in wettest year (8 years) 73 (1885)
,, ,, driest ,, ,, 18 (1880)

2. Leh—Elevation 11,503 feet

	Temperature.					Hum.	Cloud.	Rainfall.		Barometer.	
				M. Range.							
	Mean.	Mean Max.	Mean Min.	Daily.	Month.	Mean.	Mean.	Ins.	Days.	Mean.	Daily Range.
January .	18	35	11	25	44	61	6·4	0·2	4	19·64	·07
February .	19	35	9	26	48	61	6·7	0·2	5	·58	·08
March . .	31	46	20	26	50	56	6·5	0·2	2	·67	·08
April . .	41	57	31	26	43	42	6·4	0·1	1	·67	·08
May . .	47	64	36	27	44	40	6·5	0·1	2	·68	·08
June . .	56	74	44	30	50	37	4·9	0·2	1	·64	·09
July . .	62	80	51	29	44	44	5·1	0·5	3	·60	·10
August .	60	80	50	30	45	48	5·1	0·4	3	·63	·11
September .	52	72	42	30	46	44	4·0	0·2	1	·69	·11
October .	40	59	30	29	47	41	3·9	0·5	...	·73	·10
November .	30	47	20	27	43	50	4·0	0	...	·72	·09
December .	23	39	13	26	41	58	5·3	0·1	2	·70	·08
	40					49	5·4	2·7	24	19·66	

Mean highest temperature of year 90°
,, lowest ,, ,, –4°
,, annual range of temperature 94°
Highest recorded reading (10 years) 93° (1876)
Lowest ,, ,, ,, –17° (1878)

Absolute range of temperature . 110°
Rainfall of wettest year (11 years) 5·4″ (1879)
,, driest ,, ,, 0·4″ (1876)
Rainy days in wettest year (8 years) 43 (1886)
,, ,, driest ,, ,, 8 (1878)

3. Murree—Elevation 6344 feet

	Temperature.					Hum.	Cloud.	Rainfall.		Barometer.	
	Mean.	Mean Max.	Mean Min.	M. Range. Daily.	M. Range. Month.	Mean.	Mean.	Ins.	Days.	Mean.	Daily Range
January .	39	50	36	14	34	57	5·7	2·8	4	23·88	·04
February .	39	48	34	14	34	61	5·4	3·4	5	·84	·04
March . .	49	59	44	15	40	53	6·1	3·7	8	·86	·04
April . .	57	68	51	17	40	51	5·5	4·3	10	·83	·03
May . .	65	75	58	17	41	47	4·4	3·8	9	·79	·03
June . .	71	83	65	17	42	48	3·7	2·4	7	·72	·04
July . .	68	78	63	15	34	72	6·2	11·0	14	·70	·04
August . .	67	76	62	14	28	80	6·1	14·0	18	·74	·04
September .	65	74	59	15	30	69	3·6	6·1	9	·83	·04
October .	58	69	53	16	34	46	2·2	2·2	4	·92	·04
November .	49	60	45	15	33	44	2·9	1·7	3	·93	·04
December .	43	54	40	14	33	49	4·5	1·2	1	·92	·04
Year .	56					56	4·7	56·6	92	23·83	

Mean highest temperature of year 93°
,, lowest ,, ,, 24°
,, annual range of temperature 69°
Highest recorded reading (10 years) 98·7° (1880)
Lowest ,, ,, ,, 16·7° (1886)

Absolute range of temperature . 82°
Rainfall of wettest year (12 years) 71·8″ (1885)
,, driest ,, ,, 39·1″ (1880)
Rainy days in wettest year (11 years) 123 (1885)
,, ,, driest ,, ,, 74 (1880)

4. Simla—Elevation 7048 feet

	Temperature.					Hum.	Cloud.	Rainfall.		Barometer.	
	Mean.	Mean Max.	Mean Min.	M. Range. Month.	M. Range. Daily.	Mean.	Mean.	Ins.	Days.	Mean.	Daily Range.
January .	41	54	38	16	36	58	5·0	2·8	3	23·29	·06
February .	41	53	36	17	34	56	5·0	2·7	5	·24	·05
March . .	50	63	45	18	40	53	5·0	3·0	6	·27	·05
April . .	58	72	52	20	38	51	4·6	2·8	6	·26	·05
May . .	64	77	57	20	39	49	4·1	4·7	9	·22	·05
June . .	67	79	61	18	36	64	6·1	7·9	10	·15	·05
July . .	64	73	61	12	22	88	8·5	19·3	21	·12	·05
August . .	63	71	60	11	20	91	8·6	18·1	22	·16	·05
September .	61	72	58	14	23	82	6·2	6·0	12	·25	·05
October .	56	68	51	17	28	53	1·0	1·4	2	·33	·06
November .	49	61	43	18	31	50	1·5	0·3	1	·32	·06
December .	45	57	40	17	30	47	3·5	1·1	2	·31	·06
Year .	55					62	4·9	70·1	99	23·24	

Mean highest temperature of year 88°
,, lowest ,, ,, 25°
,, annual range of temperature 63°
Highest recorded reading (11 years) 94·4° (1879)
Lowest ,, ,, ,, 19·7° (1883)

Absolute range of temperature . 74·7°
Rainfall of wettest year (24 years) 94·9″ (1864)
,, driest ,, ,, 52·1″ (1867)
Rainy days in wettest year (12 years) 136 (1884-1885)
Rainy days in driest year (12 years) 74 (1877)

5 CHAKRATA—Elevation 7052 feet

	Temperature.					Hum.	Cloud.	Rainfall.		Barometer.	
	Mean.	Mean Max.	Mean Min.	M. Range. Daily.	M. Range. Month.	Mean.	Mean.	Ins.	Days.	Mean.	Daily Range.
January .	42	51	35	16	35	63	4·5	2·3	5	23·26	·05
February .	43	51	35	16	34	63	4·8	2·7	7	·22	·05
March . .	51	62	44	18	40	53	4·2	3·2	6	·25	·05
April . .	60	70	50	19	39	46	3·7	1·7	6	·24	·04
May . .	64	73	55	18	37	51	3·8	2·9	10	·20	·04
June . .	67	75	60	15	31	66	5·2	7·5	12	·13	·05
July . .	64	70	60	10	21	91	8·7	17·2	23	·11	·05
August . .	64	70	59	10	19	91	8·6	15·7	24	·15	·05
September .	63	69	58	12	21	83	6·7	5·6	12	·23	·05
October .	58	66	51	15	28	59	1·8	0·7	2	·30	·05
November .	51	59	43	16	29	56	1·5	0·2	1	·31	·05
December .	46	55	39	16	31	52	3·3	1·1	2	·30	·05
Year .	56					65	4·7	60·8	110	23·22	

Mean highest temperature of year 85°
,, lowest ,, ,, 24°
,, annual range of temperature 61°
Highest recorded reading (11 yrs.) 89·6° (1877)
Lowest ,, ,, ,, 18·7° (1876)

Absolute range of temperature . 70·9°
Rainfall of wettest year (18 years) 80·7″ (1885)
,, driest ,, ,, 42·2″ (1869)
Rainy days in wettest year (12 years) 127 (1885)
,, ,, driest ,, ,, 95 (1877)

6. RANIKHET—Elevation 6069 feet

	Temperature.					Hum.	Cloud.	Rainfall.		Barometer.	
	Mean	Mean Max.	Mean Min.	M. Range. Daily.	M. Range. Month.	Mean.	Mean.	Ins.	Days.	Mean.	Daily Range.
January .	46	55	42	13	32	61	3·8	1·9	3	24·12	·06
February .	48	55	40	15	33	60	4·6	2·0	5	·08	·06
March . .	57	65	49	16	37	52	3·8	2·2	5	·09	·06
April . .	65	73	56	17	34	45	3·4	1·3	4	·06	·06
May . .	68	76	59	17	34	52	3·8	3·0	10	·02	·05
June . .	71	78	64	14	28	64	5·6	6·0	11	23·94	·05
July . .	68	74	63	11	20	85	8·7	12·7	21	·93	·05
August . .	67	73	62	11	19	86	8·5	11.6	20	·96	·06
September .	66	73	61	12	22	80	6·1	6·0	11	24·03	·06
October .	61	69	54	15	25	61	1·6	1·1	2	·11	·06
November .	55	62	47	15	27	57	1·2	0·3	1	·15	·06
December .	49	57	43	14	28	56	2·6	1·0	2	·14	·06
Year .	60					63	4·5	49·1	95	24·05	

Mean highest temperature of year 86°
,, lowest ,, ,, 30°
,, annual range of temperature 56°
Highest recorded reading (11 years) 87·8° (1876)
Lowest ,, ,, ,, 26° (1877)

Absolute range of temperature . 61·8°
Rainfall of wettest year (16 years) 61·2″ (1874)
,, driest ,, ,, 26·4″ (1873)
Rainy days in wettest year (12 years) 107 (1884)
,, ,, driest ,, ,, 83 (1877)

7. Darjiling—Elevation (Old Observatory) 6912 feet

	Temperature.					Hum.	Cloud.	Rainfall.		Barometer.	
				M. Range.							
	Mean.	Mean Max.	Mean Min.	Daily.	Month.	Mean.	Mean.	Ins.	Days.	Mean.	Daily Range.
January .	41	50	35	15	27	78	5·7	0·7	2	23·40	·08
February .	44	53	37	16	30	78	5·9	1·2	4	·38	·07
March . .	50	61	44	17	32	71	5·3	2·4	5	·38	·07
April . .	56	67	50	17	29	77	6·1	3·7	11	·37	·08
May . .	59	68	53	15	27	84	7·2	7·1	18	·35	·07
June . .	62	69	58	11	21	91	8·7	24·1	24	·28	·06
July . .	63	70	59	11	19	92	8·7	30·5	28	·28	·06
August .	63	70	59	11	19	92	8·5	26·0	26	·32	·07
September .	61	69	57	12	21	91	8·1	17·8	22	·38	·08
October .	57	66	51	15	26	81	5·2	6·4	7	·44	·08
November .	50	60	43	17	28	75	4·4	0·2	1	·47	·08
December .	44	54	37	17	27	73	4·3	0·2	1	·45	·08
Year .	54					82	6·5	120·3	149	23·38	

Mean highest temperature of year 78°
,, lowest ,, ,, 30°
,, annual range of temperature 48°
Highest recorded reading (12 years) 84·2 (1877)
Lowest ,, ,, ,, 26° (1874)

Absolute range of temperature . 58·2°
Rainfall of wettest year (21 years) 160·9″ (1879)
,, driest ,, ,, 77·2″ (1873)
Rainy days in wettest year (14 years) 171 (1870)
,, ,, driest ,, ,, 124 (1878)

8. Darjiling—Elevation (New Observatory) 7421 feet (4 years)

	Temperature.					Hum.	Cloud.	Rainfall.		Barometer.	
				M. Range.							
	Mean.	Mean Max.	Mean Min.	Daily.	Month.	Mean.	Mean.	Ins.	Days.	Mean.	Daily Range.
January .	40	45	36	9	21	83	4·3	0·7	3	23·00	·08
February .	39	45	33	12	29	82	4·6	1·3	5	22·92	·08
March . .	47	56	43	13	27	76	4·7	1·7	4	·96	·08
April . .	53	62	47	15	29	75	6·0	5·3	9	·93	·08
May . .	54	62	49	13	27	88	7·2	7·7	18	·92	·08
June . .	56	65	55	10	20	92	8·4	28·4	21	·87	·07
July . .	61	67	57	10	16	94	8·6	28·5	27	·86	·06
August .	61	66	56	10	16	94	9·0	28·5	28	·89	·07
September .	59	65	55	10	19	94	8·1	16·9	21	·99	·07
October .	54	61	49	12	24	87	5·9	7·5	6	23·05	·07
November .	48	56	43	13	23	79	3·7	0·1	...	·04	·08
December .	43	50	38	12	23	81	3·8	0·5	1	·02	·08
Year .	51					85		127·1	143	22·95	

Mean highest temperature of year 72°
,, lowest ,, ,, 25°
,, annual range of temperature 47°
Highest recorded reading (5 years) 72·1°
Lowest ,, ,, ,, 10·9° (1883)

Absolute range of temperature . 52·2°
Rainfall of wettest year (5 years) 144·2″ (1882)
,, driest ,, ,, 107·5″ (1884)
Rainy days in wettest year ,, 151 (1885)
,, driest ,, ,, 124 (1883)

9. Shillong—Elevation 4792 feet (4 years)

	Temperature.			M. Range.		Hum.	Cloud.	Rainfall.	
	Mean.	Mean Max.	Mean Min.	Daily.	Month.	Mean.	Mean.	Ins.	Days.
January	51	62	42	20	31	70	2·5	0·4	1
February	54	64	46	18	30	65	3·4	0·8	4
March	62	71	54	17	32	59	3·1	2·0	5
April	65	74	58	16	29	65	5·2	3·7	11
May	68	76	63	13	22	77	6·6	10·0	21
June	69	76	65	11	18	84	7·8	17·0	24
July	69	76	66	10	16	87	8·6	14·0	22
August	69	76	65	11	17	88	8·5	14·4	23
September	67	74	64	10	20	89	8·1	15·4	22
October	63	72	58	14	25	86	6·2	6·2	13
November	57	66	49	17	30	76	3·4	1·0	3
December	51	61	42	19	31	75	3·2	0·4	1
Year	62					77	5·6	85·3	150

Mean highest temperature of year 82°
,, lowest ,, ,, 34°
,, annual range of temperature 48°
Highest recorded reading (4 years) 84° (1872)
Lowest ,, ,, ,, 32·5° (1872)

Absolute range of temperature . 51·5°
Rainfall of wettest year (20 years) 121·2″ (1867)
,, driest ,, ,, 53·6″ (1873)
Rainy days in wettest year (18 years) 191 (1883, 1884)
Rainy days in driest year (18 years) 100 (1868)

10. Pachmarhi—Elevation 3528 feet

	Temperature.			M. Range.		Hum.	Cloud.	Rainfall.		Barometer.	
	Mean.	Mean Max.	Mean Min.	Daily.	Month.	Mean.	Mean.	Ins.	Days.	Mean.	Daily Range.
January	58	72	47	25	41	52	2·2	0·5	2	26·55	·09
February	62	75	51	24	42	43	2·1	0·3	2	·52	·09
March	72	85	60	25	44	32	2·1	0·4	1	·49	·10
April	79	92	69	23	36	26	2·2	0·3	2	·42	·10
May	83	94	74	20	34	32	2·8	0·6	4	·34	·09
June	78	86	72	14	30	65	6·6	10·8	14	·24	·08
July	71	76	67	9	18	89	8·7	28·8	25	·21	·07
August	70	75	67	8	16	90	8·4	18·2	23	·26	·07
September	70	77	66	11	20	83	7·3	15·1	17	·34	·09
October	67	78	61	17	32	61	2·9	1·9	5	·48	·09
November	60	73	50	23	36	53	1·9	0·4	1	·54	·09
December	56	70	45	25	38	54	2·0	0·7	2	·56	·09
Year	69					57	4·1	78·0	98	·41	

Mean highest temperature of year 100°
,, lowest ,, ,, 35°
,, annual range of temperature 65°
Highest recorded reading (11 yrs.) 102·4° (1881)
Lowest ,, ,, ,, 30° (1878)

Absolute range of temperature . 72·4°
Rainfall of wettest year (16 years) 106·1″ (1871)
,, driest ,, ,, 62·0″ (1880, 1885)
Rainy days in wettest year (11 years) 115 (1878)
,, ,, driest ,, ,, 70 (1876)

11. Chikalda—Elevation 3656 feet

	Temperature.					Hum.	Cloud.	Rainfall.		Barometer.	
	Mean.	Mean Max.	Mean Min.	M. Range. Daily.	M. Range. Month.	Mean.	Mean.	Ins.	Days.	Mean.	Daily Range.
January .	64	73	56	17	31	52	1·7	0·5	1	26·43	·10
February .	67	77	59	18	35	43	1·3	0·1	1	·39	·10
March . .	75	86	67	19	35	34	1·8	0·4	1	·37	·11
April . .	81	92	72	20	31	27	2·0	0·1	2	·29	·11
May . .	82	94	73	21	33	35	2·4	0·5	3	·22	·10
June . .	75	84	69	15	30	69	6·4	11·9	15	·13	·08
July . .	69	74	66	8	19	91	8·9	17·8	24	·11	·08
August . .	68	73	65	8	15	92	8·7	16·6	22	·14	·08
September .	68	74	65	9	19	89	7·5	12·3	19	·21	·09
October .	69	77	64	13	22	63	3·0	4·6	6	·35	·09
November .	65	74	59	15	25	55	1·9	0·5	2	·40	·10
December .	62	72	55	17	27	53	1·9	1·3	2	·43	·10
Year .	70					59	4·0	66·6	98	26·29	

Mean highest temperature of year 99°
„ lowest „ „ 46°
„ annual range of temperature 53°
Highest recorded reading (11 years) 103° (1876)
Lowest „ „ „ 39·5° (1880)

Absolute range of temperature . 63·5°
Rainfall of wettest year (15 years) 97·1″ (1867)
„ driest „ „ 39·3″ (1866)
Rainy days in wettest year (11 yrs.) 121 (1884)
„ „ driest „ „ 80 (1876)

12. Mount Abu—Elevation 3945 feet

	Temperature.					Hum.	Cloud.	Rainfall.		Barometer.	
	Mean.	Mean Max.	Mean Min.	M. Range. Daily.	M. Range. Month.	Mean.	Mean.	Ins.	Days.	Mean.	Daily Range.
January .	58	68	52	16	34	39	2·6	0·2	1	26·15	·07
February .	60	69	53	16	36	34	3·0	0·4	1	·10	·07
March . .	69	79	62	17	36	30	2·9	0·1	1	·08	·08
April . .	75	85	67	18	32	29	2·6	0	1	·03	·08
May . .	79	89	71	18	33	35	2·0	1·0	3	25·96	·07
June . .	75	84	69	15	27	65	5·5	5·1	8	·86	·06
July . .	70	75	66	9	25	85	8·7	22·2	22	·82	·06
August . .	68	73	65	8	20	87	8·4	22·5	22	·86	·06
September .	69	76	65	11	21	76	6·4	9·1	15	·96	·07
October .	70	79	64	15	27	45	2·4	2·1	2	26·09	·07
November .	64	74	57	17	28	36	1·1	0·2	1	·13	·08
December .	59	70	53	17	30	41	1·9	0·2	1	·14	·07
Year .	68					50	4·0	63·1	78	26·02	

Mean highest temperature of year 96°
„ lowest „ „ 39°
„ annual range of temperature 57°
Highest recorded reading (10 years) 101° (1881)
Lowest „ „ „ 32·8° (1878)

Absolute range of temperature . 68·2°
Rainfall of wettest year (27 years) 123″ (1862)
„ driest „ „ 19·2″ (1877)
Rainy days in wettest year (8 years) 88 (1884)
„ „ driest „ „ 66 (1885)

13. Ootacamund—Elevation 7252 feet (13 Months)[1]

	Temperature.					Hum.	Cloud.	Rainfall.		Barometer.	
	Mean.	Mean Max.	Mean Min.	M. Range. Daily.	M. Range. Month.	Mean.	Mean.	Ins.	Days.	Mean.	Daily Range.
January	48	66	35	31	44	55	1·9	0·5	1	23·20	·06
February	51	68	39	29	42	61	4·2	0·2	...	·22	·07
March	55	71	44	27	45	46	3·3	1·2	1	·23	·08
April	58	72	51	21	29	64	5·6	3·9	11	·20	·08
May	59	71	53	18	31	70	6·4	6·2	16	·16	·09
June	56	65	53	12	23	84	8·7	6·0	17	·13	·06
July	55	62	52	10	20	84	9·2	5·6	23	·14	·05
August	55	63	51	12	24	82	8·6	4·2	19	·14	·07
September	55	63	50	13	25	79	8·0	3·7	13	·16	·07
October	54	64	50	14	24	89	8·7	9·8	22	·21	·09
November	53	63	47	16	35	90	6·7	2·9	16	·20	·07
December	51	64	42	22	35	71	4·9	1·6	4	·18	·07
Year	55					71	6·3	45·8	143	23·18	

Highest recorded temperature . . . 77·3°
Lowest „ „ . . . 25·3°
Range during year . . . . 52°

Rainfall of wettest year (10 years) 58·4″ (1880)
„ driest „ „ 33·7″ (1867)

[1] The rainfall is the mean of 11 years.

14. Wellington—Elevation 6200 feet

	Temperature.					Hum.	Cloud.	Rainfall.		Barometer.	
	Mean.	Mean Max.	Mean Min.	M. Range. Daily.	M. Range. Month.	Mean.	Mean.	Ins.	Days.	Mean.	Daily Range.
January	55	66	45	21	34	70	3·3	0·8	3	24·26	·07
February	58	70	46	24	37	59	2·8	0·3	1	·26	·08
March	63	74	53	21	31	60	2·9	2·0	6	·27	·08
April	65	76	56	20	26	63	4·4	2·9	7	·23	·08
May	66	76	59	17	26	69	5·6	4·1	14	·20	·07
June	63	72	58	14	22	75	7·4	3·6	15	·16	·06
July	63	71	58	13	22	73	7·5	3·2	14	·16	·06
August	62	71	57	14	22	77	7·4	4·0	16	·17	·07
September	62	71	56	15	25	75	6·9	4·7	13	·19	·07
October	61	68	55	13	22	84	7·3	9·8	18	·21	·08
November	58	66	53	13	28	84	7·1	8·5	16	·24	·08
December	56	65	48	17	34	78	5·6	4·1	10	·26	·07
Year	61					72	5·7	48·0	133	·22	

Mean highest temperature of year 80°
„ lowest „ „ 37°
Mean annual range of temperature 43°
Highest recorded reading (7 years) 80·9° (1886)
Lowest „ „ „ 34·2° (1885)

Absolute range of temperature . 46·7°
Rainfall of wettest year (10 years) 63·91″ (1885)
„ driest „ „ 40·77″ (1876)
Rainy days in wettest year „ 158 (1880)
„ „ driest „ „ 75 (1876)

15. Newara Eliya—Elevation 6240 feet

	Temperature.					Hum.	Cloud.	Rainfall.		Barometer.	
	Mean.	Mean Max.	Mean Min.	M. Range. Daily.	M. Range. Month.	Mean.	Mean.	Ins.	Days.	Mean.	Daily Range.
January .	57	67	48	19	31	81	5·0	5·6	11	24·01	?
February .	57	69	45	24	38	73	3·8	2·5	6	·02	?
March . .	59	70	46	24	36	73	4·0	3·0	8	·03	?
April . .	60	71	48	23	33	78	5·0	5·6	12	·01	?
May . .	61	70	52	18	30	83	5·7	8·2	14	23·99	?
June . .	59	65	53	12	22	88	8·0	14·4	23	·98	?
July . .	58	64	52	12	23	88	7·5	13·7	22	·98	?
August . .	59	65	52	13	23	86	7·2	9·4	22	·98	?
September .	59	66	52	14	24	86	6·6	8·9	18	·99	?
October .	59	67	51	16	26	87	6·4	10·5	23	24·00	?
November .	59	67	50	17	28	85	5·8	9·2	19	·02	?
December .	58	66	49	17	30	83	5·6	7·9	17	·01	?
Year .	59					83	5·9	98·9	195	24·00	?

Mean highest temperature of year 77°
„ lowest „ „ 35°
„ annual range of temperature 42°
Highest recorded reading (11 years) 79° (1877)
Lowest „ „ (9 years) 32° (1883)

Absolute range of temperature . 47°
Rainfall of wettest year (17 years) 129·3″ (1882)
„ driest „ „ 75·4″ (1870)
Rainy days in wettest year (11 years) 229 (1877)
„ „ driest „ „ 177 (1881)

16. Peshawar—Elevation 1110 feet

	Temperature.					Hum.	Cloud.	Rainfall.		Barometer.	
	Mean.	Mean Max.	Mean Min.	M. Range. Daily.	M. Range. Month.	Mean.	Mean.	Ins.	Days.	Mean.	Daily Range.
January .	50	65	39	26	40	62	4·5	1·6	3	28·96	·09
February .	52	66	40	26	43	57	4·1	1·2	3	·91	·10
March . .	62	76	51	25	47	57	4·8	1·8	4	·83	·10
April . .	70	85	60	25	46	55	4·5	2·0	5	·69	·10
May . .	80	96	68	28	48	43	3·0	0·7	3	·56	·10
June . .	89	105	76	29	47	39	2·2	0·3	1	·39	·10
July . .	89	103	78	25	41	51	2·6	1·7	3	·37	·11
August . .	87	100	78	22	38	59	2·7	2·0	3	·43	·11
September .	81	95	70	25	39	56	1·7	0·8	2	·59	·11
October .	71	86	57	29	46	48	1·5	0·2	1	·79	·10
November .	58	76	44	32	50	55	1·9	0·6	2	·91	·10
December .	51	67	38	29	44	62	3·3	0·6	2	·97	·10
Year .	70					54	3·1	13·5	32	28·70	

Mean highest temperature of year 115°
„ lowest „ „ 29°
„ annual range of temperature 86°
Highest recorded reading (10 years) 119° (1880)
Lowest „ „ „ 24·7° (1879)

Absolute range of temperature . 94·3°
Rainfall of wettest year (27 years) 27·9″ (1877)
„ driest „ „ 5·0″ (1880)
Rainy days in wettest year (12 years) 51 (1885)
„ „ driest „ „ 19 (1875)

17. Rawalpindi—Elevation 1652 feet

	Temperature.					Hum.	Cloud.	Rainfall.		Barometer.	
				M. Range.							
	Mean.	Mean Max.	Mean Min.	Daily.	Month.	Mean.	Mean.	Ins.	Days.	Mean.	Daily Range.
January .	49	64	38	26	40	72	4·3	2·4	5	28·37	·08
February .	52	64	40	24	41	70	4·2	2·0	5	·31	·08
March . .	61	76	50	26	47	64	4·3	1·9	5	·25	·08
April . .	71	86	59	27	49	58	3·8	2·3	6	·13	·08
May . .	81	96	67	29	49	51	2·8	1·6	5	·02	·08
June . .	89	103	74	29	48	49	2·1	1·7	4	27·88	·09
July . .	87	98	76	22	42	65	3·9	7·4	9	·86	·10
August . .	84	95	75	20	35	72	3·7	7·3	11	·94	·09
September .	80	94	69	25	38	67	2·0	3·2	6	28·06	·08
October .	70	87	56	31	48	56	1·2	0·6	3	·23	·08
November .	57	76	42	34	50	61	1·8	0·9	2	·34	·08
December .	50	67	36	31	43	67	3·0	1·1	3	·39	·08
Year .	69					63	3·1	32·4	64	28·15	

Mean highest temperature of year 114°
,, lowest ,, ,, 20°
,, annual range of temperature 85°
Highest recorded reading (11 years) 118° (1880)
Lowest ,, ,, ,, 23·9° (1878)

Absolute range of temperature . 94·1°
Rainfall of wettest year (26 years) 46″ (1875)
,, driest ,, ,, 20·3″ (1880)
Rainy days in wettest year (12 years) 76 (1885)
,, ,, driest ,, ,, 43 (1878)

18. Sialkote—Elevation 830 feet

	Temperature.					Hum.	Cloud.	Rainfall.		Barometer.	
				M. Range.							
	Mean.	Mean Max.	Mean Min.	Daily.	Month.	Mean.	Mean.	Ins.	Days.	Mean.	Daily Range.
January .	52	67	43	24	38	70	4·6	1·4	2	29·21	·08
February	56	70	45	25	41	68	4·3	1·8	4	·14	·08
March . .	66	82	55	27	51	58	4·5	1·9	3	·05	·08
April . .	77	93	66	27	49	43	3·9	1·6	3	28·92	·08
May . .	85	101	73	28	47	38	3·0	1·2	3	·81	·08
June . .	91	105	81	24	44	44	2·4	3·2	4	·66	·09
July . .	87	98	80	18	40	66	4·8	11·6	9	·66	·08
August . .	85	96	78	18	34	71	4·7	9·1	9	·73	·09
September .	83	95	74	21	36	66	2·7	3·2	4	·85	·09
October .	75	92	62	30	45	53	1·1	0·6	1	29·03	·09
November .	62	80	49	31	47	56	1·9	0·4	1	·15	·08
December .	53	70	42	28	40	69	3·1	0·8	1	·20	·08
Year .	73					59	3·4	36·8	44	28·95	

Mean highest temperature of year 117°
,, lowest ,, ,, 34°
,, annual range of temperature 83°
Highest recorded reading (10 yrs.) 121·3° (1877)
Lowest ,, ,, (9 years) 31·6° (1878)

Absolute range of temperature . 89·7°
Rainfall of wettest year (27 years) 54·9″ (1862)
,, driest ,, ,, 20·2″ (1880)
Rainy days in wettest year (12 years) 59 (1884)
,, ,, driest ,, ,, 35 (1876, 1880)

19. Lahore—Elevation 732 feet

	Temperature.					Hum.	Cloud.	Rainfall.		Barometer.	
	Mean.	Mean Max.	Mean Min.	M. Range. Daily.	M. Range. Month.	Mean.	Mean.	Ins.	Days.	Mean.	Daily Range.
January .	54	68	43	25	39	60	3·2	0·7	2	29·34	·08
February .	59	71	46	25	32	57	3·4	1·1	3	·28	·08
March . .	69	83	57	26	48	48	3·1	1·1	2	·18	·08
April . .	81	95	66	29	48	37	2·5	0·6	3	·05	·09
May . .	88	102	73	29	48	33	2·4	0·9	3	28·92	·08
June . .	93	107	81	26	45	37	2·9	1·8	3	·77	·09
July . .	89	99	81	18	39	58	4·2	7·4	7	·78	·09
August . .	88	98	80	18	32	61	3·7	4·6	6	·85	·09
September .	85	97	75	22	34	55	1·7	2·4	4	·97	·09
October . .	77	93	62	31	46	46	0·8	0·6	1	29·15	·08
November .	64	81	49	32	48	47	1·2	0·2	1	·30	·08
December .	55	71	43	28	41	56	2·0	0·5	2	·35	·08
Year .	75					50	2·6	21·9	37	29·08	

Mean highest temperature of year 117°
,, lowest ,, ,, 34°
,, annual range of temperature 83°
Highest recorded reading (11 yrs.) 120·3° (1879)
Lowest ,, ,, ,, 29·8° (1878)

Absolute range of temperature . 90·5°
Rainfall of wettest year (27 years) 37·8″ (1875)
,, driest ,, ,, 8·7″ (1871)
Rainy days in wettest year (12 yrs.) 47 (1881)
,, ,, driest ,, ,, 26 (1875)

20. Ludhiana—Elevation 812 feet

	Temperature.					Hum.	Cloud.	Rainfall.		Barometer.	
	Mean.	Mean Max.	Mean Min.	M. Range. Daily.	M. Range. Month.	Mean.	Mean.	Ins.	Days.	Mean.	Daily Range.
January .	52	69	43	26	40	65	3·7	1·3	3	29·23	·10
February .	57	71	46	25	43	64	3·6	1·2	3	·17	·09
March . .	68	84	57	27	48	55	3·8	1·6	3	·07	·09
April . .	78	96	66	30	48	50	2·8	0·9	2	28·94	·10
May . .	85	103	72	31	47	47	2·2	1·2	3	·83	·09
June . .	91	107	80	27	45	49	2·9	2·3	4	·69	·10
July . .	86	99	80	19	36	68	5·7	9·7	9	·69	·10
August . .	86	96	80	16	32	70	5·2	6·5	7	·76	·10
September .	83	95	75	20	31	67	2·8	4·3	4	·88	·10
October . .	75	92	64	28	43	54	0·8	1·2	1	29·05	·09
November .	63	80	50	30	46	52	1·3	0·1	...	·18	·09
December .	54	71	43	28	41	61	2·6	0·9	1	·24	·09
Year .	73					59	3·1	31·2	40	28·98	

Mean highest temperature of year 116°
,, lowest ,, ,, 35°
,, annual range of temperature 81°
Highest recorded reading (11 yrs.) 118·3° (1881)
Lowest ,, ,, (10 years) 32°[1] (1878)

Absolute range of temperature 86·3°
Rainfall of wettest year (27 years) 46·2″ (1872)
,, driest ,, ,, 15·6″ (1876)
Rainy days in wettest year (12 years) 55 (1885)
,, ,, driest ,, ,, 24 (1876)

[1] A reading of 24° is recorded in December 1879, but it appears to be of very doubtful validity.

21. Delhi—Elevation 718 feet

	Temperature.					Hum.	Cloud.	Rainfall.		Barometer.	
	Mean.	Mean Max.	Mean Min.	M. Range. Daily.	M. Range. Month.	Mean.	Mean.	Ins.	Days.	Mean.	Daily Range.
January .	59	71	48	23	38	57	3·4	1·0	2	29·32	·10
February .	62	75	52	23	42	50	3·0	0·5	2	·26	·10
March . .	74	88	62	26	49	43	3·1	0·7	2	·16	·11
April . .	84	99	72	27	44	33	2·7	0·4	1	·02	·12
May . .	89	104	78	26	44	39	2·2	0·7	3	28·92	·11
June . .	93	105	84	21	40	48	3·6	3·4	4	·78	·11
July . .	87	95	81	14	28	68	6·0	8·5	10	·78	·09
August . .	86	94	80	14	25	68	5·7	6·9	9	·85	·10
September .	84	94	77	17	27	65	3·5	4·5	5	·96	·11
October .	79	92	68	24	38	49	0·8	0·5	1	29·14	·10
November .	68	83	56	27	43	45	0·7	0·1	...	·27	·10
December .	60	74	49	25	37	52	1·8	0·4	1	·33	·10
Year .	77					51	3·0	27·6	40	29·07	

Mean highest temperature of year 116°
„ lowest „ „ 40°
„ annual range of temperature 76°
Highest recorded reading (11 yrs.) 117·6° (1878)
Lowest „ „ „ 36·3° (1878)

Absolute range of temperature . 81·3°
Rainfall of wettest year (36 years) 48·5″ (1860)
„ driest „ „ 8·1″ (1868)
Rainy days in wettest year (12 years) 56 (1885)
„ „ driest „ „ 27 (1877)

22. Sirsa—Elevation 662 feet

	Temperature.					Hum.	Cloud.	Rainfall.		Barometer.	
	Mean.	Mean Max.	Mean Min.	M. Range. Daily.	M. Range. Month.	Mean.	Mean.	Ins.	Days.	Mean.	Daily Range.
January .	56	71	42	29	46	55	3·8	0·7	2	29·39	·09
February .	60	75	46	29	49	49	4·0	0·3	1	·33	·10
March . .	71	88	57	31	55	44	4·3	0·5	1	·22	·10
April . .	82	100	67	33	49	36	3·6	0·4	2	·08	·10
May . .	88	105	75	30	49	37	2·9	0·6	3	28·97	·10
June . .	93	107	83	24	43	42	3·5	2·3	3	·83	·10
July . .	89	100	80	20	36	60	5·4	3·7	7	·82	·10
August . .	88	99	80	19	30	60	5·4	3·6	5	·89	·10
September .	85	97	75	24	37	56	3·1	2·0	5	29·02	·10
October .	78	95	63	32	49	39	1·0	0·3	1	·20	·10
November .	65	84	48	36	52	40	1·2	...	...	·33	·09
December .	57	75	41	34	46	51	2·6	0·3	1	·39	·09
Year .	76					47	3·4	14·7	31	29·12	

Mean highest temperature of year 116°
„ lowest „ „ 34°
„ annual range of temperature 82°
Highest recorded reading (11 yrs.) 118·3° (1879)
Lowest „ „ (9 years) 29·0° (1878)

Absolute range of temperature . 88·4°
Rainfall of wettest year (33 years) 24·6″ (1863)
„ driest „ „ 6·2″ (1880)
Rainy days in wettest year (12 years) 49 (1885)
„ „ driest „ „ 15 (1880)

23. Dera Ishmail Khan—Elevation 573 feet

	Temperature.					Hum.	Cloud.	Rainfall.		Barometer.	
				M. Range.							
	Mean.	Mean Max.	Mean Min.	Daily.	Month.	Mean.	Mean.	Ins.	Days.	Mean.	Daily Range.
January .	52	70	41	29	45	55	3·1	0·4	2	29·49	·09
February .	56	72	44	28	46	50	2·9	0·7	3	·44	·09
March . .	67	83	55	28	50	49	3·2	0·9	3	·34	·10
April . .	77	92	65	27	48	45	2·9	0·8	3	·20	·10
May . .	87	101	74	27	49	39	1·8	0·4	2	·07	·10
June . .	93	107	80	27	44	44	1·6	0·6	2	28·90	·10
July . .	91	103	82	21	41	59	2·3	1·8	4	·89	·09
August . .	90	101	81	20	36	62	2·0	1·6	3	·95	·09
September .	86	100	75	25	39	56	1·0	0·6	2	29·10	·09
October .	75	93	62	31	48	46	0·5	0·1	1	·29	·09
November .	62	81	47	34	52	48	1·4	0·1	1	·44	·08
December .	54	72	40	32	46	52	2·3	0·3	1	·51	·09
Year .	74					50	2·1	8·3	27	29·22	

Mean highest temperature of year 117°
,, lowest ,, ,, 31°
,, annual range of temperature 86°
Highest recorded reading (11 yrs.) 121·5° (1882)
Lowest ,, ,, ,, 26° (1878)

Absolute range of temperature . 95·5°
Rainfall of wettest year (24 years) 16·2″ (1878)
,, driest ,, ,, 4·4″ (1880)
Rainy days in wettest year (12 years) 35 (1884-1885)
Rainy days in driest year (12 years) 12 (1880)

24. Mooltan—Elevation 420 feet

	Temperature.					Hum.	Cloud.	Rainfall.		Barometer.	
				M. Range.							
	Mean.	Mean Max.	Mean Min.	Daily.	Month.	Mean.	Mean.	Ins.	Days.	Mean.	Daily Range.
January .	54	70	43	27	42	58	2·6	0·4	1	29·64	·10
February .	58	74	46	28	45	52	2·6	0·3	1	·59	·10
March . .	70	85	58	27	50	50	2·4	0·5	1	·48	·11
April . .	80	96	66	30	49	45	1·8	0·3	1	·34	·12
May . .	89	104	76	28	48	45	1·3	0·5	1	·20	·10
June . .	94	107	82	25	40	51	1·0	0·4	2	·05	·09
July . .	92	103	83	20	34	59	2·3	2·2	3	·03	·09
August . .	89	100	82	18	31	64	2·2	1·3	3	·10	·09
September .	87	99	77	22	34	62	1·1	0·8	1	·24	·10
October .	77	94	64	30	45	55	0·3	0·1	...	·44	·09
November .	66	82	52	30	48	53	0·7	0·1	...	·59	·10
December .	56	73	43	30	44	56	1·8	0·3	1	·66	·10
Year .	76					54	1·7	7·2	15	29·36	

Mean highest temperature of year 114°
,, lowest ,, ,, 34°
,, annual range of temperature 80°
Highest recorded reading (11 yrs.) 118·4° (1876)
Lowest ,, ,, ,, 29° (1878)

Absolute range of temperature . 89·4°
Rainfall of wettest year (25 years) 15·5″ (1877)
,, driest ,, ,, 1·9″ (1871)
Rainy days in wettest year (12 years) 22 (1884)
,, ,, driest ,, ,, 9 (1875, 1883)

25. Jacobabad—Elevation 186 feet

	Temperature.					Hum.	Cloud.	Rainfall.		Barometer.	
	Mean.	Mean Max.	Mean Min.	M. Range. Daily.	M. Range. Month.	Mean.	Mean.	Ins.	Days.	Mean.	Daily Range.
January .	57	74	43	31	48	46	2·4	0·2	1	29·89	·12
February .	62	77	48	29	51	39	3·2	0·2	2	·82	·12
March . .	74	90	60	30	54	41	3·3	0·3	1	·70	·13
April . .	83	99	68	31	51	38	3·1	0·2	1	·56	·13
May . .	91	108	76	32	50	36	1·4	0·1	1	·42	·12
June . .	96	111	83	28	44	42	1·1	0·1	...	·26	·12
July . .	94	107	83	24	37	53	2·2	1·4	3	·24	·12
August . .	91	103	81	22	33	58	2·4	1·4	4	·33	·11
September .	88	101	76	25	39	55	0·9	0·3	...	·47	·11
October .	78	96	63	33	51	46	0·4	...	...	·67	·11
November .	65	84	50	34	52	45	0·9	0·1	...	·82	·11
December .	58	75	43	32	47	48	1·8	0·1	...	·90	·12
Year .	78					46	1·9	4·4	13	29·59	

Mean highest temperature of year 118°
,, lowest ,, ,, 32°
Mean annual range of temperature 86°
Highest recorded reading (9 yrs.) 120·9° (1882)
Lowest ,, ,, ,, 29·2° (1881)

Absolute range of temperature . 91·7°
Rainfall of wettest year (26 years) 12·1″ (1869)
,, driest ,, ,, 0·7″ (1881)
Rainy days in wettest year (9 years) 35 (1878)
,, ,, driest ,, ,, 5 (1880)

26. Hyderabad—Elevation 94 feet

	Temperature.					Hum.	Cloud.	Rainfall.		Barometer.	
	Mean.	Mean Max.	Mean Min.	M. Range. Daily.	M. Range. Month.	Mean.	Mean.	Ins.	Days.	Mean.	Daily Range.
January .	63	77	51	26	45	50	1·9	0·2	1	29·96	·12
February .	66	80	54	26	48	43	2·6	0·1	1	·90	·12
March . .	78	93	64	29	51	42	1·8	0·1	...	·78	·12
April . .	85	101	71	30	44	43	1·3	0·2	...	·66	·11
May . .	91	106	78	28	44	45	0·7	0·1	1	·53	·11
June . .	91	103	81	22	34	53	1·5	0·4	1	·37	·09
July . .	88	99	80	19	33	61	2·9	2·8	5	·36	·09
August . .	86	95	78	17	29	65	3·1	3·2	5	·44	·08
September .	86	97	77	20	31	63	1·7	0·8	2	·58	·09
October .	83	97	71	26	42	46	0·3	...	...	·75	·10
November .	72	87	59	28	46	48	0·6	0·1	...	·89	·10
December .	64	78	52	26	40	48	1·1	...	...	·97	·11
Year .	79					51	1·6	8·0	16	29·68	

Mean highest temperature of year 115°
,, lowest ,, ,, 40°
,, annual range of temperature 75°
Highest recorded reading (8 years) 121° (1886)
Lowest ,, ,, ,, 38·3° (1878)

Absolute range of temperature . 82·7°
Rainfall of wettest year (21 years) 20·2″ (1869)
,, driest ,, ,, 1·9″ (1868)
Rainy days in wettest year (9 years) 30 (1878)
,, ,, driest ,, , 5 (1880)

27. Kurrachee—Elevation 49 feet

	Temperature.					Hum.	Cloud.	Rainfall.		Barometer.	
	Mean.	Mean Max.	Mean Min.	M. Range. Daily.	M. Range. Month.	Mean.	Mean.	Ins.	Days.	Mean.	Daily Range.
January .	65	77	54	23	37	57	2·6	0·6	1	30·02	·11
February .	68	79	57	22	37	59	3·1	0·3	2	29·98	·10
March . .	76	86	67	19	39	67	3·3	0·2	1	·88	·10
April . .	80	89	72	17	32	68	2·3	0·2	1	·79	·09
May . .	85	93	79	14	29	74	2·2	0·1	...	·65	·08
June . .	87	93	82	11	20	74	4·5	0·2	1	·52	·07
July . .	84	90	81	9	18	78	7·1	3·1	6	·49	·07
August . .	82	88	79	9	16	78	6·8	1·8	4	·57	·07
September .	82	88	77	11	19	76	4·5	0·9	2	·69	·08
October .	80	91	72	19	35	65	1·0	0·1	...	·86	·09
November .	72	86	62	24	40	56	1·2	0·1	1	·96	·10
December .	67	79	56	23	35	57	2·1	0·2	...	30·03	·10
Year .	77					67	3·4	7·8	19	29·79	

Mean highest temperature of year 107°
" lowest " " 45°
" annual range of temperature 62°
Highest recorded reading (11 yrs.) 117·6° (1879)
Lowest " " (10 years) 41° (1886)

Absolute range of temperature . 76·6°
Rainfall of wettest year (31 years) 28·0″ (1869)
" driest " " 0·5″ (1871)
Rainy days in wettest year (12 years) 30 (1878)
" " driest " " 9 (1880)

28. Bickanir—Elevation 744 feet

	Temperature.					Hum.	Cloud.	Rainfall.		Barometer.	
	Mean.	Mean Max.	Mean Min.	M. Range. Daily.	M. Range. Month.	Mean.	Mean.	Ins.	Days.	Mean.	Daily Range.
January .	61	74	49	25	38	37	3·2	0·3	2	29·30	·10
February .	64	76	53	23	42	39	3·8	0·2	2	·23	·10
March . .	77	90	66	24	48	36	3·9	0·1	1	·12	·10
April . .	87	98	76	22	40	32	3·2	0·2	1	·00	·11
May . .	93	106	82	24	41	34	2·8	1·6	4	28·87	·10
June . .	95	106	86	20	35	42	2·8	1·8	4	·75	·10
July . .	90	100	83	17	33	55	5·6	3·9	8	·72	·09
August . .	87	97	80	17	27	61	6·0	2·7	7	·80	·10
September .	87	97	79	18	27	58	4·0	1·5	4	·92	·10
October .	84	96	73	23	36	41	1·5	0·1	1	29·10	·10
November .	71	83	59	24	40	35	0·8	0·1	1	·23	·10
December .	62	75	52	23	35	35	2·4	...	1	·29	·10
Year .	80					42	3·3	12·5	36	29·03	

Mean highest temperature of year 114°
" lowest " " 41°
" annual range of temperature 73°
Highest recorded reading (8 years) 117·4° (1879)
Lowest " " (7 years) 35° (1878)

Absolute range of temperature . 82·4°
Rainfall of wettest year (9 years) 17·9″ (1878)
" driest " " 7·7″ (1885)
Rainy days in wettest year (8 years) 43 (1879, 1881)
Rainy days in driest year (8 years) 30 (1883)

29. Jeypore—Elevation 1431 feet

	Temperature.					Hum.	Cloud.	Rainfall.		Barometer.	
	Mean.	Mean Max.	Mean Min.	M. Range. Daily.	M. Range. Month.	Mean.	Mean.	Ins.	Days.	Mean.	Daily Range.
January .	61	75	49	26	43	53	2·5	0·3	2	28·57	·10
February .	64	78	50	28	48	45	2·7	0·2	1	·54	·11
March . .	76	89	61	28	51	39	2·8	0·2	2	·45	·11
April . .	85	99	69	30	45	28	2·4	0·2	1	·35	·11
May . .	90	104	75	29	46	35	2·6	0·7	5	·25	·11
June . .	91	102	80	22	36	52	4·2	3·3	6	·13	·10
July . .	84	91	77	14	28	77	7·3	9·0	15	·12	·09
August . .	82	91	75	16	25	73	6·6	6·1	12	·19	·09
September .	83	92	73	19	30	67	4·1	3·1	8	·29	·10
October .	78	94	64	30	42	44	1·6	0·2	1	·46	·10
November .	69	84	52	32	46	42	0·7	0·1	1	·56	·10
December .	62	77	48	29	40	52	1·8	0·4	2	·60	·10
Year .	77					51	3·3	23·8	56	28·38	

Mean highest temperature of year 112°
,, lowest ,, ,, 38°
,, annual range of temperature 74°
Highest recorded reading (6 years) 115·8° (1886)
Lowest ,, ,, ,, 34·5° (1886)

Absolute range of temperature . 81·3°
Rainfall of wettest year (18 years) 35·9″ (1879)
,, driest ,, ,, 10·6″ (1868)
Rainy days in wettest year (10 years) 72 (1885)
,, ,, driest ,, ,, 33 (1880)

30. Ajmere—Elevation 1611 feet

	Temperature.					Hum.	Cloud.	Rainfall.		Barometer.	
	Mean.	Mean Max.	Mean Min.	M. Range. Daily.	M. Range. Month.	Mean.	Mean.	Ins.	Days.	Mean.	Daily Range.
January .	58	75	45	30	47	51	2·2	0·2	1	28·42	·10
February .	61	76	46	30	51	46	2·3	0·3	2	·37	·10
March . .	72	90	58	32	53	40	2·4	0·4	1	·29	·10
April . .	83	98	69	29	47	35	2·1	0·1	1	·19	·10
May . .	89	103	77	26	45	39	1·8	0·7	4	·08	·10
June . .	88	101	80	21	39	49	3·5	2·5	4	27·98	·09
July . .	82	93	77	16	30	69	6·3	6·9	13	·95	·08
August . .	80	89	75	14	25	74	6·1	7·3	11	28·02	·08
September .	81	91	72	19	31	68	3·5	3·4	7	·11	·09
October .	75	93	63	30	43	50	0·9	0·3	1	·30	·09
November .	66	84	49	35	49	45	0·8	0·1	1	·39	·09
December .	59	77	44	33	46	50	1·4	0·3	1	·44	·10
Year .	74					51	2·8	22·5	47	28·21	

Mean highest temperature of year 112°
,, lowest ,, ,, 34°
,, annual range of temperature 78°
Highest recorded reading (10 yrs.) 119·2° (1879)
Lowest ,, ,, ,, 30·1° (1878)

Absolute range of temperature . 89·1°
Rainfall of wettest year (24 years) 36·4″ (1875)
,, driest ,, ,, 9·3″ (1868)
Rainy days in wettest year (11 years) 55 (1875, 1881)
Rainy days in driest year (11 years) 35 (1877)

31. Deesa—Elevation 465 feet

	Temperature.					Hum.	Cloud.	Rainfall.		Barometer.	
				M. Range.							
	Mean.	Mean Max.	Mean Min.	Daily.	Month.	Mean.	Mean.	Ins.	Days.	Mean.	Daily Range.
January .	67	82	50	32	46	38	2·0	0·1	...	29·55	·12
February .	71	84	54	30	49	31	2·2	0·2	1	·50	·12
March . .	81	95	64	31	48	31	2·3	0·1	...	·43	·13
April . .	88	101	71	30	43	28	1·7	0·1	1	·33	·13
May . .	92	105	78	27	42	40	1·4	0·2	1	·23	·13
June . .	90	101	81	20	32	53	4·5	2·2	4	·13	·12
July . .	83	91	78	13	28	74	7·8	9·8	13	·10	·10
August . .	82	89	76	13	21	73	7·5	8·5	12	·18	·10
September .	82	91	74	17	27	69	5·2	3·3	8	·28	·11
October .	80	93	66	27	39	44	1·6	0·8	1	·42	·11
November .	74	89	57	32	43	35	1·0	0·1	...	·51	·11
December .	69	84	52	32	43	38	1·6	0·1	1	·55	·12
Year .	80					46	3·2	25·5	42	29·35	

Mean highest temperature of year 112°
,, lowest ,, ,, 40°
,, annual range of temperature 72°
Highest recorded reading (11 yrs.) 118·6° (1886)
Lowest ,, ,, ,, 34·1° (1880)

Absolute range of temperature . 84·5°
Rainfall of wettest year (28 years) 51·4″ (1884)
,, driest ,, ,, 11·7″ (1877)
Rainy days in wettest year (12 years) 65 (1881)
,, ,, driest ,, ,, 20 (1877)

32. Dehra—Elevation 2232 feet

	Temperature.					Hum.	Cloud.	Rainfall.		Barometer.	
				M. Range.							
	Mean.	Mean Max.	Mean Min.	Daily.	Month.	Mean.	Mean.	Ins.	Days.	Mean.	Daily Range.
January .	55	66	45	21	34	?	3·9	2·2	3	27·75	·08
February .	57	69	46	23	38	?	4·1	1·9	4	·70	·08
March . .	66	79	55	24	42	?	3·5	1·4	4	·64	·08
April . .	77	88	64	24	42	?	3·0	0·7	2	·55	·09
May . .	82	93	69	24	40	?	3·0	1·6	6	·47	·09
June . .	84	94	75	19	37	?	4·3	8·6	10	·35	·08
July . .	80	86	75	11	23	?	7·7	25·0	22	·35	·07
August . .	79	85	74	11	20	?	7·4	24·8	22	·41	·08
September .	78	85	71	14	23	?	4·8	9·3	13	·50	·09
October .	71	82	61	21	32	?	1·0	0·8	3	·65	·09
November .	62	74	51	23	34	?	1·1	0·1	1	·74	·09
December .	56	68	45	23	33	?	2·5	0·6	2	·77	·09
Year .	71					?	3·9	77·0	92	27·57	

Mean highest temperature of year 105°
,, lowest ,, ,, 36°
,, annual range of temperature 69°
Highest recorded reading (10 years) 106·8° (1876, 1879)
Lowest recorded reading (10 yrs.) 33·9° (1876)

Absolute range of temperature . 72·9°
Rainfall of wettest year (38 yrs.) 119·9″ (1885)
,, driest ,, ,, 35·1″ (1848)
Rainy days in wettest year (11 years) 114 (1875)
,, ,, driest ,, ,, 60 (1877)

33. Roorkee—Elevation 887 feet

	Temperature.					Hum.	Cloud.	Rainfall.		Barometer.	
	Mean.	Mean Max.	Mean Min.	M. Range. Daily.	M. Range. Month.	Mean.	Mean.	Ins.	Days.	Mean.	Daily Range.
January .	56	70	44	26	41	64	3·3	2·0	3	29·12	·10
February .	60	74	47	27	44	60	3·4	1·4	4	·06	·09
March . .	70	85	56	29	50	49	2·5	1·0	3	28·97	·10
April . .	82	97	66	31	49	34	2·3	0·4	1	·85	·11
May . .	88	101	73	28	45	36	1·9	1·2	3	·75	·11
June . .	90	102	79	23	42	51	3·3	5·1	7	·62	·11
July . .	85	93	78	15	28	76	6·4	12·5	15	·63	·09
August . .	84	92	77	15	25	76	6·2	12·3	14	·69	·10
September .	83	93	74	19	30	72	4·2	5·1	7	·80	·11
October .	75	89	62	27	43	59	0·8	0·6	1	·96	·11
November .	64	81	49	34	44	56	0·8	0·2	1	29·09	·10
December .	57	72	43	29	40	62	2·2	0·4	2	·14	·10
Year .	75					58	3·1	42·2	61	28·89	

Mean highest temperature of year 113°
,, lowest ,, ,, 35°
,, annual range of temperature 78°
Highest recorded reading (10 yrs.) 115·5° (1878)
Lowest ,, ,, ,, 30·9° (1879)

Absolute range of temperature . 84·6°
Rainfall of wettest year (33 years) 74·3″ (1853)
,, driest ,, ,, 24·1″ (1877)
Rainy days in wettest year (18 years) 88 (1874)
,, ,, driest ,, ,, 26 (1877)

34. Meerut—Elevation 737 feet

	Temperature.					Hum.	Cloud.	Rainfall.		Barometer.	
	Mean.	Mean Max.	Mean Min.	M. Range. Daily.	M. Range. Month.	Mean.	Mean.	Ins.	Days.	Mean.	Daily Range.
January .	57	71	44	27	42	63	2·9	1·0	2	29·28	·10
February .	62	75	47	28	45	57	2·7	0·7	2	·23	·09
March . .	73	88	58	30	52	50	2·6	0·7	2	·12	·10
April . .	83	98	68	30	45	40	2·1	0·4	1	·00	·10
May . .	89	101	74	27	43	41	1·9	0·8	3	28·89	·11
June . .	92	102	82	20	38	48	3·2	3·6	5	·76	·11
July . .	86	94	80	14	27	71	6·7	9·2	11	·77	·09
August . .	85	93	79	14	24	73	6·3	7·2	11	·83	·10
September .	83	93	76	17	28	68	4·1	4·0	6	·94	·10
October .	76	91	63	28	42	56	0·7	0·5	1	29·12	·10
November .	65	82	50	32	45	54	0·6	0·1	...	·25	·10
December .	58	73	44	29	40	62	1·7	0·3	1	·30	·10
Year .	76					57	2·9	28·5	45	29·04	

Mean highest temperature of year 112°
,, lowest ,, ,, 37°
,, annual range of temperature 75°
Highest recorded reading (9 years) 115° (1880)
Lowest ,, ,, ,, 33·5° (1878)

Absolute range of temperature . 81·5°
Rainfall of wettest year (39 years) 45·3″ (1880)
,, driest ,, ,, 15·6″ (1868)
Rainy days in wettest year (11 yrs.) 62 (1884)
,, ,, driest ,, ,, 29 (1876, 1877)

35. AGRA—Elevation 555 feet

	Temperature.					Hum.	Cloud.	Rainfall.		Barometer.	
	Mean.	Mean Max.	Mean Min.	M. Range. Daily.	M. Range. Month.	Mean.	Mean.	Ins.	Days.	Mean.	Daily Range.
January .	60	74	48	26	40	56	2·0	0·5	2	29·49	·11
February .	65	78	52	26	46	47	2·4	0·3	1	·43	·11
March . .	77	92	63	29	51	38	1·9	0·2	2	·33	·12
April . .	88	101	73	28	44	29	1·5	0·2	1	·19	·12
May . .	94	106	80	26	42	32	1·3	0·7	3	·08	·12
June . .	95	105	85	20	38	44	3·6	2·9	5	28·96	·11
July . .	87	94	81	13	26	72	6·4	9·8	13	·96	·10
August. .	85	93	80	13	24	73	6·3	6·7	10	29·03	·10
September .	84	94	77	17	27	67	3·9	4·3	7	·14	·11
October .	80	93	69	24	40	48	0·8	0·4	1	·32	·11
November .	70	84	55	29	44	43	0·5	...	...	·44	·11
December .	62	75	48	27	39	53	1·3	0·2	1	·50	·11
Year .	79					50	2·7	26·2	46	29·24	

Mean highest temperature of year 116°
„ lowest „ „ 40°
„ annual range of temperature 76°
Highest recorded reading (10 yrs.) 120·3° (1878)
Lowest „ „ „ 36·3° (1878)

Absolute range of temperature . 84°
Rainfall of wettest year (40 years) 46·5″ (1873)
„ driest „ „ 10·0″ (1877)
Rainy days in wettest year (14 yrs.) 61 (1879)
„ „ driest „ „ 19 (1877)

36. LUCKNOW—Elevation 369 feet

	Temperature.					Hum.	Cloud.	Rainfall.		Barometer.	
	Mean.	Mean Max.	Mean Min.	M. Range. Daily.	M. Range. Month.	Mean.	Mean.	Ins.	Days.	Mean.	Daily Range.
January .	61	74	46	28	42	59	2·7	0·8	2	29·68	·11
February .	66	78	50	28	47	52	3·0	0·3	2	·61	·11
March . .	77	91	61	30	52	43	2·7	0·3	1	·50	·11
April . .	87	101	71	30	47	35	2·1	0·1	1	·37	·11
May . .	92	104	76	28	44	42	2·0	0·9	2	·27	·12
June . .	92	103	82	21	38	54	4·7	5·0	7	·15	·11
July . .	86	93	80	13	26	74	7·7	10·8	16	·15	·10
August. .	85	92	79	13	23	77	7·4	10·4	15	·22	·10
September .	85	93	77	16	27	72	5·4	7·1	9	·31	·11
October .	79	91	65	26	42	60	1·5	1·4	1	·50	·11
November .	69	83	52	31	44	53	0·8	...	...	·63	·11
December .	61	75	45	30	41	58	1·8	0·5	1	·69	·11
Year .	78					57	3·5	37·6	57	29·42	

Mean highest temperature of year 114°
„ lowest „ „ 38°
„ annual range of temperature 76°
Highest recorded reading (10 yrs.) 117·3° (1878)
Lowest „ „ „ 34·4° (1878)

Absolute range of temperature . 82·9°
Rainfall of wettest year (20 years) 64·9″ (1871)
„ driest „ „ 14·4″ (1877)
Rainy days in wettest year (14 years) 77 (1874)
„ „ driest „ „ 27 (1876)

37. Allahabad—Elevation 307 feet

	Temperature.					Hum.	Cloud.	Rainfall.		Barometer.	
	Mean.	Mean Max.	Mean Min.	M. Range. Daily.	M. Range. Month.	Mean.	Mean.	Ins.	Days.	Mean.	Daily Range.
January .	61	74	48	26	40	68	1·8	0·8	2	29·74	·12
February .	66	80	51	29	47	58	2·1	0·4	2	·67	·12
March . .	78	93	62	31	51	43	1·8	0·4	1	·56	·13
April . .	88	104	72	32	47	33	1·4	0·2	1	·43	·13
May . .	92	106	78	28	44	40	1·6	0·3	2	·33	·12
June . .	91	104	83	21	38	55	4·5	4·6	7	·20	·11
July . .	85	93	80	13	25	81	7·6	11·9	17	·21	·10
August . .	84	91	79	12	21	82	7·3	9·6	18	·28	·11
September .	83	92	77	15	24	80	5·2	6·7	11	·38	·12
October .	78	89	67	22	38	69	1·6	2·3	3	·56	·11
November .	68	82	54	28	41	64	0·9	0·2	...	·68	·11
December .	61	74	47	27	39	69	1·6	0·2	1	·75	·11
Year .	78					62	3·1	37·6	65	29·48	

Mean highest temperature of year 116°
,, lowest ,, ,, 40°
,, annual range of temperature 76°
Highest recorded reading (10 yrs.) 117·2° (1884)
Lowest ,, ,, ,, 36° (1878)

Absolute range of temperature . 81·2°
Rainfall of wettest year (39 years) 61·0″ (1854)
,, driest ,, ,, 15·7″ (1864)
Rainy days in wettest year (14 years) 80 (1879)
,, ,, driest ,, ,, 31 (1877)

38. Benares—Elevation 267 feet

	Temperature.					Hum.	Cloud.	Rainfall.		Barometer.	
	Mean.	Mean Max.	Mean Min.	M. Range. Daily.	M. Range. Month.	Mean.	Mean.	Ins.	Days.	Mean.	Daily Range.
January .	61	75	48	27	41	62	2·1	0·7	2	29·78	·12
February .	66	81	51	30	47	54	2·3	0·5	2	·71	·12
March . .	77	93	62	31	51	41	2·1	0·4	2	·61	·13
April . .	87	103	72	31	45	34	1·7	0·2	1	·48	·13
May . .	91	105	79	26	42	40	1·9	0·5	2	·37	·12
June . .	91	102	83	19	35	58	5·0	5·0	7	·25	·11
July . .	85	93	80	13	25	80	7·6	12·8	19	·26	·10
August .	84	91	78	13	21	82	7·4	10·7	17	·31	·10
September .	83	92	78	14	25	78	5·5	6·5	12	·42	·11
October .	78	90	68	22	38	66	2·3	2·1	3	·59	·11
November .	68	83	55	28	41	60	1·1	0·1	...	·73	·11
December .	61	75	48	27	39	63	1·6	0·1	1	·80	·11
Year .	78					60	3·4	39·6	68	29·53	

Mean highest temperature of year 114°
,, lowest ,, ,, 40°
,, annual range of temperature 74°
Highest recorded reading (10 yrs.) 117·3° (1878, 1884)
Lowest ,, ,, ,, 36·4° (1878)

Absolute range of temperature 80·9°
Rainfall of wettest year (39 years) 65·3″ (1874)
,, driest ,, ,, 19·4″ (1864)
Rainy days in wettest year (18 years) 91 (1874)
,, ,, driest ,, ,, 43 (1868)

39. Neemuch—Elevation 1639 feet

	Temperature.					Hum.	Cloud.	Rainfall.		Barometer.	
				M. Range.							
	Mean.	Mean Max.	Mean Min.	Daily.	Month.	Mean.	Mean.	Ins.	Days.	Mean.	Daily Range.
January .	62	78	49	29	45	40	2·2	0·1	1	28·37	·12
February .	65	80	51	29	49	35	2·9	0·2	1	·32	·12
March . .	76	92	62	30	50	30	2·5	0·1	1	·25	·12
April . .	84	99	70	29	42	28	2·1	0·1	...	·16	·12
May . .	88	103	76	27	42	31	2·2	0·5	3	·06	·11
June . .	86	98	77	21	38	50	5·0	3·9	8	27·96	·10
July . .	79	87	74	13	26	74	8·1	11·2	15	·94	·08
August . .	78	85	72	13	20	76	7·8	10·4	14	28·01	·09
September .	78	87	71	16	26	72	5·9	5·5	11	·09	·10
October .	76	90	65	25	37	47	2·1	1·0	2	·26	·10
November .	68	83	54	29	42	35	0·7	...	...	·34	·11
December .	63	77	49	28	41	41	1·7	0·2	1	·38	·11
Year .	75					47	3·6	33·2	57	28·18	

Mean highest temperature of year 111°
" lowest " " 38°
" annual range of temperature 73°
Highest recorded reading (9 yrs.) 113·8° (1886)
Lowest " " " 31·2° (1886)

Absolute range of temperature . 82·6°
Rainfall of wettest year (20 years) 48·6″ (1864)
" driest " " 16·7″ (1868)
Rainy days in wettest year (8 years) 68 (1884)
" " driest " " 45 (1880)

40. Indore—Elevation 1823 feet

	Temperature.					Hum.	Cloud.	Rainfall.		Barometer.	
				M. Range.							
	Mean.	Mean Max.	Mean Min.	Daily.	Month.	Mean.	Mean.	Ins.	Days.	Mean.	Daily Range.
January .	64	79	50	29	43	44	1·9	0·4	1	28·17	·12
February .	67	82	52	30	47	37	1·8	0·3	1	·13	·12
March . .	76	93	60	33	50	29	2·0	...	...	·07	·13
April . .	83	100	69	31	44	24	1·8	0·1	...	27·98	·12
May . .	87	102	75	27	39	32	2·2	0·6	3	·90	·12
June . .	83	96	75	21	33	60	6·3	6·8	11	·81	·11
July . .	77	84	72	12	23	80	8·4	10·4	20	·79	·09
August . .	76	83	71	12	19	81	8·4	7·8	19	·85	·09
September .	76	84	71	13	24	79	6·9	8·1	15	·93	·10
October .	74	87	64	23	35	56	2·9	1·2	4	28·07	·10
November .	66	81	54	27	40	43	1·4	0·2	1	·15	·11
December .	62	78	49	29	40	45	1·7	0·2	1	·18	·11
Year .	74					51	3·8	36·1	76	28·00	

Mean highest temperature of year 108°
" lowest " " 40°
" annual range of temperature 68°
Highest recorded reading (9 yrs.) 109·8° (1881)
Lowest " " " 36·2° (1886)

Absolute range of temperature . 73·6°
Rainfall of wettest year (19 years) 52·5″ (1870)
" driest " " 25·1″ (1877)
Rainy days in wettest year (8 years) 84 (1881)
" " driest " " 62 (1880)

41. Jhansi—Elevation 855 feet

	Temperature.					Hum.	Cloud.	Rainfall.		Barometer.	
				M. Range.							
	Mean.	Mean Max.	Mean Min.	Daily.	Month.	Mean.	Mean.	Ins.	Days.	Mean.	Daily Range.
January .	63	76	51	25	40	49	0·7	0·5	1	29·17	·10
February .	68	80	54	26	45	41	0·7	0·2	1	·12	·10
March . .	79	93	65	28	49	34	0·3	0·4	1	·03	·11
April . .	89	102	73	29	41	29	0·4	0·1	1	28·90	·12
May . .	95	107	81	26	41	31	0·6	0·3	2	·79	·11
June . .	93	104	84	20	38	47	1·7	4·0	7	·68	·10
July . .	84	92	79	13	37	74	4·3	13·6	15	·67	·09
August . .	83	90	78	12	22	76	3·9	10·5	14	·74	·09
September .	83	91	76	15	27	70	2·0	5·2	9	·84	·09
October .	81	92	68	24	36	48	0·3	0·8	1	29·01	·09
November .	73	84	57	27	41	39	0·3	...	...	·12	·09
December .	65	77	50	27	39	45	0·3	0·2	1	·18	·09
Year .	79					49	1·3	35·8	53	28·94	

Mean highest temperature of year 115°
,, lowest ,, ,, 43°
,, annual range of temperature 72°
Highest recorded reading (10 yrs.) 117·9° (1878)
Lowest ,, ,, ,, 40·7° (1878)

Absolute range of temperature . 77·2°
Rainfall of wettest year (27 yrs.) 53·8″ (1881)
,, driest ,, ,, 13·3″ (1868)
Rainy days in wettest year (11 yrs.) 75 (1884-1885)
Rainy days in driest year (11 yrs.) 24 (1877)

42. Saugor—Elevation 1769 feet

	Temperature.					Hum.	Cloud.	Rainfall.		Barometer.	
				M. Range.							
	Mean.	Mean Max.	Mean Min.	Daily.	Month.	Mean	Mean.	Ins.	Days.	Mean.	Daily Range.
January .	63	77	51	26	40	46	1·4	0·6	2	28·24	·10
February .	67	81	54	27	43	39	1·5	0·5	1	·18	·11
March . .	78	93	64	29	41	32	1·3	0·2	...	·11	·11
April . .	85	101	72	29	39	28	1·4	0·2	1	·02	·11
May . .	89	104	76	28	39	28	2·0	0·6	3	27·92	·11
June . .	85	100	77	23	37	50	4·1	6·3	11	·82	·10
July . .	78	86	74	12	24	77	6·1	16·8	19	·80	·09
August . .	77	85	73	12	20	83	5·9	11·2	18	·86	·09
September .	77	87	71	16	26	78	4·3	7·3	12	·95	·10
October .	75	87	66	21	33	55	1·4	1·3	2	28·12	·10
November .	69	82	56	26	36	42	1·0	0·4	...	·20	·10
December .	64	76	51	25	38	45	1·0	0·7	1	·24	·10
Year .	76					50	2·6	46·1	70	28·04	

Mean highest temperature of year 110°
,, lowest ,, ,, 42°
,, annual range of temperature 68°
Highest recorded reading (8 yrs.) 111·4° (1881)
Lowest ,, ,, ,, 40·1° (1883)

Absolute range of temperature . 71·3°
Rainfall of wettest year (31 years) 70·1″ (1867)
,, driest ,, ,, 22·1″ (1848)
Rainy days in wettest year (11 yrs.) 106 (1884)
,, ,, driest ,, ,, 56 (1879, 1880)

43. Sutna—Elevation 1040 feet

	Temperature.					Hum.	Cloud.	Rainfall.		Barometer.	
				M. Range.							
	Mean.	Mean Max.	Mean Min.	Daily.	Month.	Mean.	Mean.	Ins.	Days.	Mean.	Daily Range.
January .	61	75	47	28	44	49	2·1	0·7	2	28·98	·12
February .	65	79	51	28	47	39	2·3	0·5	2	·93	·12
March . .	76	92	61	31	50	29	2·1	0·3	2	·83	·12
April . .	85	100	71	29	43	23	2·2	0·1	2	·71	·13
May . .	90	104	77	27	41	28	2·8	0·4	4	·60	·12
June . .	89	99	81	18	36	49	5·1	5·8	10	·49	·11
July . .	82	87	77	10	25	76	7·1	15·4	18	·48	·09
August . .	81	86	76	10	20	77	6·8	11·4	16	·55	·10
September .	81	87	75	12	23	72	5·3	5·1	11	·65	·11
October .	76	87	65	22	38	55	2·2	1·9	2	·83	·10
November .	66	81	52	29	43	44	1·2	0·2	1	·94	·11
December .	60	75	45	30	42	47	1·6	0·4	1	·99	·11
Year .	76					49	3·4	42·2	71	28·75	

Mean highest temperature of year 111°
,, lowest ,, ,, 38°
,, annual range of temperature 73°
Highest recorded reading (10 yrs.) 115·0° (1886)
Lowest ,, ,, ,, 34·2° (1878)

Absolute range of temperature . 80·8°
Rainfall of wettest year (15 years) 63·7″ (1882)
,, driest ,, ,, 21·8″ (1880)
Rainy days in wettest year (10 years) 86 (1885)
,, ,, driest ,, ,, 57 (1880)

44. Jubbulpore—Elevation 1341 feet

	Temperature.					Hum.	Cloud.	Rainfall.		Barometer.	
				M. Range.							
	Mean.	Mean Max.	Mean Min.	Daily.	Month.	Mean.	Mean.	Ins.	Days.	Mean.	Daily Range.
January .	62	78	48	30	46	59	2·2	0·6	2	28·67	·12
February .	66	82	52	30	50	50	2·3	0·5	1	·62	·12
March . .	76	93	62	31	51	38	2·4	0·5	1	·54	·13
April . .	85	101	71	30	43	29	2·2	0·2	1	·44	·14
May . .	90	104	78	26	41	30	3·3	0·5	4	·34	·13
June . .	86	97	79	18	37	59	6·5	8·5	13	·23	·11
July . .	79	86	75	11	23	81	8·7	18·6	22	·22	·09
August . .	78	85	74	11	19	82	8·2	13·8	19	·29	·10
September .	79	87	73	14	23	79	6·4	8·3	12	·36	·11
October .	74	87	64	23	39	65	2·7	1·5	3	·53	·11
November .	66	81	52	29	44	56	1·7	0·3	1	·65	·11
December .	61	76	46	30	45	58	1·9	0·3	1	·68	·12
Year .	75					57	4·0	53·6	80	28·46	

Mean highest temperature of year 111°
,, lowest ,, ,, 36°
,, annual range of temperature 75°
Highest recorded reading (10 yrs.) 112·5° (1883)
Lowest ,, ,, ,, 32° (1879

Absolute range of temperature . 80·5°
Rainfall of wettest year (43 yrs.) 94·8″ (1884)
,, driest ,, ,, 28·8″ (1868)
Rainy days in wettest year (13 yrs.) 107 (1885)
,, driest ,, ,, 69 (1873)

45. PATNA—Elevation 183 feet

	TEMPERATURE.					HUM.	CLOUD.	RAINFALL.		BAROMETER.	
				M. Range.							
	Mean.	Mean Max.	Mean Min.	Daily.	Month.	Mean.	Mean.	Ins.	Days.	Mean.	Daily Range.
January .	61	73	49	24	36	67	2·4	0·7	2	29·87	·13
February .	66	78	52	26	43	58	2·6	0·5	2	·80	·13
March . .	78	92	64	28	47	46	2·9	0·3	2	·68	·13
April . .	87	101	73	28	42	41	2·2	0·3	1	·56	·14
May . .	89	100	77	23	41	55	2·8	1·6	4	·47	·14
June . .	88	98	81	17	32	68	6·2	7·1	10	·35	·12
July . .	85	91	80	11	23	81	8·0	11·0	17	·36	·10
August . .	84	90	80	10	19	82	7·9	10·1	16	·42	·11
September .	84	91	79	12	21	80	6·9	7·9	12	·52	·12
October .	80	88	72	16	31	70	3·6	2·9	4	·68	·11
November .	70	82	59	23	37	61	1·7	0·2	...	·82	·12
December .	62	74	50	24	35	65	2·0	0·2	1	·88	·12
Year .	78					65	4·1	42·8	71	29·62	

Mean highest temperature of year 110°
,, lowest ,, ,, 42°
,, annual range of temperature 68°
Highest recorded reading (9 yrs.) 112·4° (1884)
Lowest ,, ,, (10 yrs.) 36·4° (1878)

Absolute range of temperature . 76°
Rainfall of wettest year (35 yrs.) 65·5″ (1886)
,, driest ,, ,, 21·8″ (1843)
Rainy days in wettest year (18 yrs.) 87 (1880)
,, ,, driest ,, ,, 51 (1882)

46. DURBHANGA—Elevation 166 feet

	TEMPERATURE.					HUM.	CLOUD.	RAINFALL.		BAROMETER.	
				M. Range.							
	Mean.	Mean Max.	Mean Min.	Daily.	Month.	Mean.	Mean.	Ins.	Days.	Mean.	Daily Range.
January .	62	72	52	20	30	71	1·3	0·4	1	29·88	·13
February	65	76	53	23	36	63	1·8	0·5	2	·81	·13
March . .	76	87	63	24	41	54	1·6	0·3	1	·69	·13
April . .	84	96	71	25	39	50	1·3	0·6	2	·57	·14
May . .	85	95	75	20	36	65	2·1	2·3	5	·51	·13
June . .	85	92	79	13	27	77	4·8	7·8	10	·38	·11
July . .	84	89	80	9	18	83	6·4	11·8	14	·38	·10
August . .	84	89	79	10	20	84	6·2	10·7	16	·44	·11
September .	84	88	79	9	19	83	5·7	9·9	11	·54	·12
October .	79	86	73	13	24	77	3·2	2·8	3	·71	·12
November .	71	81	62	19	31	69	0·9	0·1	...	·82	·12
December .	63	74	54	20	31	75	1·2	0·2	1	·89	·12
Year .	77					71	3·0	47·4	66	29·64	

Mean highest temperature of year 105°
,, lowest ,, ,, 45°
,, annual range of temperature 60°
Highest recorded reading (8 years) 107·4° (1879)
Lowest ,, ,, (9 years) 38·3° (1878)
Absolute range of temperature . 69·1°

Rainfall of wettest year (16 years) 78·7″ (1871)
,, driest ,, ,, 23·9″ (1873)
Rainy days in wettest year (15 years) 86 (1871)
,, ,, driest ,, ,, 50 (1872, 1873)

47. PURNEAH—Elevation 125 feet

	Temperature.					Hum.	Cloud.	Rainfall.		Barometer.	
				M. Range.							
	Mean.	Mean Max.	Mean Min.	Daily.	Month.	Mean.	Mean.	Ins.	Days.	Mean.	Daily Range.
January .	62	75	48	27	39	67	1·3	0·3	1	29·91	·13
February .	66	80	51	29	47	61	1·3	0·6	2	·84	·13
March . .	76	91	61	30	48	55	1·6	0·4	1	·73	·14
April . .	84	98	69	29	44	55	1·6	1·3	3	·61	·14
May . .	83	95	73	22	40	69	3·5	3·3	6	·56	·13
June . .	84	93	77	16	30	79	6·0	11·2	13	·44	·11
July . .	84	91	79	12	20	85	7·2	14·3	21	·43	·10
August . .	84	90	79	11	21	85	7·2	14·5	21	·48	·11
September .	83	90	78	12	22	83	6·1	14·4	14	·59	·12
October .	79	88	71	17	31	77	3·4	4·0	5	·74	·12
November .	71	83	59	24	36	69	1·0	0·1	...	·84	·12
December .	63	77	50	27	38	70	1·2	0·2	1	·91	·13
Year .	77					71	3·5	64·6	88	29·67	

Mean highest temperature of year 105°
,, lowest ,, ,, 40°
,, annual range of temperature 65°
Highest recorded reading (9 years) 109·7° (1879)
Lowest ,, ,, (10 years) 36·9° (1878)

Absolute range of temperature . 72·8°
Rainfall of wettest year (16 years) 131·4″ (1879)
,, driest ,, ,, 38·4″ (1883)
Rainy days in wettest year (15 years) 112 (1874)
,, ,, driest ,, ,, 68 (1872)

48. GYA—Elevation 375 feet

	Temperature.					Hum.	Cloud.	Rainfall.		Barometer.	
				M. Range.							
	Mean.	Mean Max.	Mean Min.	Daily.	Month.	Mean.	Mean.	Ins.	Days.	Mean.	Daily Range.
January .	64	76	51	25	38	54	1·8	0·7	2	29·68	·11
February .	69	81	55	26	43	45	2·4	0·5	2	·60	·11
March . .	81	94	66	28	46	37	2·2	0·4	2	·50	·12
April . .	90	103	74	29	43	37	1·6	0·3	1	·38	·13
May . .	92	104	78	26	42	49	2·5	1·2	4	·29	·13
June . .	89	100	80	20	36	63	6·6	6·2	10	·17	·11
July . .	84	92	79	13	24	79	8·4	11·9	19	·18	·10
August . .	84	91	79	12	20	81	8·5	9·9	17	·24	·10
September .	84	91	78	13	22	78	7·0	7·0	12	·33	·11
October .	80	89	71	18	32	65	3·4	2·5	4	·50	·10
November .	72	83	59	24	38	55	1·7	0·2	1	·63	·10
December .	64	76	51	25	36	57	1·8	0·2	1	·69	·11
Year .	79					58	4·0	41·0	75	29·43	

Mean highest temperature of year 113°
,, lowest ,, ,, 43°
,, annual range of temperature 70°
Highest recorded reading (11 yrs.) 116·2° (1878)
Lowest ,, ,, ,, 39·6° (1878)

Absolute range of temperature . 76·6°
Rainfall of wettest year (27 years) 67·4″ (1886)
,, driest ,, ,, 20·6″ (1854)
Rainy days in wettest year (16 yrs.) 99 (1885)
,, ,, driest ,, ,, 54 (1873)

49. Hazaribagh—Elevation 2007 feet

	Temperature.					Hum.	Cloud.	Rainfall.		Barometer.	
	Mean.	Mean Max.	Mean Min.	M. Range. Daily.	M. Range. Month.	Mean.	Mean.	Ins.	Days.	Mean.	Daily Range.
January .	61	73	51	22	36	50	2·2	0·4	2	27·98	·10
February .	65	77	54	23	39	43	2·2	0·8	3	·93	·10
March . .	75	88	64	24	43	35	2·5	0·7	3	·87	·10
April . .	84	97	72	25	39	30	2·6	0·4	2	·77	·11
May . .	85	98	74	24	39	42	3·3	1·6	7	·69	·10
June . .	82	92	76	16	32	65	7·3	8·3	15	·58	·08
July . .	78	85	74	11	20	85	9·0	12·6	23	·57	·08
August . .	78	84	73	11	18	85	8·6	12·7	22	·62	·09
September .	78	85	72	13	20	82	7·7	8·0	17	·71	·10
October .	74	83	66	17	28	65	4·1	3·4	7	·86	·09
November .	67	77	57	20	31	53	2·1	0·3	1	·97	·09
December .	61	71	50	21	33	49	2·0	0·2	1	28·00	·10
Year .	74					57	4·5	49·4	103	27·80	

Mean highest temperature of year 106°
,, lowest ,, ,, 43°
,, annual range of temperature 63°
Highest recorded reading (10 yrs.) 109·3° (1884)
Lowest ,, ,, ,, 39·2° (1878)

Absolute range of temperature . 70·1°
Rainfall of wettest year (24 years) 62·2″ (1885)
,, driest ,, ,, 35·1″ (1872)
Rainy days in wettest year (19 yrs.) 134 (1874)
,, ,, driest ,, ,, 73 (1869)

50. Berhampore—Elevation 66 feet

	Temperature.					Hum.	Cloud.	Rainfall.		Barometer.	
	Mean.	Mean Max.	Mean Min.	M. Range. Daily.	M. Range. Month.	Mean.	Mean.	Ins.	Days.	Mean.	Daily Range.
January .	65	78	53	25	36	69	2·1	0·4	1	29·97	·13
February .	69	82	56	26	42	61	2·1	1·0	2	·90	·14
March . .	78	94	66	28	47	57	2·8	1·0	2	·79	·15
April . .	85	101	74	27	42	59	3·2	1·9	4	·68	·15
May . .	85	96	76	20	40	71	5·3	4·8	9	·60	·14
June . .	84	93	78	15	27	83	8·2	9·7	14	·48	·11
July . .	83	90	79	11	20	86	8·8	10·3	18	·49	·10
August . .	83	89	79	10	18	87	8·8	10·8	19	·55	·11
September .	83	89	78	11	20	86	8·2	9·8	15	·64	·12
October .	80	88	74	14	24	78	4·8	5·3	6	·79	·12
November .	73	82	64	18	32	73	2·5	0·3	...	·91	·13
December .	66	77	55	22	33	72	2·0	0·1	1	·97	·12
Year .	78					74	4·9	55·4	91	29·73	

Mean highest temperature of year 109°
,, lowest ,, ,, 46°
,, annual range of temperature 63°
Highest recorded reading (11 yrs.) 112·4° (1876)
Lowest ,, ,, ,, 43·1° (1876, 1886)

Absolute range of temperature . 69·3°
Rainfall of wettest year (31 years) 73″ (1861)
,, driest ,, ,, 31·5″ (1873)
Rainy days in wettest year (18 yrs.) 113 (1871)
,, ,, driest ,, ,, 72 (1874)

51. BURDWAN—Elevation 99 feet

	Temperature.					Hum.	Cloud.	Rainfall.		Barometer.	
				M. Range.							
	Mean.	Mean Max.	Mean Min.	Daily.	Month.	Mean.	Mean.	Ins.	Days.	Mean.	Daily Range.
January .	66	79	55	24	37	64	1·7	0·3	1	29·95	·14
February .	70	83	58	25	42	58	2·1	0·8	3	·88	·14
March . .	80	94	69	25	44	55	2·2	1·2	3	·76	·14
April . .	86	100	75	25	39	55	2·5	2·4	5	·65	·15
May . .	85	97	77	20	36	67	4·0	5·0	10	·58	·13
June . .	84	94	78	16	29	78	7·4	9·8	16	·47	·11
July . .	84	90	79	11	19	83	8·2	12·3	20	·46	·10
August . .	83	89	79	10	17	84	8·2	12·4	23	·52	·11
September .	83	90	79	11	18	83	7·2	8·2	16	·61	·12
October .	80	88	75	13	25	76	4·3	4·6	6	·76	·11
November .	73	83	64	19	31	66	2·6	0·5	1	·88	·12
December .	66	77	56	21	34	64	2·0	0·2	1	·94	·13
Year .	78					69	4·4	57·7	105	29·71	

Mean highest temperature of year 108°
,, lowest ,, ,, 47°
,, annual range of temperature 61°
Highest recorded reading (11 yrs.) 110·7° (1879, 1885)
Lowest recorded reading ,, 44·2° (1878, 1886)

Absolute range of temperature . 66·5°
Rainfall of wettest year (26 years) 99·3″ (1861)
,, driest ,, ,, 40·5″ (1870)
Rainy days in wettest year (16 yrs.) 131 (1880)
,, ,, driest ,, ,, 80 (1873)

52. CALCUTTA—Elevation 21 feet

	Temperature.					Hum.	Cloud.	Rainfall.		Barometer.	
				M. Range.							
	Mean.	Mean Max.	Mean Min.	Daily.	Month.	Mean.	Mean.	Ins.	Days.	Mean.	Daily Range.
January .	65	77	55	22	33	71	1·4	0·4	2	30·02	·13
February .	70	82	61	21	37	69	2·2	1·0	3	29·95	·13
March . .	79	91	70	21	38	69	2·2	1·3	3	·86	·14
April . .	85	96	76	20	33	71	2·8	2·3	5	·75	·14
May . .	85	94	77	17	30	76	4·4	5·6	10	·66	·12
June . .	84	91	79	12	25	84	7·0	11·8	18	·55	·10
July . .	83	88	78	10	17	87	8·1	13·0	24	·54	·09
August .	82	87	78	9	16	89	8·2	13·9	24	·60	·10
September .	82	87	78	9	17	88	7·1	10·0	18	·68	·11
October .	80	87	75	12	23	83	4·2	5·4	8	·83	·11
November .	72	82	64	18	29	74	2·4	0·6	2	·96	·12
December .	65	76	56	20	31	72	2·0	0·3	1	30·02	·13
Year .	78					78	4·3	65·6	118	29·78	

Mean highest temperature of year 102°
,, lowest ,, ,, 48°
,, annual range of temperature 54°
Highest recorded reading (8 years) 105·3° (1879, 1885)
Lowest recorded reading (8 years) 45° (1878)

Absolute range of temperature . 60·3°
Rainfall of wettest year (57 years) 93·3″ (1871)
,, driest ,, ,, 43·6″ (1837)
Rainy days in wettest year (34 yrs.) 154 (1861)
,, ,, driest ,, ,, 72 (1853)

53. JESSORE—Elevation 33 feet

	Temperature.					Hum.	Cloud.	Rainfall.		Barometer.	
	Mean.	Mean Max.	Mean Min.	M. Range. Daily.	M. Range. Month.	Mean.	Mean.	Ins.	Days.	Mean.	Daily Range.
January .	65	78	53	25	37	68	1·7	0·5	1	30·00	·12
February .	70	83	57	26	45	63	2·1	0·9	3	29·94	·13
March . .	79	93	69	24	44	62	2·8	2·0	5	·84	·14
April . .	83	97	75	22	39	67	3·5	4·7	8	·74	·14
May . .	84	94	76	18	32	76	5·1	7·5	12	·67	·13
June . .	83	91	79	12	23	84	7·8	13·5	18	·55	·10
July . .	83	89	79	10	18	87	8·4	11·5	23	·54	·10
August .	82	89	79	10	17	87	7·9	11·6	23	·60	·10
September .	82	89	78	11	19	86	7·5	9·1	17	·69	·11
October .	80	89	75	14	25	81	4·7	5·6	8	·83	·11
November .	73	83	64	19	33	72	2·4	0·9	2	·95	·11
December .	66	78	55	23	35	70	1·7	0·2	1	30·00	·12
Year .	77					75	4·6	68·0	121	29·78	

Mean highest temperature of year 104°
,, lowest ,, ,, 44°
,, annual range of temperature 60°
Highest recorded reading (10 yrs.) 107·5° (1879)
Lowest ,, ,, ,, 39·4° (1884)

Absolute range of temperature . 68·1°
Rainfall of wettest year (26 yrs.) 100·4″ (1861)
,, driest ,, ,, 40·2″ (1851)
Rainy days in wettest year (18 yrs.) 131 (1880)
,, ,, driest ,, ,, 90 (1869)

54. DACCA—Elevation 22 feet

	Temperature.					Hum.	Cloud.	Rainfall.		Barometer.	
	Mean.	Mean Max.	Mean Min.	M. Range. Daily.	M. Range. Month.	Mean.	Mean.	Ins.	Days.	Mean.	Daily Range.
January .	66	78	55	23	34	69	1·6	0·3	1	30·01	·12
February	71	82	57	25	39	65	2·0	1·1	3	29·95	·13
March . .	79	90	69	21	38	68	3·0	2·5	4	·86	·13
April . .	83	93	74	19	34	73	3·8	5·8	8	·76	·13
May . .	83	92	75	17	29	80	5·3	9·2	12	·70	·11
June . .	83	90	78	12	20	87	7·6	13·3	17	·58	·09
July . .	83	90	79	11	16	87	7·9	12·8	21	·57	·09
August . .	83	89	80	9	16	87	7·8	12·4	20	·62	·09
September .	83	88	79	9	18	86	6·9	10·2	15	·71	·11
October .	81	88	75	13	23	81	4·1	5·2	6	·84	·11
November .	75	84	66	18	30	72	2·0	0·7	2	·95	·11
December .	68	78	57	21	30	71	1·4	0·2	1	30·01	·12
Year .	78					77	4·5	73·7	110	29·80	

Mean highest temperature of year 100°
,, lowest ,, ,, 48°
,, annual range of temperature 52°
Highest recorded reading (10 yrs.) 105·6° (1879)
Lowest ,, ,, ,, 45·2° (1878)

Absolute range of temperature . 60·4°
Rainfall of wettest year (32 yrs.) 112·6″ (1880)
,, driest ,, ,, 52·2″ (1874)
Rainy days in wettest year (18 yrs.) 140 (1874)
,, ,, driest ,, . ,, 83 (1872)

55. Chittagong—Elevation 87 feet

	Temperature.					Hum.	Cloud.	Rainfall.		Barometer.	
				M. Range.							
	Mean.	Mean Max.	Mean Min.	Daily.	Month.	Mean.	Mean.	Ins.	Days.	Mean.	Daily Range.
January .	67	78	55	23	34	72	1·3	0·4	1	29·95	·11
February .	71	81	58	23	36	70	1·8	1·2	2	·91	·11
March .	77	86	67	19	34	74	2·9	1·9	4	·83	·11
April . .	81	89	73	16	26	77	4·0	4·6	6	·75	·11
May . .	82	89	75	14	27	80	5·0	9·2	12	·67	·10
June . .	81	86	76	10	19	86	7·4	23·8	19	·57	·09
July . .	81	86	76	10	18	87	7·5	22·2	24	·56	·08
August .	81	86	76	10	16	87	7·2	20·5	23	·60	·09
September .	81	87	76	11	18	86	6·5	14·1	18	·68	·11
October .	80	86	73	13	23	83	4·4	5·7	9	·79	·11
November .	74	83	66	17	29	79	2·7	1·6	3	·88	·11
December .	68	78	58	20	31	75	2·1	0·6	1	·94	·11
Year .	77					80	4·4	105·8	122	29·76	

Mean highest temperature of year 94°
,, lowest ,, ,, 48°
,, annual range of temperature 46°
Highest recorded reading (10 yrs.) 98·7° (1878)
Lowest ,, ,, ,, 45·4° (1878)

Absolute range of temperature . 53·3°
Rainfall of wettest year (26 yrs.) 154·1″ (1868)
,, driest ,, ,, 83·1″ (1874)
Rainy days in wettest year (19 yrs.) 146 (1881)
,, ,, driest ,, ,, 99 (1867)

56. Cuttack—Elevation 80 feet

	Temperature.					Hum.	Cloud.	Rainfall.		Barometer.	
				M. Range.							
	Mean.	Mean Max.	Mean Min.	Daily.	Month.	Mean.	Mean.	Ins.	Days.	Mean.	Daily Range.
January .	72	85	60	25	38	65	1·7	0·4	1	29·96	·13
February .	76	90	65	25	40	63	1·9	0·6	2	·89	·14
March . .	83	98	73	25	39	62	2·3	1·1	2	·81	·15
April . .	88	102	78	24	37	62	3·0	1·5	3	·70	·14
May . .	89	100	79	21	35	65	3·8	3·2	8	·61	·13
June . .	86	95	80	15	30	74	6·6	10·7	15	·50	·10
July . .	83	90	78	12	21	81	7·5	12·6	22	·50	·09
August .	83	90	78	12	20	81	7·0	11·2	21	·55	·10
September .	83	90	78	12	19	82	6·3	9·8	19	·64	·11
October .	81	90	74	16	28	75	4·2	5·8	10	·78	·11
November .	75	85	65	20	33	68	2·3	1·0	2	·91	·11
December .	70	82	58	24	34	66	1·7	0·5	1	·97	·12
Year .	81					70	4·0	58·4	106	29·74	

Mean highest temperature of year 110°
,, lowest ,, ,, 51°
,, annual range of temperature 59°
Highest recorded reading (10 yrs.) 117·7° (1879)
Lowest ,, ,, ,, 49·1° (1879)

Absolute range of temperature . 68·6°
Rainfall of wettest year (23 years) 91·9″ (1875)
,, driest ,, ,, 41·1″ (1877)
Rainy days in wettest year (18 years) 124 (1880)
,, ,, driest ,, ,, 80 (1876)

57. Dhubri—Elevation 115 feet

	Temperature.					Hum.	Cloud.	Rainfall.		Barometer.	
	Mean.	Mean Max.	Mean Min.	M. Range. Daily.	M. Range. Month.	Mean.	Mean.	Ins.	Days.	Mean.	Daily Range.
January .	63	74	54	20	29	77	1·5	0·3	1	29·94	·15
February .	65	76	54	22	35	65	1·7	0·5	1	·88	·15
March . .	75	87	65	22	39	61	2·4	1·9	3	·77	·16
April . .	79	89	71	18	32	68	2·9	5·1	8	·67	·15
May . .	78	86	73	13	29	80	4·6	15·6	16	·61	·13
June . .	80	86	77	9	20	87	7·1	26·4	21	·50	·12
July . .	82	88	79	9	17	85	6·6	15·4	18	·48	·11
August . .	81	87	78	9	16	86	7·3	11·4	18	·53	·12
September .	80	85	77	8	19	88	6·6	12·4	15	·63	·13
October .	78	84	73	11	21	81	2·2	4·3	6	·78	·13
November .	71	79	63	16	25	75	1·3	...	...	·86	·13
December .	64	74	56	18	27	78	1·2	0·2	1	·93	·14
Year .	75					78	3·8	93·5	108	29·71	

Mean highest temperature of year 98°
,, lowest ,, ,, 48°
,, annual range of temperature 50°
Highest recorded reading (5 yrs.) 100·8° (1885)
Lowest ,, ,, ,, 45·3° (1886)
Absolute range of temperature . 55·5°

Rainfall of wettest year (14 yrs.) 156·3″ (1878)
,, driest ,, ,, 69″ (1877, 1881)
Rainy days in wettest year (8 years) 129 (1881)
,, ,, driest ,, ,, 81 (1878)

58. Sibsagar—Elevation 333 feet

	Temperature.					Hum.	Cloud.	Rainfall.		Barometer.	
	Mean.	Mean Max.	Mean Min.	M. Range. Daily.	M. Range. Month.	Mean.	Mean.	Ins.	Days.	Mean.	Daily Range.
January .	59	70	49	21	32	85	5·4	1·1	7	29·74	·14
February .	62	73	53	20	36	81	6·2	2·2	9	·67	·14
March . .	69	79	59	20	37	79	6·4	4·4	14	·59	·14
April . .	74	82	66	16	30	81	7·3	9·9	18	·50	·14
May . .	78	86	71	15	30	82	8·2	11·0	19	·43	·13
June . .	83	90	76	14	25	83	8·6	14·1	20	·31	·12
July . .	84	91	78	13	23	83	8·8	15·6	22	·28	·13
August . .	83	90	78	12	23	84	8·7	16·0	21	·33	·13
September .	82	89	76	13	24	85	8·2	11·7	18	·43	·14
October .	77	85	70	15	29	85	6·9	5·2	10	·57	·14
November .	69	79	60	19	31	84	4·8	1·3	3	·67	·14
December .	61	72	51	21	31	85	4·6	0·6	3	·73	·15
Year .	73					83	7·0	93·1	164	29·52	

Mean highest temperature of year 99°
,, lowest ,, ,, 42°
,, annual range of temperature 57°
Highest recorded reading (10 yrs.) 102·1° (1883)
Lowest ,, ,, ,, 40° (1878)

Absolute range of temperature . 62·1°
Rainfall of wettest year (30 yrs.) 121·7″ (1871)
,, driest ,, ,, 72·2″ (1866)
Rainy days in wettest year (17 yrs.) 195 (1874)
,, ,, driest ,, ,, 137 (1870)

59. Silchar—Elevation 104 feet

	Temperature.					Hum.	Cloud.	Rainfall.		Barometer.	
				M. Range.							
	Mean.	Mean Max.	Mean Min.	Daily.	Month.	Mean.	Mean.	Ins.	Days.	Mean.	Daily Range.
January .	64	77	52	25	35	75	2·7	0·6	2	29·94	·13
February .	67	79	55	24	38	70	3·1	2·6	6	·89	·13
March .	73	85	63	22	36	72	4·0	7·9	12	·81	·14
April . .	78	87	69	18	30	76	5·1	13·0	17	·72	·14
May . .	80	87	72	15	30	81	6·1	15·7	19	·65	·13
June . .	82	89	76	13	24	85	7·5	19·1	22	·54	·11
July . .	82	89	77	12	21	85	7·6	20·6	25	·53	·12
August .	82	89	77	12	22	86	7·6	18·2	25	·58	·13
September .	82	89	76	13	24	84	6·9	14·2	19	·66	·13
October .	80	88	72	16	28	81	4·6	6·4	9	·78	·13
November .	73	84	64	20	32	77	3·2	1·0	2	·88	·12
December .	66	78	55	23	36	76	2·8	0·7	1	·94	·13
Year .	76					79	5·1	120·0	159	29·74	

Mean highest temperature of year 99°
,, lowest ,, ,, 45°
,, annual range of temperature 54°
Highest recorded reading (11 yrs.) 101·8° (1886)
Lowest ,, ,, ,, 43° (1880)

Absolute range of temperature . 58·8°
Rainfall of wettest year (29 yrs.) 188·2″ (1866)
,, driest ,, ,, 65·3″ (1868)
Rainy days in wettest year (18 yrs.) 187 (1874)
,, ,, driest ,, ,, 137 (1882)

60. Seoni—Elevation 2030 feet

	Temperature.					Hum.	Cloud.	Rainfall.		Barometer.	
				M. Range.							
	Mean.	Mean Max.	Mean Min.	Daily.	Month.	Mean.	Mean.	Ins.	Days.	Mean.	Daily Range.
January .	64	80	51	29	42	52	3·6	0·7	2	27·96	
February .	68	84	55	29	43	45	3·7	0·9	3	·92	
March . .	77	94	64	30	48	34	3·4	0·5	2	·86	
April . .	84	100	71	29	41	31	3·7	0·6	2	·78	
May . .	87	103	76	27	40	34	3·5	0·9	6	·69	
June . .	82	94	75	19	34	62	4·1	10·1	16	·59	
July . .	76	85	73	12	22	81	4·7	15·0	23	·57	
August . .	76	84	72	12	20	80	4·4	11·0	21	·62	
September .	76	86	71	15	24	76	3·9	8·2	16	·70	
October .	73	86	65	21	36	59	3·9	1·9	4	·85	
November .	66	81	55	26	40	50	3·7	0·4	1	·94	
December .	63	77	50	27	40	51	3·6	0·7	2	·98	
Year .	74					55	3·8	50·9	98	27·79	

Mean highest temperature of year 109°
,, lowest ,, ,, 41°
,, annual range of temperature 68°
Highest recorded reading (9 yrs.) 110·6°
Lowest ,, ,, ,, 36·4°

Absolute range of temperature . 74·2°
Rainfall of wettest year (27 years) 70·0″ (1884)
,, driest ,, ,, 31·4″ (1868)
Rainy days in wettest year (13 yrs.) 130 (1885)
,, ,, driest ,, ,, 80 (1873)

61. Raipur—Elevation 960 feet

	Temperature.					Hum.	Cloud.	Rainfall.		Barometer.	
				M. Range.							
	Mean.	Mean Max.	Mean Min.	Daily.	Month.	Mean.	Mean.	Ins.	Days.	Mean.	Daily Range.
January .	67	81	55	26	39	54	1·3	0·3	1	29·05	·12
February .	71	86	61	25	42	49	1·2	0·3	1	28·99	·13
March . .	80	97	69	28	45	40	1·6	0·6	1	·90	·14
April . .	88	103	76	27	40	36	2·2	0·6	2	·79	·14
May . .	92	105	80	25	41	36	3·2	0·9	4	·69	·14
June . .	86	97	78	19	38	63	6·3	10·4	12	·60	·11
July . .	79	87	75	12	24	82	8·1	14·8	18	·60	·10
August . .	79	87	75	12	20	82	7·6	12·1	18	·66	·10
September .	80	88	75	13	21	79	6·4	7·7	13	·73	·11
October .	77	88	68	20	32	67	3·0	2·2	4	·90	·12
November .	70	82	60	22	39	59	1·9	0·8	1	29·00	·11
December .	66	78	54	24	36	56	1·5	0·3	1	·05	·12
Year .	78					59	3·7	51·0	76	28·83	

Mean highest temperature of year 112°
,, lowest ,, ,, 46°
,, annual range of temperature 66°
Highest recorded reading (10 yrs.) 115·6° (1878)
Lowest ,, ,, ,, 41·4° (1883)

Absolute range of temperature . 74·2°
Rainfall of wettest year (21 yrs.) 85·9″ (1884)
,, driest ,, ,, 35·2″ (1873)
Rainy days in wettest year (13 yrs.) 96 (1885)
,, ,, driest ,, ,, 61 (1873)

62. Nagpur—Elevation 1025 feet

	Temperature.					Hum.	Cloud.	Rainfall.		Barometer.	
				M. Range.							
	Mean.	Mean Max.	Mean Min.	Daily.	Month.	Mean.	Mean.	Ins.	Days.	Mean.	Daily Range.
January .	69	83	55	28	42	51	2·2	0·6	1	28·97	·14
February .	73	89	59	30	46	42	2·0	0·4	2	·91	·14
March . .	82	99	67	32	49	32	2·5	0·6	2	·83	·15
April . .	89	105	75	30	42	28	3·1	0·5	2	·73	·15
May . .	93	108	80	28	42	30	4·1	0·8	4	·64	·14
June . .	86	98	78	20	38	60	7·1	8·8	15	·56	·11
July . .	79	88	75	13	25	80	8·6	13·3	20	·56	·10
August . .	79	88	75	13	23	78	8·3	8·9	18	·61	·11
September .	79	89	73	16	24	76	7·4	7·8	15	·67	·12
October .	77	90	68	22	35	60	4·0	2·3	3	·82	·12
November .	71	84	59	25	39	52	2·6	0·4	1	·94	·12
December .	67	80	54	26	38	52	2·4	0·5	1	·98	·12
Year .	79					53	4·5	44·9	84	28·77	

Mean highest temperature of year 115°
,, lowest ,, ,, 46°
,, annual range of temperature 69°
Highest recorded reading (10 yrs.) 117·7° (1883, 1885)
Lowest recorded reading (10 yrs.) 43·1° (1883)

Absolute range of temperature . 74·6°
Rainfall of wettest year (39 years) 65·3″ (1831)
,, driest ,, ,, 25·5″ (1868)
Rainy days in wettest year (13 yrs.) 119 (1884)
,, ,, driest ,, ,, 61 (1876)

63. Akola—Elevation 930 feet

	Temperature.					Hum.	Cloud.	Rainfall.		Barometer.	
	Mean.	Mean Max.	Mean Min.	M. Range. Daily.	M. Range. Month.	Mean.	Mean.	Ins.	Days.	Mean.	Daily Range.
January .	68	85	53	32	47	47	1·3	0·5	1	29·07	·14
February .	73	90	57	33	50	36	1·5	0·1	1	·01	·14
March .	82	99	66	33	51	28	1·2	0·4	1	28·94	·15
April .	89	105	74	31	45	22	1·3	0·1	1	·84	·14
May .	93	107	81	26	40	27	1·9	0·2	2	·76	·14
June .	86	98	78	20	37	56	6·2	5·2	11	·69	·12
July .	80	89	74	15	26	74	8·5	7·8	17	·69	·10
August .	79	88	73	15	24	74	8·0	6·7	13	·74	·11
September .	78	88	72	16	26	75	6·4	5·8	12	·81	·12
October .	77	90	66	24	39	61	2·5	2·2	4	·94	·12
November .	70	86	56	30	45	51	1·3	0·4	1	29·03	·13
December .	66	82	51	31	43	51	1·5	0·8	2	·08	·13
Year .	78					50	3·5	30·2	66	28·89	

Mean highest temperature of year 113°
,, lowest ,, ,, 41°
,, annual range of temperature 72°
Highest recorded reading (10 yrs.) 115·5° (1881)
Lowest ,, ,, ,, 36·7° (1878)

Absolute range of temperature . 78·8°
Rainfall of wettest year (25 yrs.) 48·9″ (1884)
,, driest ,, ,, 10·4″ (1866)
Rainy days in wettest year (13 yrs.) 77 (1878)
,, ,, driest ,, ,, 47 (1876)

64. Surat—Elevation 36 feet

	Temperature.					Hum.	Cloud.	Rainfall.		Barometer.	
	Mean.	Mean Max.	Mean Min.	M. Range. Daily.	M. Range. Month.	Mean.	Mean.	Ins.	Days.	Mean.	Daily Range.
January .	70	87	56	31	43	53	0·8	...	...	29·98	·12
February .	72	89	58	31	49	50	0·8	...	...	·94	·13
March .	79	96	65	31	48	51	0·9	...	...	·88	·13
April .	84	100	72	28	41	50	1·0	...	...	·80	·13
May .	86	98	77	21	37	59	1·6	0·2	1	·73	·12
June .	85	94	79	15	27	68	5·4	9·4	11	·63	·10
July .	81	88	78	10	21	80	8·0	13·8	20	·61	·08
August .	81	88	77	11	18	79	6·8	9·8	17	·67	·09
September .	81	88	76	12	21	79	6·0	7·3	12	·74	·11
October .	80	91	70	21	33	67	2·4	1·4	4	·85	·11
November .	75	90	62	28	39	55	0·7	0·1	...	·92	·11
December .	71	86	58	28	41	55	1·1	0·1	1	·97	·12
Year .	79					62	3·0	42·1	66	29·81	

Mean highest temperature of year 109°
,, lowest ,, ,, 48°
,, annual range of temperature 61°
Highest recorded reading (9 yrs.) 113·2° (1881)
Lowest ,, ,, (8 yrs.) 45·3° (1879)

Absolute range of temperature . 67·9°
Rainfall of wettest year (18 yrs.) 89·3″ (1878)
,, driest ,, ,, 16·7″ (1877)
Rainy days in wettest year (8 yrs.) 84 (1878)
,, ,, driest ,, ,, 48 (1880)

65. Bombay—Elevation 37 feet

	Temperature.					Hum.	Cloud.	Rainfall.		Barometer.	
	Mean.	Mean Max.	Mean Min.	M. Range. Daily.	M. Range. Month.	Mean.	Mean.	Ins.	Days.	Mean.	Daily Range.
January .	74	82	68	14	24	70	1·5	0·1	1	29·95	·12
February .	75	82	69	13	26	69	1·3	...	1	·92	·12
March . .	79	85	74	11	22	73	1·8	...	...	·88	·12
April . .	82	88	77	11	17	75	2·3	...	...	·81	·11
May . .	85	90	80	10	15	75	4·1	0·5	2	·77	·10
June . .	83	88	80	8	17	82	7·9	20·8	20	·67	·08
July . .	81	85	77	8	13	87	9·1	24·7	29	·67	·06
August . .	80	84	77	7	13	87	8·8	15·1	26	·73	·08
September .	80	84	76	8	13	86	7·5	10·8	21	·78	·09
October .	81	87	76	11	18	81	4·3	1·8	7	·84	·11
November .	80	86	73	13	22	71	2·2	0·5	1	·91	·11
December .	76	84	70	14	23	70	1·8	0·1	...	·95	·12
Year .	80					77	4·4	74·4	108	29·82	

Mean highest temperature of year 95°
,, lowest ,, ,, 61°
,, annual range of temperature 34°
Highest recorded reading (37 yrs.) 100·2° (1857)
Lowest ,, ,, ,, 53·3° (1847)

Absolute range of temperature . 46·9°
Rainfall of wettest year (40 yrs.) 114·9″ (1849)
,, driest ,, ,, 40·6″ (1871)
Rainy days in wettest year (11 yrs.) 123 (1883)
,, ,, driest ,, ,, 82 (1876)

66. Mangalore—Elevation 52 feet

	Temperature.					Hum.	Cloud.	Rainfall.		Barometer.	
	Mean.	Mean Max.	Mean Min.	M. Range. Daily.	M. Range. Month.	Mean.	Mean.	Ins.	Days.	Mean.	Daily Range.
January .	76	88	69	19	30	62	1·5	0·2	...	29·91	·12
February .	78	87	71	16	26	69	1·4	0·1	...	·89	·13
March . .	81	89	76	13	19	73	2·0	0·1	...	·87	·13
April . .	83	91	79	12	19	73	2·8	2·0	3	·81	·11
May . .	83	90	79	11	19	74	4·4	8·1	8	·78	·09
June . .	78	85	75	10	17	86	7·9	37·8	27	·78	·07
July . .	77	83	74	9	15	89	7·7	37·9	30	·79	·07
August .	77	83	74	9	14	89	7·3	23·1	28	·80	·08
September .	77	83	74	9	14	87	6·5	11·3	22	·84	·09
October .	78	85	74	11	17	83	5·7	8·0	15	·84	·10
November .	78	87	73	14	25	76	4·0	1·9	7	·84	·11
December .	77	88	71	17	26	67	2·6	0·5	2	·88	·11
Year .	79					77	4·5	131·0	142	29·84	

Mean highest temperature of year 94°
,, lowest ,, ,, 63°
,, annual range of temperature 31°
Highest recorded reading (7 years) 97·4° (1886)
Lowest ,, ,, (6 years) 62·2° (1882, 1884)

Absolute range of temperature . 35·2°
Rainfall of wettest year (28 yrs.) 182·3″ (1878)
,, driest ,, ,, 95·5″ (1881)
Rainy days in wettest year (9 years) 156 (1880)
,, ,, driest ,, ,, 122 (1877)

67. Mercara—Elevation 3695 feet

	Temperature.					Hum.	Cloud.	Rainfall.		Barometer.	
	Mean.	Mean Max.	Mean Min.	M. Range. Daily.	M. Range. Month.	Mean.	Mean.	Ins.	Days.	Mean.	Daily Range.
January .	64	75	55	20	30	68	2·7	0·3	1	26·31	·10
February .	67	81	57	24	36	60	2·5	...	...	·31	·11
March .	70	84	61	23	32	64	3·6	1·1	4	·29	·11
April . .	72	85	64	21	29	73	5·2	2·2	8	·25	·10
May . .	70	81	65	16	26	80	6·8	6·1	13	·23	·08
June . . .	66	72	64	8	18	93	9·6	25·8	26	·21	·06
July . .	64	69	62	7	14	97	9·9	42·1	29	·22	·06
August .	64	70	62	8	16	95	9·5	25·7	27	·22	·07
September .	64	72	62	10	19	93	9·2	12·3	23	·25	·08
October .	66	75	62	13	22	88	7·9	8·0	17	·26	·09
November .	65	74	61	13	24	84	6·9	2·6	10	·28	·09
December .	64	74	57	17	26	77	4·8	0·6	3	·30	·09
Year .	67					81	6·5	126·8	161	26·29	

Mean highest temperature of year 96°
,, lowest ,, ,, 49°
,, annual range of temperature 41°
Highest recorded reading (6 years) 91·1° (1886)
Lowest ,, ,, ,, 47·6° (1884)

Absolute range of temperature . 43·5°
Rainfall of wettest year (23 years) 227·1″ (1882)
,, driest ,, ,, 94·2″ (1875)
Rainy days in wettest year (7 years) 184 (1880)
,, ,, driest ,, ,, 148 (1882)

68. Cochin—Elevation 11 feet

	Temperature.					Hum.	Cloud.	Rainfall.		Barometer.	
	Mean.	Mean Max.	Mean Min.	M. Range. Daily.	M. Range. Month.	Mean.	Mean.	Ins.	Days.	Mean.	Daily Range.
January .	79	89	71	18	25	70	2·2	0·9	1	29·93	·13
February .	80	90	73	17	24	72	2·4	0·7	2	·92	·15
March .	83	91	76	15	21	75	3·1	2·1	5	·91	·14
April . .	84	91	78	13	20	77	4·0	4·4	8	·86	·13
May . .	82	89	77	12	21	81	5·7	12·7	17	·84	·11
June . .	78	85	74	11	17	88	7·4	30·7	27	·86	·09
July . .	77	83	74	9	16	88	7·2	22·7	27	·87	·09
August .	79	83	74	9	16	87	6·2	12·4	22	·88	·10
September .	78	84	74	10	15	86	5·5	9·4	19	·90	·11
October .	79	86	74	12	17	84	5·9	12·1	19	·89	·12
November .	79	87	74	13	20	81	4·8	5·1	12	·90	·13
December .	79	88	72	16	23	75	3·3	1·9	5	·92	·13
Year .	80					80	4·8	115·1	164	29·89	

Mean highest temperature of year 95°
,, lowest ,, ,, 67°
,, annual range of temperature 28°
Highest recorded reading (10 yrs.) 99·5° (1879)
Lowest ,, ,, ,, 63·6° (1876)

Absolute range of temperature . 35·9°
Rainfall of wettest year (23 yrs.) 227·1″ (1882)
,, driest ,, ,, 94·2″ (1875)
Rainy days in wettest year (11 yrs.) 189 (1880)
,, ,, driest ,, ,, 122 (1876)

69. MALAGAON—Elevation 1430 feet

	Temperature.					Hum.	Cloud.	Rainfall.		Barometer.	
				M. Range.							
	Mean.	Mean Max.	Mean Min.	Daily.	Month.	Mean.	Mean.	Ins.	Days.	Mean.	Daily Range.
January .	68	86	52	34	49	41	1·7	0·2	1	28·56	·14
February .	72	90	55	35	51	38	1·3	0·2	1	·51	·14
March . .	80	98	64	34	52	31	1·9	...	...	·46	·14
April . .	85	102	70	32	45	28	2·1	0·3	1	·37	·13
May . .	88	104	75	29	41	39	2·2	0·7	2	·30	·13
June . .	83	95	75	20	33	63	5·7	5·0	10	·22	·11
July . .	79	87	73	14	23	72	7·8	4·2	15	·22	·09
August . .	78	86	72	14	23	72	7·4	4·3	13	·27	·09
September .	77	86	70	16	26	72	6·8	6·4	15	·34	·11
October .	76	89	65	24	38	57	3·3	2·4	6	·45	·12
November .	70	86	56	30	43	47	2·0	0·4	1	·53	·12
December .	66	83	51	32	45	47	2·1	0·6	1	·56	·13
Year .	76					51	3·7	24·7	66	28·40	

Mean highest temperature of year 109°
,, lowest ,, ,, 40°
,, annual range of temperature 69°
Highest recorded reading (8 yrs.) 110·8° (1883)
Lowest ,, ,, ,, 36·1° (1879)

Absolute range of temperature . 74·7°
Rainfall of wettest year (17 yrs.) 38·4″ (1884)
,, driest ,, ,, 15·7″ (1876)
Rainy days in wettest year (8 yrs.) 81 (1884)
,, ,, driest ,, ,, 48 (1885)

70. POONA—Elevation 1849 feet

	Temperature.					Hum.	Cloud.	Rainfall.		Barometer.	
				M. Range.							
	Mean.	Mean Max.	Mean Min.	Daily.	Month.	Mean.	Mean.	Ins.	Days.	Mean.	Daily Range.
January .	72	86	54	32	49	41	1·0	0·2	1	28·10	·14
February .	76	90	56	34	50	33	1·7	...	...	·09	·14
March . .	83	98	64	34	47	29	2·5	0·2	1	·05	·14
April . .	86	101	69	32	43	31	2·4	0·6	2	27·99	·14
May . .	85	100	71	29	41	42	3·0	1·6	2	·94	·12
June . .	79	89	72	17	27	69	7·6	5·6	13	·88	·09
July . .	75	81	70	11	19	79	9·0	6·6	21	·87	·07
August .	75	83	69	14	22	79	8·7	4·1	20	·91	·08
September .	75	83	68	15	23	77	8·1	4·3	15	·97	·10
October .	78	87	66	21	33	58	4·9	4·1	8	28·05	·12
November .	75	85	59	26	41	46	2·8	0·8	2	·10	·13
December .	72	83	54	29	41	41	2·7	0·2	1	·14	·13
Year .	78					52	4·6	28·3	86	28·01	

Mean highest temperature of year 106°
,, lowest ,, ,, 44°
,, annual range of temperature 62°
Highest recorded reading (8 yrs.) 109·2° (1886)
Lowest ,, ,, ,, 40·8° (1881)

Absolute range of temperature . 68·4°
Rainfall of wettest year (43 years) 56·9″ (1861)
,, driest ,, ,, 14·2″ (1844)
Rainy days in wettest year (11 yrs.) 94 (1883)
,, ,, driest ,, ,, 71 (1876)

71. Sholapur—Elevation 1590 feet

	Temperature.					Hum.	Cloud.	Rainfall.		Barometer.	
				M. Range.							
	Mean.	Mean Max.	Mean Min.	Daily.	Month.	Mean.	Mean.	Ins.	Days.	Mean.	Daily Range.
January .	72	87	58	29	42	42	2·2	...	...	28·40	·13
February .	77	94	62	32	43	34	1·9	0·1	1	·35	·14
March . .	83	100	68	32	43	29	2·5	0·3	1	·30	·14
April . .	88	105	74	31	39	26	3·2	0·7	3	·21	·15
May . .	89	104	76	28	39	34	4·0	1·2	4	·15	·14
June . .	82	95	72	23	35	56	7·1	4·6	14	·12	·12
July . .	79	90	71	19	28	65	8·1	4·3	16	·12	·10
August .	78	89	71	18	27	66	8·1	6·0	17	·15	·11
September .	77	88	70	18	26	70	8·3	7·5	14	·20	·12
October .	77	89	68	21	32	60	5·9	3·7	8	·29	·12
November .	73	87	63	24	39	53	3·7	0·7	3	·35	·12
December .	70	85	58	27	39	49	3·0	0·4	2	·39	·12
Year . .	79					49	4·8	29·5	83	28·25	

Mean highest temperature of year 110°
„ lowest „ „ 47°
„ annual range of temperature 63°
Highest recorded reading (8 yrs.) 112·1° (1884)
Lowest „ „ „ 42·8° (1884)

Absolute range of temperature . 69·3°
Rainfall of wettest year (32 yrs.) 68·0″ (1878)
„ driest „ „ 10·6″ (1876)
Rainy days in wettest year (8 yrs.) 99 (1878)
„ „ driest „ „ 58 (1881)

72. Belgaum—Elevation 2550 feet

	Temperature.					Hum.	Cloud	Rainfall.		Barometer.	
				M. Range.							
	Mean.	Mean. Max.	Mean. Min.	Daily	Month.	Mean.	Mean.	Ins.	Days.	Mean.	Daily Range.
January .	72	83	58	25	36	47	1·7	0·1	...	27·44	·13
February .	76	89	60	29	41	39	1·5	...	...	·41	·13
March . .	80	95	65	30	49	41	2·1	0·5	3	·38	·14
April . .	81	97	67	30	40	50	2·6	2·0	6	·32	·13
May . .	80	94	68	26	38	59	3·4	2·8	7	·29	·12
June . .	74	82	68	14	26	80	7·7	9·3	21	·25	·08
July . .	71	76	67	9	16	89	8·8	15·2	28	·25	·07
August. .	70	77	66	11	18	88	8·5	9·0	25	·27	·08
September .	71	78	65	13	21	86	7·7	3·7	22	·32	·10
October .	74	83	65	18	27	70	5·5	4·7	13	·36	·12
November .	73	82	62	20	31	57	3·7	1·2	5	·40	·12
December .	71	82	59	23	33	50	2·9	0·3	3	·43	·12
Year .	74					63	4·7	48·8	133	27·34	

Mean highest temperature of year 102°
„ lowest „ „ 51°
„ annual range of temperature 51°
Highest recorded reading (10 yrs.) 109° (1879)
Lowest „ „ „ 47·1° (1878)

Absolute range of temperature . 61·9°
Rainfall of wettest year (42 yrs.) 73·4″ (1882)
„ driest „ „ 35·2″ (1880)
Rainy days in wettest year (11 yrs.) 174 (1883)
„ „ driest „ „ 91 (1876)

73.—Secunderabad—Elevation 1787 feet

	Temperature.					Hum.	Cloud.	Rainfall.		Barometer.	
				M. Range.							
	Mean.	Mean. Max.	Mean. Min.	Daily.	Month.	Mean.	Mean.	Ins.	Days.	Mean.	Daily Range.
January .	70	84	57	27	37	54	1·7	0·3	1	28·19	·14
February .	76	91	62	29	40	45	1·4	0·2	...	·15	·15
March . .	82	98	69	29	40	39	1·7	0·7	3	·10	·15
April . .	87	102	75	27	40	36	1·9	0·7	4	·03	·15
May . .	89	103	78	25	39	37	2·6	1·4	5	27·95	·15
June . .	82	94	74	20	34	58	5·4	3·7	12	·91	·13
July . .	77	86	71	15	26	72	6·7	6·0	17	·91	·11
August . .	77	86	70	16	25	72	6·0	5·7	18	·94	·12
September .	76	86	70	16	24	75	5·9	5·2	15	·98	·13
October .	76	87	67	20	31	66	4·0	3·3	8	28·07	·13
November .	72	83	61	22	35	60	2·9	0·8	4	·15	·14
December .	69	81	56	25	37	57	2·2	0·3	2	·19	·14
Year .	78					56	3·5	28·3	89	28·05	

Mean highest temperature of year 109°
,, lowest ,, ,, 48°
,, annual range of temperature 61°
Highest recorded reading (9 yrs.) 110·4° (1883)
Lowest ,, ,, ,, 41·7° (1879)

Absolute range of temperature . 68·7°
Rainfall of wettest year (44 yrs.) 44·6″ (1878)
,, driest ,, ,, 12·9″ (1859)
Rainy days in wettest year (11 yrs.) 116 (1883)
,, ,, driest ,, ,, 57 (1876)

74. Bellary—Elevation 1455 feet

	Temperature.					Hum.	Cloud.	Rainfall.		Barometer.	
				M. Range.							
	Mean.	Mean Max.	Mean Min.	Daily.	Month.	Mean.	Mean.	Ins.	Days.	Mean.	Daily Range.
January .	73	88	61	27	38	50	2·0	0·1	...	28·48	·14
February .	77	91	65	29	41	37	1·5	...	...	·44	·15
March . .	85	101	72	29	40	33	2·3	0·6	1	·39	·15
April . .	89	104	77	27	37	35	3·1	0·8	3	·32	·15
May . .	88	103	77	26	38	42	4·8	1·8	5	·27	·15
June . .	83	95	76	19	30	58	7·4	1·8	6	·26	·12
July . .	81	92	75	17	27	62	8·0	1·3	8	·27	·11
August .	81	92	74	18	26	63	7·4	2·3	8	·29	·12
September .	80	91	73	18	27	64	7·3	3·7	9	·32	·13
October .	79	90	71	19	30	64	6·3	3·9	9	·38	·14
November .	75	87	66	21	34	61	4·7	1·0	5	·44	·14
December .	73	86	61	25	35	57	3·4	0·3	1	·48	·14
Year .	80					52	4·9	17·6	55	28·36	

Mean highest temperature of year 108°
,, lowest ,, ,, 54°
,, annual range of temperature 54°
Highest recorded reading (11 yrs.) 110·3° (1882)
Lowest ,, ,, (9 yrs.) 50·8° (1879)

Absolute range of temperature . 59·5°
Rainfall of wettest year (33 yrs.) 30·2″ (1885)
,, driest ,, ,, 7·2″ (1876)
Rainy days in wettest year (11 yrs.) 69 (1879)
,, ,, driest ,, ,, 33 (1876)

75. Bangalore—Elevation 2981 feet

	Temperature.					Hum.	Cloud.	Rainfall.		Barometer.	
				M. Range.							
	Mean.	Mean Max.	Mean Min.	Daily.	Month.	Mean.	Mean.	Ins.	Days.	Mean.	Daily Range.
January .	67	79	56	23	33	60	2·8	0·2	1	27·04	·12
February .	72	85	59	26	36	52	2·3	0·1	...	·02	·13
March . .	77	90	64	26	37	49	2·6	0·6	2	26·99	·13
April . .	80	93	69	24	32	52	3·5	1·3	3	·94	·13
May . .	79	92	69	23	33	61	4·5	5·0	10	·90	·12
June . .	74	85	67	18	27	73	5·7	3·2	13	·88	·10
July . .	72	83	66	17	24	77	6·2	4·0	14	·89	·09
August .	72	83	66	17	24	77	6·1	5·9	17	·90	·11
September .	72	82	65	17	25	76	5·9	6·3	13	·93	·12
October .	72	82	65	17	24	75	5·7	6·4	13	·96	·12
November .	70	79	62	17	27	73	4·9	1·9	7	27·00	·12
December .	67	78	59	19	29	68	4·0	0·7	3	·04	·12
Year .	73					66	4·5	35·6	96	26·96	

Mean highest temperature of year 98°
,, lowest ,, ,, 51°
,, annual range of temperature 47°
Highest recorded reading (10 yrs.) 99·4° (1882)
Lowest ,, ,, ,, 45·8° (1877)

Absolute range of temperature . 53·6°
Rainfall of wettest year (51 yrs.) 56·7″ (1874)
,, driest ,, ,, 16·0″ (1838)
Rainy days in wettest year (11 yrs.) 149 (1878)
,, ,, driest ,, ,, 73 (1876)

76. Masulipatam—Elevation 10 feet

	Temperature.					Hum.	Cloud.	Rainfall.		Barometer.	
				M. Range.							
	Mean.	Mean Max.	Mean Min.	Daily.	Month.	Mean.	Mean.	Ins.	Days.	Mean.	Daily Range.
January .	74	83	66	17	27	73	3·3	0·3	1	30·01	·12
February .	77	87	68	19	29	74	3·2	0·1	1	29·96	·12
March . .	81	92	73	19	30	74	3·5	0·3	1	·90	·13
April . .	85	95	77	18	30	75	3·9	0·1	1	·81	·13
May . .	88	99	81	18	36	72	4·8	1·7	4	·71	·13
June . .	87	98	81	17	34	67	6·6	4·4	8	·65	·12
July . .	84	94	79	15	27	72	6·9	5·6	15	·67	·12
August .	84	92	78	14	25	74	6·6	6·0	16	·70	·12
September .	83	91	78	13	23	76	6·3	6·5	16	·74	·13
October .	81	89	76	13	24	78	5·5	8·8	12	·84	·12
November .	77	85	71	14	24	78	5·1	4·0	6	·94	·12
December .	74	83	67	16	24	75	4·2	0·7	2	30·00	·12
Year .	81					74	5·0	38·5	83	29·83	

Mean highest temperature of year 109°
,, lowest ,, ,, 59°
,, annual range of temperature 50°
Highest recorded reading (11 yrs.) 115·9° (1878)
Lowest ,, ,, (10 yrs.) 58·1° (1883)

Absolute range of temperature . 57·8°
Rainfall of wettest year (24 yrs.) 69·4″ (1878)
,, driest ,, ,, 25·5″ (1868)
Rainy days in wettest year (12 yrs.) 101 (1875)
,, ,, driest ,, ,, 61 (1876)

77. Madras—Elevation 22 feet

	Temperature.					Hum.	Cloud.	Rainfall.		Barometer.	
				M. Range.							
	Mean.	Mean Max.	Mean Min.	Daily.	Month.	Mean.	Mean.	Ins.	Days.	Mean.	Daily Range.
January .	76	85	68	17	26	72	4·1	1·0	3	29·99	·12
February .	77	87	68	19	28	71	2·8	0·3	1	·97	·13
March . .	81	90	72	18	29	73	2·5	0·4	1	·91	·14
April . .	85	93	77	16	27	72	2·9	0·6	1	·82	·13
May . .	87	98	81	17	32	67	3·9	2·2	3	·74	·13
June . .	88	99	81	18	31	61	6·5	2·1	10	·70	·12
July . .	86	97	79	18	28	64	7·1	3·8	14	·72	·12
August . .	85	95	77	18	27	69	6·4	4·4	14	·75	·13
September .	84	94	77	17	28	70	6·2	4·7	11	·77	·13
October .	81	89	75	14	25	77	6·2	10·8	14	·84	·13
November .	78	85	72	13	25	79	6·3	13·7	14	·92	·12
December .	76	83	70	13	23	77	5·4	5·1	9	·98	·11
Year .	82					71	5·0	49·1	95	29·84	

Mean highest temperature of year 108°
,, lowest ,, ,, 60°
,, annual range of temperature 48°
Highest recorded reading (27 yrs.) 112·9° (1880)
Lowest ,, ,, ,, 57·6° (1876)

Absolute range of temperature . 55·3°
Rainfall of wettest year (74 yrs.) 88·4″ (1827)
,, driest ,, ,, 18·5″ (1832)
Rainy days in wettest year (26 yrs.) 119 (1847)
,, ,, driest ,, ,, 73 (1876)

78. Trichinopoly—Elevation 275 feet

	Temperature.					Hum.	Cloud.	Rainfall.		Barometer.	
				M. Range.							
	Mean.	Mean Max.	Mean Min.	Daily.	Month.	Mean.	Mean.	Ins.	Days.	Mean.	Daily Range.
January .	76	87	67	20	30	68	4·2	1·0	1	29·71	·14
February .	78	92	68	24	34	62	2·9	0·5	...	·70	·16
March . .	83	98	73	25	36	58	3·1	0·7	1	·66	·16
April . .	87	101	78	23	33	54	4·4	1·8	1	·57	·16
May . .	88	102	78	24	33	56	5·6	3·8	5	·52	·15
June . .	86	99	78	21	29	57	7·4	1·3	3	·50	·12
July . .	85	98	78	20	27	57	7·7	2·2	3	·51	·12
August . .	84	96	76	20	28	61	7·8	4·4	7	·53	·13
September .	83	95	75	20	27	63	7·4	5·3	8	·55	·14
October .	81	91	74	17	26	73	7·8	7·8	11	·59	·14
November .	79	87	71	16	26	76	7·4	5·2	9	·65	·13
December .	76	85	69	16	27	74	5·9	3·1	6	·70	·13
Year .	82					63	6·0	37·1	55	29·60	

Mean highest temperature of year 106°
,, lowest ,, ,, 60°
,, annual range of temperature 46°
Highest recorded reading (12 yrs.) 107·9° (1876, 1881)
Lowest recorded reading (12 yrs.) 56·6° (1884)

Absolute range of temperature . 51·3°
Rainfall of wettest year (33 yrs.) 95·3″ (1863)
,, driest ,, ,, 17·1″ (1845)
Rainy days in wettest year (11 years) 70 (1877)
,, ,, driest ,, ,, 31 (1876)

79. COIMBATORE—Elevation 1348 feet

	Temperature.					Hum.	Cloud.	Rainfall.		Barometer.	
				M. Range.							
	Mean.	Mean Max.	Mean Min.	Daily.	Month.	Mean.	Mean.	Ins.	Days.	Mean.	Daily Range.
January .	74	87	64	23	34	59	3·4	0·3	1	28·61	·15
February .	77	92	65	27	37	52	2·4	0·1	1	·58	·17
March . .	81	97	70	27	36	54	2·9	0·6	3	·55	·17
April . .	83	98	74	24	32	58	4·2	1·8	4	·49	·16
May . .	81	95	74	21	32	65	5·4	2·6	10	·45	·14
June . .	78	90	72	18	28	71	6·6	1·8	9	·44	·10
July . .	77	89	71	18	25	72	6·5	1·3	9	·46	·09
August . .	77	88	71	17	25	73	6·1	1·2	8	·47	·11
September .	77	90	71	19	26	71	5·7	1·2	6	·49	·13
October .	77	88	71	17	26	75	6·3	5·7	15	·51	·15
November .	76	86	69	17	27	74	6·1	3·4	13	·56	·14
December .	74	85	66	19	31	68	4·9	1·1	6	·59	·14
Year .	78					66	5·0	21·1	85	28·52	

Mean highest temperature of year 102°
,, lowest ,, ,, 56°
,, annual range of temperature 46°
Highest recorded reading (11 yrs.) 103·8° (1879)
Lowest ,, ,, ,, 54° (1876)

Absolute range of temperature . 49·8°
Rainfall of wettest year (24 yrs.) 29·0″ (1880)
,, driest ,, ,, 12·5″ (1867)
Rainy days in wettest year (11 years) 118 (1882)
,, ,, driest ,, ,, 55 (1876)

80. MADURA—Elevation 448 feet

	Temperature.					Hum.	Cloud.	Rainfall.		Barometer.	
				M. Range.							
	Mean.	Mean Max.	Mean Min.	Daily.	Month.	Mean.	Mean.	Ins.	Days.	Mean.	Daily Range.
January .	77	88	68	20	29	67	4·2	0·7	1	29·50	·16
February .	79	93	68	25	33	61	3·9	0·4	1	·48	·17
March . .	83	97	72	25	35	60	4·0	0·6	1	·44	·17
April . .	86	101	76	25	33	59	4·6	2·0	2	·37	·17
May . .	86	101	76	25	33	61	5·1	2·8	5	·32	·15
June . .	85	99	76	23	30	59	5·5	1·6	3	·31	·13
July . .	84	99	76	23	30	59	5·3	1·7	3	·33	·12
August . .	84	97	75	22	30	63	5·3	4·7	7	·34	·13
September .	83	97	75	22	29	63	5·0	4·5	8	·36	·14
October .	81	92	73	19	27	73	5·4	8·7	14	·40	·14
November .	79	87	72	15	25	75	5·6	5·1	9	·44	·15
December .	77	86	70	16	25	74	5·1	2·2	5	·48	·15
Year .	82					65	4·9	35·0	59	29·40	

Mean highest temperature of year 105°
,, lowest ,, ,, 62°
,, annual range of temperature 43°
Highest recorded reading (12 yrs.) 106·7° (1880)
Lowest ,, ,, ,, 59·8° (1876)

Absolute range of temperature . 46·9°
Rainfall of wettest year (24 yrs.) 54·0″ [1] (1877)
,, driest ,, ,, 20·7″ (1876)
Rainy days in wettest year (11 yrs.) 75 (1885)
,, ,, driest ,, ,, 42 (1876)

[1] A much higher rainfall, viz. 78·0 inches, is given in the tables for 1853, but as a register kept in the suburb of Pasumalai by the missionaries gives only 31 inches in the same year, I cannot accept the former as valid.

81. Trincomalee—Elevation 175 feet

	Temperature.					Hum.	Cloud.	Rainfall.		Barometer.	
	Mean.	Mean Max.	Mean Min.	M. Range. Daily.	M. Range. Month.	Mean.	Mean.	Ins.	Days.	Mean.	Daily Range.
January .	78	83	74	9	15	77	5·8	6·2	10	29·88	·10
February .	80	85	75	10	17	74	4·5	2·4	3	·87	·11
March . .	82	88	75	13	20	75	4·1	1·3	4	·84	·12
April . .	85	92	77	15	23	70	4·2	1·6	4	·77	·12
May . .	85	94	77	17	26	67	5·1	2·2	6	·71	·11
June . .	85	93	76	17	25	67	6·2	1·9	4	·70	·09
July . .	85	93	76	17	24	65	5·7	2·2	2	·71	·10
August . .	84	93	75	18	25	68	5·7	4·2	9	·72	·10
September .	83	92	75	17	24	69	5·7	4·6	7	·74	·12
October .	81	89	75	14	23	74	6·0	8·9	16	·77	·12
November .	79	85	74	11	18	82	6·2	13·1	19	·82	·11
December .	78	83	74	9	15	81	6·4	13·2	20	·85	·11
Year .	82					72	5·5	61·8	104	29·78	

Mean highest temperature of year 99°
,, lowest ,, ,, 68°
,, annual range of temperature 31°
Highest recorded reading (11 yrs.) 101·4° (1886)
Lowest ,, ,, ,, 65·3° (1876)

Absolute range of temperature . 36·1°
Rainfall of wettest year (16 yrs.) 78·5″ (1883)
,, driest ,, ,, 48·5″ (1879)
Rainy days in wettest year (12 yrs.) 128 (1879)
,, ,, driest ,, ,, 78 (1884)

82. Kandy—Elevation 1696 feet.

	Temperature.					Hum.	Cloud.	Rainfall.		Barometer.	
	Mean.	Mean Max.	Mean Min.	M Range. Daily.	M Range. Month.	Mean.	Mean.	Ins.	Days.	Mean.	Daily Range.
January .	74	80	69	11	21	76	5·1	5·0	10	28·22	·10
February .	76	82	68	14	24	71	4·3	2·6	4	·21	·12
March . .	79	86	71	15	23	69	4·8	3·1	9	·20	·12
April . .	79	87	71	16	22	74	5·3	6·9	14	·16	·11
May . .	79	86	73	13	20	76	5·7	6·7	13	·14	·09
June . .	77	84	71	13	17	81	6·8	9·1	21	·15	·08
July . .	76	80	71	9	16	80	6·5	7·8	21	·16	·08
August .	76	81	71	10	19	79	6·5	6·2	19	·16	·09
September .	76	83	70	13	21	79	6·3	5·8	16	·18	·10
October .	76	82	69	13	21	80	6·4	11·3	23	·18	·10
November .	75	82	68	14	22	80	5·9	11·6	18	·19	·10
December .	75	81	68	13	23	80	5·9	8·4	17	·20	·10
Year .	76					77	5·8	84·5	185	28·18	

Mean highest temperature of year 90°
,, lowest ,, ,, 60°
,, annual range of temperature 30°
Highest recorded reading (5 yrs.[1]) 94·6° (1886)
Lowest ,, ,, ,, 54° (1886)

Absolute range of temperature . 40·6°
Rainfall of wettest year (17 yrs.) 111·1″ (1882)
,, driest ,, ,, 67″ (1884)
Rainy days in wettest year (12 years) 213 (1877)
,, ,, driest ,, ,, 160 (1884)

1 From 1876 to 1882 the maxima and minima temperatures recorded at Kandy are evidently untrustworthy, owing apparently to want of proper exposure. These have therefore been rejected. According to Sir Emerson Tennant's description, the diurnal range of temperature in the cooler months is much greater than that shown by the above table.

83. Colombo—Elevation 40 feet

	Temperature.					Hum.	Cloud.	Rainfall.		Barometer.	
	Mean.	Mean Max.	Mean Min.	M. Range. Daily.	M. Range. Month.	Mean.	Mean.	Ins.	Days.	Mean.	Daily Range.
January .	79	86	72	14	21	74	4·9	3·0	6	29·88	·12
February .	80	87	73	14	22	73	4·2	1·7	4	·88	·13
March . .	82	88	75	13	18	76	4·7	5·5	9	·87	·13
April . .	83	89	77	12	18	77	5·6	8·8	13	·82	·12
May. . .	83	87	78	9	18	79	6·8	13·2	20	·81	·10
June . .	82	85	78	7	14	81	7·6	8·2	17	·82	·08
July . .	81	85	77	8	14	80	6·8	5·5	12	·83	·08
August .	81	84	77	7	13	80	7·0	4·5	13	·84	·09
September .	81	85	77	8	13	80	6·9	4·9	14	·86	·11
October .	81	85	76	9	15	80	6·8	12·9	21	·86	·11
November .	80	85	74	11	16	79	6·2	12·7	17	·86	·11
December .	80	85	73	12	19	76	5·9	6·4	13	·87	·12
Year .	81					78	6·1	87·3	159	29·85	

Mean highest temperature of year 93°
„ lowest „ „ 68°
„ annual range of temperature 25°
Highest recorded reading (12 yrs.) 95·8° (1885)
Lowest „ „ „ 65·8° (1880)

Absolute range of temperature . 30°
Rainfall of wettest year (17 yrs.) 139·7″ (1878)
„ driest „ „ 57·0″ (1874)
Rainy days in wettest year (12 years) 195 (1885)
„ „ driest „ „ 125 (1876)

84. Akyab—Elevation 20 feet

	Temperature.					Hum.	Cloud.	Rainfall.		Barometer.	
	Mean.	Mean Max.	Mean Min.	M. Range. Daily.	M. Range. Month.	Mean.	Mean.	Ins.	Days.	Mean.	Daily Range.
January .	69	82	59	23	34	72	1·6	0·1	...	30·00	·11
February .	73	85	60	25	37	70	1·1	0·2	...	29·95	·12
March .	79	89	68	21	34	73	1·9	0·5	1	·90	·12
April . .	84	92	75	17	26	74	3·0	1·6	2	·83	·12
May . .	84	91	77	14	25	78	5·3	12·2	14	·76	·10
June . .	82	86	77	9	18	87	7·0	51·6	27	·68	·09
July . .	81	84	77	7	15	89	8·5	51·0	29	·67	·08
August .	81	85	77	8	15	89	8·0	38·6	28	·71	·09
September .	82	87	77	10	16	86	6·9	23·0	22	·76	·11
October .	81	88	76	12	19	83	4·8	12·4	12	·85	·11
November .	78	85	71	14	25	81	3·2	3·9	3	·94	·10
December .	72	82	64	18	30	77	2·5	0·6	2	·99	·11
Year .	79					80	4·5	195·7	140	29·84	

Mean highest temperature of year 96°
„ lowest „ „ 51°
„ annual range of temperature 45°
Highest recorded reading (12 years) 99·1° (1882)
Lowest „ „ „ 47·3° (1878)

Absolute range of temperature . 51·8°
Rainfall of wettest year (28 yrs.) 254·1″ (1850)
„ driest „ „ 142″ (1874)
Rainy days in wettest year (17 years) 156 (1879)
„ „ driest „ „ 118 (1869)

85. MANDALAY—Elevation 250 (?) feet

	Temperature.					Hum.	Cloud.	Rainfall.		Barometer.	
				M. Range.							
	Mean.	Mean Max.	Mean Min.	Daily.	Month.	Mean.	Mean.	Ins.	Days.	Mean.	Daily Range.
January .	71	84	59	25	33	64		...	...	29·75	·15
February .	74	90	60	30	42	54		...	...	·70	·16
March .	83	98	69	29	43	43		...	...	·59	·17
April . .	91	102	81	21	35	49		0·8	4	·50	·19
May . .	90	101	82	19	34	53		4·6*	12*	·48	·17
June . .	86	94	79	15	28	66		5·1*	11*	·45	·14
July . .	84	93	79	14	27	67	8·4	2·9*	10*	·44	·14
August .	82	92	78	14	21	73	9·1	2·4*	10*	·48	·13
September .	83	90	78	12	22	77	8·6	4·7*	10*	·50	·14
October .	[81]	...	...	...	...	[76]	...	4·2	6	[29·61]	...
November .	75	84	68	16	26	75	3·7	2·6	7	·72	·12
December .	72	84	62	22	30	69	1·4	0·2	2	·76	·14
Year .	81 ?					64 ?		27·5	72	29·58 ?	

Highest recorded temperature . . . . 106·1° (1879)
Lowest ,, ,, ... 52·9° (1879)
Annual range . . . . 53·2°

N.B.—The values for October in brackets are interpolated.

* For the five months May to September, the rainfall data are the means of the two years 1879 and 1886.

86. THYET MYO—Elevation 134 feet

	Temperature.					Hum.	Cloud.	Rainfall.		Barometer.	
				M. Range.							
	Mean.	Mean Max.	Mean Min.	Daily.	Month.	Mean.	Mean.	Ins.	Days.	Mean.	Daily Range.
January .	68	85	54	31	46	68	1·8	...	...	29·85	·14
February .	73	92	56	36	48	58	1·0	...	...	·79	·15
March. .	82	100	66	34	48	56	1·2	0·1	...	·73	·16
April . .	87	103	75	28	38	56	2·5	0·7	2	·64	·16
May . .	86	99	78	21	34	67	4·8	5·3	10	·61	·14
June . .	81	90	76	14	23	82	6·4	7·9	20	·58	·10
July . .	80	89	76	13	19	84	6·6	8·0	22	·57	·11
August .	80	89	76	13	18	84	6·3	8·5	20	·59	·11
September .	81	90	76	14	20	82	5·9	7·8	18	·65	·12
October .	81	90	75	15	22	80	5·3	4·9	10	·73	·12
November .	76	87	69	18	31	78	3·8	2·3	4	·80	·12
December .	71	84	60	24	36	72	3·0	...	1	·85	·13
Year .	79					72	4·0	45·5	107	29·70	

Mean highest temperature of year 108°
,, lowest ,, ,, 45°
,, annual range of temperature 63°
Highest recorded reading (7 yrs.) 110·2° (1881)
Lowest ,, ,, ,, 40·3° (1878)

Absolute range of temperature . 69·9°
Rainfall of wettest year (17 years) 62·6″ (1877)
,, driest ,, ,, 33·6″ (1884)
Rainy days in wettest year (9 years) 133 (1879)
,, ,, driest ,, ,, 106 (1878)

87. TOUNGHOO—Elevation 181 feet

	Temperature.					Hum.	Cloud.	Rainfall.		Barometer.	
	Mean.	Mean Max.	Mean Min.	M. Range. Daily.	M. Range. Month.	Mean.	Mean.	Ins.	Days.	Mean.	Daily Range.
January .	70	86	57	29	39	70	0·9	...	...	29·83	·13
February .	74	91	59	32	43	64	0·6	0·2	...	·76	·14
March .	80	98	67	31	44	61	0·5	...	...	·70	·16
April . .	85	100	74	26	35	62	1·8	1·5	5	·63	·17
May . .	83	97	76	21	32	70	4·7	6·6	12	·59	·14
June . .	79	88	75	13	21	84	7·7	13·4	24	·57	·10
July . .	78	87	75	12	19	87	8·0	17·5	26	·56	·10
August .	78	87	75	12	19	88	8·2	18·1	27	·58	·10
September .	80	89	75	14	19	85	6·7	11·8	22	·62	·12
October .	80	90	75	15	20	82	4·3	7·4	14	·69	·12
November .	77	87	71	16	28	80	2·8	1·4	4	·76	·11
December .	72	85	63	22	32	78	2·1	0·2	1	·80	·11
Year .	78					76	4·0	78·1	135	29·67	

Mean highest temperature of year 104°
,, lowest ,, ,, 51°
,, annual range of temperature 53°
Highest recorded reading (9 yrs.) 106·5° (1881)
Lowest ,, ,, ,, 47·1° (1883, 1884)

Absolute range of temperature . 59·4°
Rainfall of wettest year (17 yrs.) 101·3″ (1871)
,, driest ,, ,, 61″ (1873, 1874)
Rainy days in wettest year (9 yrs.) 154 (1884)
,, ,, driest ,, ,, 91 (1878)

88. RANGOON—Elevation 41 feet

	Temperature.					Hum.	Cloud.	Rainfall.		Barometer.	
	Mean.	Mean Max.	Mean Min.	M. Range. Daily.	M. Range. Month.	Mean.	Mean.	Ins.	Days.	Mean.	Daily Range.
January .	75	88	64	24	35	66	2·4	0·2	1	29·96	·13
February .	77	93	65	28	38	62	2·2	0·1	1	·91	·14
March . .	81	97	71	26	37	64	2·2	0·1	...	·87	·15
April . .	84	98	76	22	31	68	3·6	1·8	4	·80	·15
May . .	83	93	77	16	27	76	6·5	10·9	16	·75	·12
June . .	79	86	77	9	18	88	8·8	18·4	28	·73	·09
July . .	78	85	76	9	16	90	9·0	21·3	29	·73	·08
August . .	78	85	76	9	16	91	9·0	18.6	27	·75	·09
September .	78	85	76	9	16	89	8·3	16·0	26	·79	·11
October .	80	88	76	12	18	85	6·2	8·1	14	·84	·12
November .	78	87	73	14	24	80	4·5	3·4	6	·89	·11
December .	76	87	68	19	29	74	3·2	0·1	1	·94	·12
Year .	79					78	5·5	99·0	153	29·83	

Mean highest temperature of year 104°
,, lowest ,, ,, 58°
,, annual range of temperature 46°
Highest recorded reading (11 yrs.) 106·7° (1877)
Lowest ,, ,, ,, 55·7° (1883)

Absolute range of temperature . 51°
Rainfall of wettest year (17 yrs.) 143·4″ (1871)
,, driest ,, ,, 69·1″ (1874)
Rainy days in wettest year (10 yrs.) 161 (1880, 1882)
Rainy days in driest year (10 yrs.) 132 (1878)

89. Moulmein—Elevation 94 feet

	Temperature.					Hum.	Cloud.	Rainfall.		Barometer.	
				M. Range.							
	Mean.	Mean Max.	Mean Min.	Daily.	Month.	Mean.	Mean.	Ins.	Days.	Mean.	Daily Range.
January .	75	89	64	25	34	63	0·8	...	...	29·89	·12
February .	77	92	66	26	35	63	0·7	0·1	...	·84	·12
March . .	81	94	72	22	31	66	0·6	0·1	...	·81	·13
April . .	83	95	76	19	27	69	2·4	3·0	7	·75	·13
May . .	82	91	76	15	24	76	5·0	19·7	18	·70	·12
June . .	78	84	75	9	18	89	8·2	38·4	28	·68	·09
July . .	77	83	74	9	15	90	8·3	43·9	29	·68	·08
August . .	77	83	74	9	18	91	8·2	43·0	28	·69	·09
September .	78	84	74	10	18	87	6·3	30·3	27	·72	·11
October .	80	88	74	14	20	81	3·2	8·4	15	·76	·12
November .	78	88	72	16	26	74	1·7	1·5	5	·80	·11
December .	76	87	67	20	30	70	1·3	0·1	1	·85	·11
Year .	78					77	3·9	188·5	158	29·77	

Mean highest temperature of year 99°
„ lowest „ „ 58°
„ annual range of temperature 41°
Highest recorded reading (10 yrs.) 101·5° (1878)
Lowest „ „ „ 54·8° (1883)

Absolute range of temperature . 46·7°
Rainfall of wettest year (37 yrs.) 279″ (1865)
„ driest „ „ 134·1″[1] (1878)
Rainy days in wettest year (10 yrs.) 175 (1879)
„ „ driest „ „ 135 (1878)

[1] Only 124·2 inches was recorded in 1850, and 124·3 inches in 1852; but up to 1865 no rain whatever was registered in the five dry months November to March, and the registers are probably incomplete.

90. Mergui—Elevation 96 feet

	Temperature.					Hum.	Cloud.	Rainfall.		Barometer.	
				M. Range.							
	Mean.	Mean Max.	Mean Min.	Daily.	Month.	Mean.	Mean.	Ins.	Days.	Mean.	Daily Range.
January .	77	89	68	21	30	73	2·7	0·5	2	29·87	·11
February .	79	90	71	19	20	75	3·1	1·4	3	·86	·12
March .	80	92	73	19	25	77	3·8	2·3	6	·83	·12
April . .	81	92	74	18	24	79	5·4	5·7	8	·78	·12
May . .	80	90	75	15	23	83	7·3	15·9	16	·75	·10
June . .	77	85	73	12	20	89	9·1	29·8	25	·74	·09
July . .	76	84	73	11	18	89	9·0	31·6	28	·74	·08
August .	77	84	73	11	18	90	8·7	28·1	26	·75	·09
September .	77	85	73	12	17	90	8·7	26·3	25	·78	·10
October .	78	87	73	14	19	86	7·3	13·5	20	·80	·11
November .	77	87	71	16	24	80	5·3	3·5	7	·82	·11
December .	76	88	69	19	27	74	2·8	0·5	2	·85	·10
Year .	78					82	6·1	159·1	168	29·80	

Mean highest temperature of year 97°
„ lowest „ „ 61°
„ annual range of temperature 36°
Highest recorded reading (9 years) 98·8° (1878)
Lowest „ „ „ 57° (1882)

Absolute range of temperature . 41·8°
Rainfall of wettest year (22 years) 213″ (1882)
„ driest „ „ 132″ (1875)
Rainy days in wettest year (9 years) 188 (1878)
„ „ driest „ „ 149 (1885)

91. Port Blair—Elevation 61 feet

	Temperature.					Hum.	Cloud.	Rainfall.		Barometer.	
	Mean.	Mean Max.	Mean Min.	M. Range. Daily.	M. Range. Month.	Mean.	Mean.	Ins.	Days.	Mean.	Daily Range.
January .	79	86	75	11	18	77	3·4	0·9	2	29·90	·10
February .	80	88	74	14	20	78	2·8	1·3	2	·89	·11
March .	82	91	76	15	21	78	2·9	0·4	1	·87	·11
April . .	83	92	78	14	20	78	4·1	2·4	5	·80	·11
May . .	81	89	78	11	19	84	5·0	15·9	20	·75	·09
June . .	81	86	78	8	16	86	7·0	17·9	25	·73	·08
July . .	80	86	77	9	15	86	7·1	16·5	26	·74	·07
August .	80	85	77	8	15	87	6·9	15·2	25	·75	·08
September .	79	85	76	9	15	88	6·9	19·6	25	·78	·09
October .	80	87	77	10	15	87	5·8	11·8	22	·81	·10
November .	80	87	77	10	16	84	5·0	9·5	15	·84	·10
December .	79	86	76	10	18	81	4·5	5·3	9	·88	·10
Year .	80					83	5·1	116·7	177	29·81	

Mean highest temperature of year 95°
,, lowest ,, ,, 69°
,, annual range of temperature 26°
Highest recorded reading (12 yrs.) 96·4° (1878, 1885)
Lowest recorded reading (12 yrs.) 66° (1875)

Absolute range of temperature . 30·4°
Rainfall of wettest year (19 yrs.) 149·8″ (1869)
,, driest ,, ,, 100·1″ (1871)
Rainy days in wettest year (14 yrs.) 206 (1882)
,, ,, driest ,, ,, 158 (1885)

92. Nancowry—Elevation 81 feet

	Temperature.					Hum.	Cloud.	Rainfall.		Barometer.	
	Mean.	Mean Max.	Mean Min.	M. Range. Daily.	M. Range. Month.	Mean.	Mean.	Ins.	Days.	Mean.	Daily Range.
January .	80	86	77	9	13	75	5·2	2·9	8	29·86	·09
February .	81	87	78	9	15	74	4·7	1·3	4	·86	·10
March .	82	89	78	11	15	74	4·6	2·1	5	·84	·10
April . .	83	89	79	10	17	76	5·3	5·3	9	·80	·10
May . .	81	88	79	9	17	79	6·6	12·1	18	·77	·09
June . .	80	87	78	9	16	82	7·3	12·7	22	·76	·08
July . .	80	86	77	9	16	82	6·9	12·6	20	·77	·07
August .	79	86	77	9	16	82	6·9	12·6	19	·78	·08
September .	79	87	77	10	16	81	6·9	10·6	19	·80	·09
October .	79	86	76	10	17	84	6·8	13·5	21	·81	·09
November .	79	86	76	10	16	82	6·5	11·8	18	·82	·09
December .	79	86	76	10	15	79	5·8	12·2	15	·84	·09
Year .	80					79	6·1	109·7	178	29·81	

Mean highest temperature of year 94°
,, lowest ,, ,, 72°
,, annual range of temperature 22°
Highest recorded reading (10 yrs.) 90·5° (1884)
Lowest ,, ,, (12 yrs.) 70·2° (1875)

Absolute range of temperature . 20·3°
Rainfall of wettest year (18 yrs.) 141·3″ (1882)
,, driest ,, ,, 79·3″ (1877)
Rainy days in wettest year (12 yrs.) 227 (1879)
,, ,, driest ,, ,, 155 (1874)

APPENDIX II

TABULAR LIST OF STORMS IN THE BAY OF BENGAL, THEIR PLACE OF ORIGIN, TRACKS, AND RATE OF PROGRESSION

Date.		Place of Origin.		Subsequent Course.		Coast Crossing.	Authority.
		N. Lat.	E. Long.	Direction.	Miles per Hour.		
April and May	28th April–2d May 1872	7°– 8°	83°–84°	N.W. to N.N.W.	4–5	Sadras	Auct.
	2d–5th May 1874	8°–10°	84°–86°	N.W.	7–8	Madras	"
	12th–20th May 1877	8°–10°	83°–84°	N.W. to N.N.W.	4–7	Pennair River	Eliot
	18th–21st May 1879	12°–14°	84°–88°	W.N.W.	?	Madras	"
	28th–29th May 1879	20°–21°	89°–91°	W. by N.	3–4	Balasore Bay	"
	29th May–2d June 1880	18°–20°	89°–91°	N.N.W.	3	Hooghly River	"
	25th–27th May 1881	20°–21°	90°–91°	N. by W.	12	Sunderbuns	"
	13th–17th May 1884	11°	89°–90°	N. to N.E.	5–8	Akyab	"
	22d–24th May 1886	10°–11°	83°–84°	W.N.W.	7	Negapatam	"
	21st–26th May 1887	16°–17°	91°–92°	N.W.	5–8	Balasore Bay	"
June	5th–9th June 1869	18°–20°	89°	N.	2	Mutlah River	Auct.
	28th June–1st July 1872	20°–21°	88°	N.W. to W.	8–10	Balasore	Willson
	25th June 1880	19°	88°–89°	W.N.W.	3	Chilka Lake	Eliot
	2d–4th June 1881	19°	89°	N.	4	Mutlah River	"
	5th June 1881	21°–22°	88°–89°	W.	?	Balasore Bay	"
	13th–14th June 1881	21°–22°	89°–90°	N.W.	6–7	Hooghly River	"
	14th June 1883	21°–22°	87°–88°	N.N.W.	?	Balasore	"
	26th–29th June 1883	20°–21°	89°–90°	N.W.	2–3	"	"
	18th–19th June 1884	20°–21°	88°–89°	W.	3–4	False Point	"
	30th June–1st July 1884	20°–21°	88°–89°	N.W.	4–5	Saugor Island	"
	16th June 1885	21°–22°	89°–90°	N.	?	Sunderbuns	"
	25th–26th June 1885	21°–22°	87°–88°	N.W.	?	Balasore	"
	14th–16th June 1886	20°	89°	N. to N.W.	7–8	Hooghly River	"

Date.		Place of Origin.		Subsequent Course.		Coast Crossing.	Authority.
		N. Lat.	E. Long.	Direction.	Miles per Hour.		
July	27th–29th July 1874	20°–21°	88°–89°	N. by W.	2–3	Saugor Island	Anct.
	10th–12th July 1877	21°	90°–91°	W. by N.	3–6	Balasore	Eliot
	20th–22d July 1878	18°–19°	88°	W.N.W. to N.W.	4½	Ganjam	,,
	13th–15th July 1880	21°	89°–90°	W. to W. by N.	4½	Balasore Bay	,,
	2d July 1881	20°–21°	88°–90°	W.	13	False Point	,,
	12th–16th July 1881	21°	90°–91°	W. by N. to N.W.	1–3	Saugor Island	,,
	11th–12th July 1882	20°–21°	89°–90°	W. by N.	4–5	False Point	,,
	17th–19th July 1882	20°–21°	89°–90°	N.W.	3	Saugor Island	,,
	23d July 1882	20°–21°	89°–90°	N.W.	6	Balasore Bay	,,
	30th–31st July 1882	19°–20°	88°–90°	W.N.W.	4	False Point	,,
	2d–4th July 1883	20°–21°	90°	N.N.W.	5	Sunderbuns	,,
	6th–7th July 1883	21°	88°–89°	W.N.W.	3	Balasore Bay	,,
	12th–14th July 1883	19°–20°	89°–90°	W.	5	Chilka Lake	,,
	26th–27th July 1883	21°–22°	90°–91°	W.N.W.	6	Saugor Island	,,
	8th–10th July 1884	20°–21°	89°	W. by N.	3–4	False Point	,,
	15th–16th July 1884	19°–20°	88°–89°	W.N.W.	8–9	Pooree	,,
	18th–21st July 1884	21°–22°	88°–89°	W. by N.	1–2	Balasore Bay	,,
	13th–15th July 1885	19°–20°	88°–90°	W.N.W.	5	False Point	,,
	22d–24th July 1885	21°	89°	N.W.	6	Hooghly River	,,
	17th–18th July 1886	20°–21°	88°–89°	W.N.W.	7	Balasore Bay	,,
August	5th–7th August 1877	21°	89°–90°	N.	6	Sunderbuns	Eliot
	13th–15th August 1877	21°	89°	N.N.W.	4	,,	,,
	18th–20th August 1877	20°–21°	90°–91°	N.N.W.	3	,,	,,
	27th–30th August 1877	18°–20°	89°–90°	N.W.	3–4	Balasore	,,
	1st–3d August 1878	19°–20°	87°–88°	W.N.W.	?	Pooree	,,
	7th–8th August 1879	21°	88°	W. by N.	?	Balasore	,,
	18th–21st August 1880	21°	89°–90°	W. by N.	3	N. of False Point	,,
	31st July–2d August 1880	20°	89°	W.N.W.	2	,,	,,

Date.		Place of Origin.		Subsequent Course.		Coast Crossing.	Authority.
		N. Lat.	E. Long.	Direction.	Miles per Hour.		
	1st–2d August 1881	20°–21°	88°–90°	W.	17 ?	False Point	Eliot
	16th–18th August 1881	21°	89°–90°	W.N.W.	13 ?	Balasore	,,
	18th–19th August 1882	21°	90°	W.N.W.	6	,,	,,
	22d–25th August 1883	18°–19°	88°–89°	W. by N.	3–4	Ganjam	,,
	1st–2d August 1884	21°	88°–89°	N. by W.	?	Saugor Island	,,
	15th–16th August 1885	21°	88°–89°	W.N.W.	?	Balasore Bay	,,
	11th–18th August 1886	16°	86°–87°	N.E.	1–2½	Pooree	,,
September	29th–30th September 1842	16°–18°	90°–91°	W.N.W.	?	,,	Piddington
	11th September 1872	21°–22°	90°–92°	?	?	Chittagong	Auct.
	19th–20th September 1872	21°–22°	90°	N. by W.	?	Sunderbuns	,,
	4th–7th September 1877	19°–20°	87°–89°	N.W.	?	False Point	Eliot
	13th–14th September 1878	19°–20°	90°–91°	W. by N.	6–7	Pooree	,,
	18th–21st September 1878	18°	89°–90°	N.N.W.	2–5	Mutlah River	,,
	12th–16th September 1879	21°	90°	W. by N.	2–3	Orissa Coast	,,
	21st–24th September 1879	19°–20°	89°–90°	W. by N.	3	,,	,,
	12th–14th September 1880	18°–19°	89°	W. by N.	9	Ganjam	,,
	21st–22d September 1880	20°	88°	W.N.W.	5	Balasore	,,
	10th–12th September 1881	20°	90°	W. by N.	7	False Point	,,
	6th–8th September 1882	19°–20°	90°	W. by N.	4	,,	,,
	14th–16th September 1882	18°	87°	W.N.W. to N.N.W.	2–4	Chilka Lake	,,
	26th–28th September 1882	15°–17°	93°–94°	N.W.	5–11	False Point	,,
	6th–8th September 1884	19°–20°	90°	W. by N.	4	,,	,,
	16th–20th September 1884	17°–19°	89°–90°	W. by N.	2–5	Vizagapatam	,,
	15th September 1885	21°–22°	88°–89°	W. by N.	?	Balasore	,,
	19th–22d September 1885	14°–15°	92°–93°	N.W.	3–15	False Point	Pedler
	17th–21st September 1886	20°–21°	87°–88°	N.N.W.	1–2	Balasore	Eliot
	27th–29th September 1886	16°	90°	N.W. to W.	3–14	Vizagapatam	,,

Date.		Place of Origin.		Subsequent Course.		Coast Crossing.	Authority.
		N. Lat.	E. Long.	Direction.	Miles per Hour.		
October .	22d–24th October 1842 . .	9°–12°	90°–91°	W. by N.	11–12	Pondicherry	Piddington
	11th–12th October 1848 . .	16°–17°	90°	N.W.	4–8	False Point	„
	21st–22d October 1851 . .	16°–18°?	86°–87°?	N.	5–6	„	„
	1st–5th October 1864 . . .	13°–14°	91°–92°	N.N.W. to N.	7–9	Hooghly River	Auct.
	26th October–1st November 1867	7°– 8°	91°–92°	N.N.W. to N.	7	Mutlah River	„
	13th–15th October 1874 . .	16°–17°	89°–90°	N.N.W.	7	Balasore	Willson
	5th–7th October 1876 . .	14°–15°	93°–94°	W.N.W.	8–9	Bimlipatam	Eliot
	26th October–1st November 1876	10°–12°	89°–91°	N. to N.E.	4–12	Backergunj	„
	13th–15th October 1878 . .	15°–16°	82°–84°	N.N.W.	5	Cocanada	„
	27th–30th October 1878 . .	13°–15°	82°–84°	N.N.W. to N.W.	4–5	„	„
	10th–14th October 1879 . .	18°–19°	94°–96°	W.N.W.	4–5	Balasore	„
	25th–27th October 1879 . .	19°–20°	92°–94°	N. by W. to N.	10	Chittagong	„
	12th–14th October 1880 . .	16°–18°	85°–88°	W. and W.N.W.	7	Vizagapatam	„
	2d–5th October 1881 . . .	16°–18°	86°–88°	N.W.	4	N. of Vizagapatam	„
	11th–14th October 1882 . .	13°–14°	85°–86°	N.	11–12	Pooree	„
	26th–31st October 1882 . .	8°– 9°	88°–90°	N.W.	6–7	Vizagapatam	„
	15th–17th October 1884 . .	9°–11°	82°–84°	W.	10	Negapatam	„
	24th–26th October 1884 . .	15°–16°	85°	N. to N.E.	8–13	Saugor Island	„
	30th October–1st November 1884	17°–19°	89°–91°	N. and N.N.E.	7	Chittagong	„
	16th–19th October 1886 . .	16°	88°–89°	W. by N.	5–6	Cocanada	„
	29th Nov.–1st Dec. 1845 . .	6°– 7°	86°–88°	W.N.W.	6	Batticaloa	Piddington
	16th–19th November 1850 . .	11°–13°	93°–95°	N.N.W.	6–7	W. of C. Negrais	„
	3d–5th November 1877 . .	16°–18°	90°–92°	N.N.E. to N. by E.	6	Chittagong	Eliot
	3d–5th November 1878 . .	13°–15°	84°–86°	N. by W. to N.N.W.	4	Vizagapatam	„
	1st–2d November 1879 . .	15°–16°	96°–98°	N. to N. by W.	3–4	Rangoon	„
	15th–19th November 1879 . .	10°–12°	86°–88°	W.N.W. to N.W.	6	Madras	„
	19th–21st November 1880 . .	8°–10°	82°–84°	W.N.W.	4–5	Negapatam	„
	10th–13th November 1881 . .	9°–11°	82°–84°	N.W.	6	Madras	„

Date.		Place of Origin.		Subsequent Course.		Coast Crossing.	Authority.
		N. Lat.	E. Long.	Direction.	Miles per Hour.		
November	21st–24th November 1882	8°–10°	88°–90°	W.N.W.	8	Sadras	Eliot
	28th–30th November 1882	9°–11°	86°–88°	W.N.W.	9–12	Cuddalore	"
	9th–14th November 1883	11°	95°	N.W. to N. and N.E.	5–11	Akyab	"
	27th Nov.–6th Dec. 1883	8°–10°	89°–91°	N.W. to N. & N.N.E.	4–9	Chittagong	"
	3d–5th November 1884	11°	83°–84°	W. by N.	3	Off Cuddalore	"
	19th–21st November 1884	10°–12°	84°–88°	W.N.W.	8	Madras	"
	1st–3d November 1885	12°–14°	83°–85°	W.N.W.	10	Pulicat Lake	"
	17th–23d November 1885	10°–12°	81°–83°	N.N.E. to N.E.	3–15	Chittagong	"
	2d–9th November 1886	10°–11°	92°	W. by N. to W.N.W.	1–8	Madras	"
	14th–23d November 1886	10°–11°	94°	N. to N.W. and W.	1–9	Vizagapatam	"
December	4th–7th December 1878	16°–17°	88°–90°	W.N.W.	5	"	"
	1st–6th December 1883	See November		N.W. to N. & N.N.E.	5–12	Chittagong	"
	12th–18th December 1884	7°– 8°	82°–84°	N.W.	2–7	Negapatam	"
	7th–9th December 1886	9°	88°	N.W. by W.	15	Madras	"

APPENDIX III

MEAN, HIGHEST, AND LOWEST RECORDED MONTHLY RAINFALL IN INCHES AT 114 STATIONS IN INDIA, CEYLON, AND BURMAH

		Jan.	Feb.	March.	April.	May.	June.	July.	August.	Sept.	Oct.	Nov.	Dec.
BALUCHISTAN													
Quetta, 9 years	Mean	1·6	1·9	2·4	1·3	0·5	0·1	0·8	0·6	0·2	0·1	...	0·4
	Max.	2·3	3·4	4·0	5·1	2·0	0·5	2·7	2·2	1·1	0·8	0·2	3·0
	Min.	0·1	0·3	0·6	0·1	...	...	...	...	...	...	...	...
PUNJAB													
Murree, 17 years	Mean	4·3	5·6	4·5	3·6	3·6	3·4	12·2	13·1	5·6	1·7	1·5	2·4
	Max.	14·2	15·9	8·5	8·9	9·8	14·0	25·1	24·5	11·6	9·5	10·2	10·1
	Min.	...	0·4	0·3	0·5	0·7	0·5	2·2	1·2	1·8	...	...	...
Kangra, 16 to 19 years	Mean	2·6	3·6	3·2	0·8	1·4	5·9	23·9	20·9	8·7	0·5	0·4	1·4
	Max.	5·5	12·4	7·2	3·3	6·9	14·6	41·5	40·7	21·6	3·6	5·0	6·9
	Min.	0·1	0·6	...	...	...	0·1	4·6	1·4	1·5	...	...	...
Simla, 25 years	Mean	2·8	2·7	3·0	2·8	4·7	7·9	19·3	18·1	6·0	1·4	0·3	1·1
	Max.	7·1	7·6	8·5	14·7	13·3	15·1	36·4	30·7	14·3	6·3	5·0	6·1
	Min.	...	...	...	...	0·1	2·2	8·8	6·5	2·1	...	...	...
Peshawar, 28 years	Mean	1·6	1·2	1·8	2·0	0·7	0·3	1·7	2·0	0·8	0·2	0·6	0·6
	Max.	5·1	5·0	5·8	7·4	3·8	3·9	4·9	11·3	7·0	1·6	8·5	3·7
	Min.	...	...	...	...	...	...	...	...	...	...	...	...

		Jan.	Feb.	March.	April.	May.	June.	July.	August.	Sept.	Oct.	Nov.	Dec.
Bannu, 25 years	Mean	0·9	0·9	1·5	1·4	0·8	1·0	2·0	3·0	0·8	0·2	0·3	0·4
	Max.	2·6	3·0	5·3	5·0	3·6	4·9	5·3	10·7	3·9	2·6	2·9	2·1
	Min.	...	...	...	...	...	...	...	...	...	...	...	...
D. I. Khan, 25 years	Mean	0·4	0·7	0·9	0·8	0·4	0·6	1·8	1·6	0·6	0·1	0·1	0·3
	Max.	2·1	3·1	3·3	2·6	2·2	4·2	8·6	6·3	4·4	1·2	1·8	2·0
	Min.	...	...	...	...	...	...	...	...	...	...	...	...
Mooltan, 25 years	Mean	0·4	0·3	0·5	0·3	0·5	0·4	2·2	1·3	0·8	0·1	0·1	0·3
	Max.	1·4	2·3	4·0	2·9	1·6	1·3	7·7	4·1	9·2	2·5	0·6	1·4
	Min.	...	...	...	...	...	...	...	...	...	...	...	...
Rawalpindi, 29 years	Mean	2·4	2·0	1·9	2·3	1·6	1·7	7·4	7·3	3·2	0·6	0·9	1·1
	Max.	7·5	5·8	12·1	7·7	8·2	5·7	16·0	15·6	10·5	2·9	10·0	4·0
	Min.	...	0·2	...	...	...	...	2·2	...	0·9	...	...	...
Jhelum, 26 years	Mean	1·6	1·2	1·7	1·1	0·8	2·0	7·0	5·4	2·4	0·6	0·3	0·8
	Max.	9·2	5·7	5·7	3·1	5·2	7·5	18·8	14·0	8·3	3·8	4·3	4·1
	Min.	...	...	...	...	...	...	2·1	1·1	...	...	...	...
Shahpur, 33 years	Mean	0·8	1·1	0·9	0·6	0·7	1·1	3·8	3·2	1·8	0·2	0·2	0·4
	Max.	2·7	9·3	3·3	2·2	3·5	4·4	10·5	12·6	9·0	1·4	3·1	2·3
	Min.	...	...	...	...	...	...	0·1	0·5	...	...	...	...
Sialkot, 29 years	Mean	1·4	1·8	1·9	1·6	1·2	3·2	11·6	9·1	3·2	0·6	0·4	0·8
	Max.	4·2	5·4	7·8	6·5	5·5	12·9	33·2	25·8	14·8	3·9	4·4	4·7
	Min.	...	...	...	...	...	...	2·0	0·7	...	...	...	...
Lahore, 30 years	Mean	0·7	1·1	1·1	0·6	0·9	1·8	7·4	4·6	2·4	0·6	0·2	0·5
	Max.	2·4	4·7	6·5	3·3	4·4	8·8	17·3	16·4	11·4	3·6	1·3	2·6
	Min.	...	...	...	...	...	...	1·1	...	...	...	...	...

		Jan.	Feb.	March.	April.	May.	June.	July.	August.	Sept.	Oct.	Nov.	Dec.
Hoshiarpur, 30 years .	Mean	1·6	1·6	1·5	0·7	1·0	3·3	10·2	9·1	4·3	0·4	0·1	1·2
	Max.	4·7	5·2	5·6	3·9	3·8	15·4	23·9	26·4	15·0	5·0	1·0	10·7
	Min.	...	...	...	...	...	0·2	2·3	0·8	...	...	...	...
Ludhiana, 27 to 31 yrs.	Mean	1·3	1·2	1·6	0·9	1·2	2·3	9·7	6·5	4·3	1·2	0·1	0·9
	Max.	3·4	4·1	7·1	4·5	4·5	8·7	26·3	15·5	16·4	9·5	1·0	6·3
	Min.	...	...	...	...	...	...	2·3	0·4	0·2	...	...	...
Umballa, 35 to 37 years	Mean	1·4	1·5	1·1	0·6	1·0	4·0	11·6	8·6	4·1	0·6	0·2	0·7
	Max.	9·6	7·1	7·3	3·9	3·6	14·8	30·5	25·9	12·8	6·0	2·3	5·2
	Min.	...	...	...	...	...	...	2·8	...	...	...	...	...
Delhi, 34 to 36 years .	Mean	1·0	0·5	0·7	0·4	0·7	3·4	8·5	6·9	4·5	0·5	0·1	0·4
	Max.	6·8	2·1	4·3	2·5	3·7	12·9	24·4	19·4	27·4	5·3	0·9	3·2
	Min.	...	...	...	...	...	...	1·1	...	0·1	...	...	...
Sirsa, 34 to 36 years .	Mean	0·7	0·3	0·5	0·4	0·6	2·3	3·7	3·6	2·0	0·3	...	0·3
	Max.	2·7	2·0	3·1	3·2	2·7	6·3	11·8	10·3	9·5	3·8	0·7	3·1
	Min.	...	...	...	...	...	0·3	0·2	...	...	...	...	...
N.W. PROVINCES													
Mussoorie, 23 to 35 yrs.	Mean	2·3	2·9	2·9	1·7	2·8	9·6	30·2	30·2	9·9	1·0	0·3	1·1
	Max.	6·7	7·7	9·2	5·6	7·6	21·4	57·0	56·6	21·9	6·5	3·2	5·6
	Min.	...	0·2	...	...	...	0·5	2·6	7·9	2·1	...	...	...
Dehra Dun, 37 to 39 yrs.	Mean	2·2	1·9	1·4	0·7	1·6	8·6	25·0	24·8	9·3	0·8	0·1	0·6
	Max.	11·1	7·2	7·1	3·5	6·2	23·5	58·1	64·5	33·9	5·1	1·5	3·8
	Min.	...	...	...	...	...	0·1	3·7	6·6	0·3	...	...	...

		Jan.	Feb.	March.	April.	May.	June.	July.	August.	Sept.	Oct.	Nov.	Dec.
Ranikhet, 16 years	Mean	1·9	2·0	2·2	1·3	3·0	6·0	12·7	11·6	6·0	1·1	0·3	1·0
	Max.	5·3	4·3	6·2	7·0	8·1	11·6	21·8	21·1	23·4	8·9	1·2	5·5
	Min.	...	...	...	...	0·2	0·6	6·0	2·8	0·7	...	...	...
Naini Tal, 36 to 38 yrs.	Mean	3·0	2·3	2·5	1·6	3·0	14·9	25·0	25·0	10·8	1·8	0·2	1·3
	Max.	7·6	8·6	12·4	6·0	8·5	34·9	54·2	46·9	30·0	12·3	1·9	11·0
	Min.	...	...	...	...	...	1·3	3·7	5·9	0·7	...	...	...
Roorkee, 32 to 33 years	Mean	2·0	1·4	1·0	0·4	1·2	5·1	12·5	12·3	5·1	0·6	0·2	0·4
	Max.	10·6	5·7	7·4	2·2	5·2	15·7	21·3	27·0	13·9	2·6	2·0	3·2
	Min.	...	...	...	...	...	...	4·2	2·0	...	...	...	...
Meerut, 37 to 39 years	Mean	1·0	0·7	0·7	0·4	0·8	3·6	9·2	7·2	4·0	0·5	0·1	0·3
	Max.	4·7	5·2	4·5	2·1	4·4	13·9	17·7	17·0	17·0	2·4	0·6	3·8
	Min.	...	...	...	...	...	...	0·8	...	...	...	...	...
Moradabad, 37 to 39 years	Mean	1·2	1·0	0·8	0·3	1·0	5·3	13·8	9·9	5·9	0·8	0·1	0·4
	Max.	5·3	5·2	3·5	2·1	4·5	17·1	35·3	21·0	16·9	7·7	2·2	5·2
	Min.	...	...	...	...	...	0·2	3·0	0·6	...	...	...	...
Bareilly, 37 to 39 years	Mean	1·0	0·8	0·7	0·3	0·9	6·0	14·5	9·3	6·7	1·8	0·1	0·3
	Max.	3·7	4·9	3·3	1·7	2·7	17·9	34·6	28·7	24·2	12·4	1·0	3·4
	Min.	...	...	...	...	...	...	1·3	0·8	...	...	...	...
Agra, 37 to 39 years	Mean	0·5	0·3	0·2	0·2	0·7	2·9	9·8	6·7	4·3	0·4	...	0·2
	Max.	2·4	2·2	2·9	1·5	4·1	8·6	18·8	18·3	13·4	4·5	0·6	2·4
	Min.	...	...	...	...	...	...	1·6	...	...	...	...	...
Lucknow, 19 to 20 yrs.	Mean	0·8	0·3	0·3	0·1	0·9	5·0	10·8	10·4	7·1	1·4	...	0·5
	Max.	4·7	1·2	1·4	0·9	4·7	15·7	19·1	23·8	18·5	9·4	0·2	3·9
	Min.	...	...	...	...	...	...	2·9	1·0	0·9	...	...	...

		Jan.	Feb.	March.	April.	May.	June.	July.	August.	Sept.	Oct.	Nov.	Dec.
Cawnpore, 37 to 39 yrs.	Mean	0·7	0·4	0·2	0·1	0·5	3·5	9·7	8·8	4·8	1·2	0·1	0·2
	Max.	3·1	2·4	1·2	1·2	2·1	11·0	21·9	26·6	11·8	6·5	1·7	2·6
	Min.	...	...	...	...	...	...	1·7	0·3	0·3	...	...	...
Allahabad, 37 to 39 yrs.	Mean	0·8	0·4	0·4	0·2	0·3	4·6	11·9	9·6	6·7	2·3	0·2	0·2
	Max.	4·2	2·7	3·5	1·4	2·5	17·9	21·9	23·3	18·7	11·3	5·3	2·9
	Min.	...	...	...	...	...	...	1·0	0·6	0·1	...	...	...
Jhansi, 26 to 27 years	Mean	0·5	0·2	0·4	0·1	0·3	4·0	13·6	10·5	5·2	0·8	...	0·2
	Max.	2·5	0·9	2·0	0·7	1·1	10·9	31·5	28·3	13·7	4·1	0·5	2·5
	Min.	...	...	...	...	...	...	0·5	0·2	0·6	...	...	...
Gorakhpur, 37 to 39 yrs.	Mean	0·7	0·5	0·4	0·3	1·5	7·7	13·3	11·8	8·8	3·0	0·2	0·1
	Max.	3·5	2·7	2·3	2·6	5·9	20·8	27·3	25·0	21·1	17·1	4·1	1·4
	Min.	...	...	...	...	...	0·5	3·5	4·1	0·7	...	...	...
Benares, 37 to 39 years	Mean	0·7	0·5	0·4	0·2	0·5	5·0	12·8	10·7	6·5	2·1	0·1	0·1
	Max.	3·4	3·1	1·8	1·5	3·6	18·8	22·5	27·1	13·9	7·0	1·7	3·0
	Min.	...	...	...	...	...	0·1	2·7	1·8	0·1	...	...	...
RAJPUTANA													
Bickanir, 9 years	Mean	0·3	0·2	0·1	0·2	1·6	1·8	3·9	2·7	1·5	0·1	0·1	...
	Max.	1·1	0·6	0·3	1·0	3·6	6·4	7·8	6·3	4·8	0·5	0·7	0·1
	Min.	...	...	...	...	0·3	...	0·1	0·1	...	...	...	...
Jeypore, 18 to 20 years	Mean	0·3	0·2	0·2	0·2	0·7	3·3	9·0	6·1	3·1	0·2	0·1	0·4
	Max.	1·2	0·9	1·4	1·0	2·0	13·6	21·5	20·8	14·7	1·5	0·5	1·5
	Min.	...	...	...	...	...	...	0·8	...	0·4	...	...	...

		Jan.	Feb.	March.	April.	May.	June.	July.	August.	Sept.	Oct.	Nov.	Dec.
Ajmere, 24 years .	Mean	0·2	0·3	0·4	0·1	0·7	2·5	6·9	7·3	3·4	0·3	0·1	0·3
	Max.	1·0	2·8	5·0	0·7	2·8	9·7	16·2	20·0	17·8	1·8	1·3	1·8
	Min.	...	...	...	...	...	...	0·4	0·2	...	...	...	...
Jodhpur, 13 years .	Mean	0·3	0·2	...	...	0·6	1·3	3·6	5·2	2·3	0·2	...	0·1
	Max.	2·1	1·0	0·2	0·4	2·1	5·0	8·1	14·6	10·6	2·6	0·4	0·6
	Min.	...	...	...	...	...	...	...	...	...	...	...	...
Abu, 27 years . .	Mean	0·2	0·4	0·1	...	1·0	5·1	22·2	22·5	9·1	2·1	0·2	0·2
	Max.	1·1	2·9	1·9	0·5	6·8	32·7	49·4	60·4	50·9	39·5	3·4	1·3
	Min.	...	...	...	...	...	...	4·9	0·7	0·4	...	...	...
CENTRAL INDIA													
Morar, 15 years . .	Mean	0·3	0·3	0·2	0·1	0·5	3·1	10·3	10·3	5·4	0·6	...	0·4
	Max.	1·6	2·1	0·9	1·3	2·1	9·0	24·2	26·9	15·3	3·9	0·2	3·2
	Min.	...	...	...	...	...	...	0·9	0·4	0·1	...	...	...
Neemuch, 19 to 20 yrs.	Mean	0·1	0·2	0·1	0·1	0·5	3·9	11·2	10·4	5·5	1·0	...	0·2
	Max.	0·4	1·6	1·7	0·9	2·7	11·6	24·3	22·3	14·5	8·2	0·5	1·9
	Min.	...	...	...	...	...	0·4	1·8	...	...	...	...	...
Indore, 19 years . .	Mean	0·4	0·3	...	0·1	0·6	6·8	10·4	7·8	8·1	1·2	0·2	0·2
	Max.	2·5	2·9	0·3	1·9	4·2	21·5	19·4	19·7	16·2	3·7	1·2	0·8
	Min.	...	...	...	...	...	2·6	2·7	3·4	0·6	...	...	...
BENGAL													
Patna, 34 to 35 years .	Mean	0·7	0·5	0·3	0·3	1·6	7·1	11·0	10·1	7·9	2·9	0·2	0·2
	Max.	3·7	2·4	2·7	1·7	5·4	18·1	23·4	18·6	26·2	7·6	1·5	1·8
	Min.	...	...	...	...	...	0·6	3·0	1·3	0·9	...	...	...

		Jan.	Feb.	March.	April.	May.	June.	July.	August.	Sept.	Oct.	Nov.	Dec.
Gya, 24 to 26 years	Mean	0·7	0·5	0·4	0·3	1·2	6·2	11·9	9·9	7·0	2·5	0·2	0·2
	Max.	2·7	2·5	2·9	2·6	11·1	14·6	20·7	20·1	16·9	12·2	3·3	3·0
	Min.	...	...	...	...	...	0·5	3·3	2·3	1·1	...	...	...
Purneah, 16 to 17 years	Mean	0·3	0·6	0·4	1·3	3·3	11·2	14·3	14·5	14·4	4·0	0·1	0·2
	Max.	1·2	2·7	1·8	3·4	4·7	22·1	25·5	28·5	61·6	9·3	0·9	1·7
	Min.	...	...	...	...	0·4	2·5	4·9	6·3	4·5	...	...	...
Bhagulpore, 32 to 34 years	Mean	0·5	0·7	0·4	0·8	0·9	8·3	11·2	10·7	7·8	4·1	0·2	0·1
	Max.	2·1	3·0	2·5	4·5	9·4	20·4	21·0	24·3	25·6	24·0	3·5	0·9
	Min.	...	...	...	...	0·1	0·6	4·8	2·7	2·2	...	...	...
Darjiling, 26 to 29 years	Mean	0·7	1·1	2·3	3·9	7·2	24·8	30·1	26·5	17·8	6·4	0·2	0·2
	Max.	3·4	3·4	17·4	9·6	13·4	46·4	53·5	43·3	28·8	27·4	1·5	1·9
	Min.	...	...	...	0·5	2·1	3·0	14·7	13·6	9·2	...	...	...
Hazaribagh, 25 to 26 years	Mean	0·4	0·8	0·7	0·4	1·6	8·3	12·6	12·7	8·0	3·4	0·3	0·2
	Max.	2·0	3·2	2·6	1·7	4·8	20·1	23·0	20·3	14·4	11·9	1·7	3·2
	Min.	...	...	...	...	...	2·2	3·9	5·0	2·0	0·2	...	...
Chybassa, 17 to 18 years	Mean	0·7	0·8	1·4	0·7	3·7	8·4	14·1	13·6	8·7	3·2	0·2	0·4
	Max.	3·0	3·7	3·7	2·1	10·7	13·6	25·9	21·2	12·5	6·0	1·2	2·6
	Min.	...	...	...	...	0·4	3·4	7·8	5·5	4·7	0·6	...	...
Burdwan, 26 to 27 yrs.	Mean	0·3	0·8	1·2	2·4	5·0	9·8	12·3	12·4	8·2	4·6	0·5	0·2
	Max.	2·3	4·1	4·2	7·8	16·3	18·2	28·3	29·6	19·7	13·1	3·4	1·2
	Min.	...	...	...	...	0·6	2·8	5·9	4·0	1·7	...	...	...
Midnapore, 19 to 22 yrs.	Mean	0·7	0·8	1·4	1·6	5·3	9·9	12·1	12·1	8·4	5·1	0·3	0·3
	Max.	4·3	3·2	7·3	5·6	15·0	22·8	22·6	30·5	17·7	15·3	2·7	3·4
	Min.	...	...	...	...	1·1	3·0	4·3	2·3	2·9	0·1	...	...

		Jan.	Feb.	March.	April.	May.	June.	July.	August.	Sept.	Oct.	Nov.	Dec.
Berhampore, 31 to 34 years	Mean	0·4	1·0	1·0	1·9	4·8	9·7	10·3	10·8	9·8	5·3	0·3	0·1
	Max.	3·2	5·4	4·0	7·9	11·0	21·7	19·4	22·5	19·0	17·0	2·7	0·8
	Min.	...	...	...	...	0·7	3·0	2·8	3·0	2·7	...	...	...
Calcutta, 58 years	Mean	0·4	1·0	1·3	2·3	5·6	11·8	13·0	13·9	10·0	5·4	0·6	0·3
	Max.	2·9	3·8	6·1	7·3	16·2	26·6	20·1	26·9	20·6	16·3	5·6	3·0
	Min.	...	...	...	...	0·1	3·0	5·0	3·8	4·0	...	...	...
Dacca, 35 to 37 years	Mean	0·3	1·0	2·5	5·8	9·2	13·3	12·8	12·4	10·2	5·2	0·7	0·2
	Max.	1·7	6·1	12·5	14·2	17·7	28·6	23·1	21·4	24·7	21·5	5·6	2·1
	Min.	...	...	...	0·5	4·1	4·5	5·1	5·1	3·7	0·2	...	...
Chittagong, 28 to 32 yrs.	Mean	0·4	1·2	1·9	4·6	9·2	23·8	22·2	20·5	14·1	5·7	1·6	0·6
	Max.	5·7	6·2	7·5	20·0	21·3	49·6	40·4	43·0	33·3	20·7	15·4	8·5
	Min.	...	...	...	...	1·3	5·7	7·6	6·6	5·5	0·5	...	...
Cuttack, 25 to 30 years	Mean	0·4	0·6	1·1	1·5	3·2	10·7	12·6	11·2	9·8	5·8	1·0	0·5
	Max.	3·0	3·4	6·1	5·8	8·5	23·1	27·2	23·4	19·3	16·2	3·5	7·6
	Min.	...	...	...	...	0·3	1·3	7·2	4·5	5·0	0·6	...	...
ASSAM													
Dibrugarh, 23 to 26 yrs.	Mean	1·4	2·5	5·7	9·9	13·4	18·3	19·7	18·0	14·3	6·1	1·4	0·9
	Max.	3·2	5·5	12·3	16·7	32·6	35·3	32·5	30·0	27·8	17·9	3·8	6·6
	Min.	...	0·3	1·8	5·5	5·4	6·6	10·0	8·2	4·4	2·4	...	...
Sibsagar, 30 to 32 years	Mean	1·1	2·2	4·4	9·9	11·0	14·1	15·6	16·0	11·7	5·2	1·3	0·6
	Max.	4·2	4·2	3·8	19·5	29·8	26·3	31·1	35·9	23·7	13·4	7·3	2·0
	Min.	...	0·6	...	3·5	4·0	4·6	7·1	9·0	4·2	1·2	...	...

		Jan.	Feb.	March.	April.	May.	June.	July.	August.	Sept.	Oct.	Nov.	Dec.
Gauhati, 35 to 39 years	Mean	0·6	0·9	2·5	5·8	10·1	12·9	12·7	11·2	8·1	3·1	0·6	0·3
	Max.	2·4	6·2	7·3	13·4	24·1	34·2	28·6	19·8	18·3	13·3	5·0	1·8
	Min.	...	...	...	2·4	...	4·7	4·5	4·5	0·1	...	...	...
Dhubri, 15 to 17 years	Mean	0·3	0·5	1·9	5·1	15·6	26·4	15·4	11·4	12·4	4·3	...	0·2
	Max.	1·3	2·2	5·5	11·2	32·3	47·1	36·2	24·8	29·2	15·5	0·3	1·5
	Min.	...	...	...	2·4	4·3	7·3	4·1	3·3	3·4	0·1	...	...
Shillong, 19 to 20 years	Mean	0·4	0·8	2·0	3·7	10·0	17·0	14·0	14·4	15·4	6·2	1·0	0·4
	Max.	1·5	3·1	5·1	5·8	18·0	36·0	29·6	30·6	26·5	20·3	7·1	1·7
	Min.	...	...	...	0·8	4·4	8·8	4·3	5·3	5·6	0·8	...	...
Silchar, 29 to 31 years	Mean	0·6	2·6	7·9	13·0	15·7	19·1	20·6	18·2	14·2	6·4	1·0	0·7
	Max.	2·3	9·6	22·1	34·7	51·7	38·7	52·2	29·6	30·2	12·4	4·5	7·3
	Min.	...	...	...	4·1	4·0	7·0	9·3	8·0	4·9	1·5	...	...
Sylhet, 30 to 34 years.	Mean	0·4	1·6	5·4	14·0	21·2	31·1	25·6	26·3	21·0	8·3	1·1	0·3
	Max.	2·9	4·6	23·7	28·1	45·9	48·5	43·3	44·9	47·3	20·4	6·0	2·6
	Min.	...	...	...	2·1	4·5	13·2	9·0	10·4	5·9	0·8	...	...
CENTRAL PROVINCES													
Saugor, 29 to 32 years	Mean	0·6	0·5	0·2	0·2	0·6	6·3	16·8	11·2	7·3	1·3	0·4	0·7
	Max.	3·1	3·2	1·7	1·6	3·3	18·2	29·8	21·4	17·8	4·8	3·9	8·1
	Min.	...	...	...	...	...	0·2	7·0	1·9	0·9	...	...	...
Jubbulpore, 41 to 43 years	Mean	0·6	0·5	0·5	0·2	0·5	8·5	18·6	13·8	8·3	1·5	0·3	0·3
	Max.	2·8	2·9	2·9	3·5	4·1	24·4	36·6	36·1	23·0	5·8	5·7	4·9
	Min.	...	...	...	...	...	0·3	5·8	0·4	0·8	..	...	...

		Jan.	Feb.	March.	April.	May.	June.	July.	August.	Sept.	Oct.	Nov.	Dec.
Hoshungabad, 28 to 33 years	Mean	0·3	0·1	0·2	...	0·5	6·5	15·5	12·9	10·4	1·2	0·4	0·6
	Max.	2·6	1·1	3·0	0·3	3·5	18·9	31·6	31·0	29·9	4·3	6·3	7·0
	Min.	...	...	...	...	...	0·1	2·1	3·0	1·0	...	...	...
Pachmarhi, 16 years	Mean	0·5	0·3	0·4	0·3	0·6	10·8	28·8	18·2	15·1	1·9	0·4	0·7
	Max.	4·0	1·6	2·2	2·8	1·9	24·5	54·8	30·3	39·6	6·9	4·2	5·5
	Min.	...	...	...	...	...	3·0	12·0	9·2	2·2	...	...	...
Raipur, 22 to 24 years	Mean	0·3	0·3	0·6	0·6	0·9	10·4	14·8	12·1	7·7	2·2	0·8	0·3
	Max.	3·5	1·6	3·8	3·6	2·7	20·6	38·9	23·7	14·0	9·7	7·4	2·1
	Min.	...	...	...	...	...	2·9	5·0	2·6	0·7	...	...	...
Nagpur, 39 to 40 years	Mean	0·6	0·4	0·6	0·5	0·8	8·8	13·3	8·9	7·8	2·3	0·4	0·5
	Max.	4·8	3·0	3·8	3·9	5·9	22·2	24·8	19·5	16·3	9·7	4·8	8·2
	Min.	...	...	...	...	...	2·8	1·6	1·8	0·1	...	...	...
					BERAR								
Akola, 25 to 26 years	Mean	0·5	0·1	0·4	0·1	0·2	5·2	7·8	6·7	5·8	2·2	0·4	0·8
	Max.	3·0	1·6	2·5	1·7	0·8	10·8	21·5	14·4	12·5	8·2	3·3	7·2
	Min.	...	...	...	...	...	...	1·3	1·0	...	...	...	...
					BOMBAY								
Jacobabad, 26 years	Mean	0·2	0·2	0·3	0·2	0·1	0·1	1·4	1·4	0·3	...	0·1	0·1
	Max.	1·0	1·9	2·6	1·1	1·2	0·7	5·4	7·9	1·7	0·3	0·9	1·1
	Min.	...	...	...	...	...	...	...	...	...	...	...	...
Kurrachee, 31 years	Mean	0·6	0·3	0·2	0·2	0·1	0·2	3·1	1·8	0·9	0·1	0·1	0·2
	Max.	2·4	1·6	1·0	4·8	1·2	2·2	13·0	11·6	8·3	1·6	1·4	1·1
	Min.	...	...	...	...	...	...	...	...	...	...	...	...

		Jan.	Feb.	March.	April.	May.	June.	July.	August.	Sept.	Oct.	Nov.	Dec.
Hyderabad, 21 to 24 years	Mean	0·2	0·1	0·1	0·2	0·1	0·4	2·8	3·2	0·8	...	0·1	...
	Max.	1·4	0·6	0·8	3·4	0·5	3·3	12·2	11·8	4·1	0·1	0·7	0·2
	Min.	...	...	...	...	...	...	...	...	...	...	...	..
Deesa, 29 to 31 years	Mean	0·1	0·2	0·1	0·1	0·2	2·2	9·8	8·5	3·3	0·8	0·1	...
	Max.	1·6	1·3	1·3	1·0	1·4	12·0	28·5	21·8	16·7	10·7	0·9	0·6
	Min.	...	...	...	...	...	...	...	...	...	...	...	...
Rajkot, 20 to 26 years	Mean	0·1	0·1	...	...	0·4	5·4	10·3	6·7	4·2	0·8	0·2	0·1
	Max.	0·4	0·7	0·2	0·2	4·3	25·4	25·0	19·0	16·8	4·9	4·1	0·6
	Min.	...	...	...	...	...	0·4	0·6	0·3	0·1	...	...	...
Surat, 18 to 24 years	Mean	...	...	...	...	0·2	9·4	13·8	9·8	7·3	1·4	0·1	0·1
	Max.	0·7	0·5	...	...	2·0	21·2	28·1	27·8	28·6	4·3	1·5	0·4
	Min.	...	...	...	...	...	0·6	2·5	1·2	0·4	...	...	...
Tanna, 19 to 30 years	Mean	...	0·1	...	...	0·4	24·8	36·2	20·1	12·8	3·3	...	...
	Max.	0·4	1·4	0·1	...	3·7	45·0	69·3	45·1	32·2	10·7	0·3	1·2
	Min.	...	...	...	...	...	9·9	17·5	8·1	2·9	...	...	...
Bombay, 39 to 70 years	Mean	0·1	...	...	...	0·5	20·8	24·7	15·1	10·8	1·8	0·5	0·1
	Max.	2·2	0·5	...	0·6	5·7	49·2	52·8	37·1	31·8	10·4	5·5	1·1
	Min.	...	...	...	...	...	3·9	4·3	3·4	1·3	...	...	...
Lanauli, 13 years	Mean	0·1	...	0·1	0·1	0·6	28·1	65·3	40·4	23·8	4·8	0·4	0·2
	Max.	0·4	0·2	1·5	0·9	3·2	54·2	91·7	60·9	47·5	9·3	1·5	2·0
	Min.	...	...	...	...	...	12·6	25·1	21·4	8·6	...	...	...
Mahableshwar, 31 to 32 years	Mean	0·4	0·1	0·4	0·9	1·4	47·3	102·1	68·6	32·9	5·8	1·1	0·4
	Max.	7·7	0·5	7·6	11·5	7·7	110·5	181·7	136·4	76·4	13·4	7·2	4·7
	Min.	...	...	...	...	...	19·7	42·5	33·9	12·2	...	...	...

		Jan.	Feb.	March.	April.	May.	June.	July.	August.	Sept.	Oct.	Nov.	Dec.
Goa, 26 to 27 years	Mean	0·1	...	...	0·1	1·6	31·8	34·1	19·9	0·6	4·4	1·1	0·1
	Max.	2·7	...	0·6	0·9	15·1	49·8	61·0	43·7	39·1	13·6	5·8	1·3
	Min.	...	...	...	...	...	14·6	9·7	6·1	1·8	...	...	...
Dhulia, 17 to 26 years	Mean	0·3	0·1	...	...	0·4	4·8	4·8	4·0	4·6	2·0	0·5	0·4
	Max.	3·0	0·7	0·1	0·3	2·0	12·0	11·6	13·8	11·7	8·7	4·7	3·2
	Min.	...	...	...	...	...	0·3	0·8	0·7	...	...	...	...
Poona, 44 years	Mean	0·2	...	0·2	0·6	1·6	5·6	6·6	4·1	4·3	4·1	0·8	0·2
	Max.	5·1	1·0	1·4	4·4	7·9	17·2	24·9	12·5	10·9	18·1	5·1	1·7
	Min.	...	...	...	...	...	...	...	0·1	...	...	...	...
Sholapur, 34 years	Mean	...	0·1	0·3	0·7	1·2	4·6	4·3	6·0	7·5	3·7	0·7	0·4
	Max.	0·9	1·7	1·5	7·6	6·8	13·3	11·8	19·0	20·3	12·6	4·7	4·0
	Min.	...	...	...	...	...	0·5	0	0·4	0·4	...	...	...
Belgaum, 34 to 35 years	Mean	0·1	...	0·5	2·0	2·8	9·3	15·2	9·0	3·7	4·7	1·2	0·3
	Max.	0·9	0·5	3·0	5·8	10·7	18·6	32·2	22·4	9·3	8·1	5·4	2·9
	Min.	...	...	...	...	...	0·9	3·1	2·3	0·1	0·2	...	...
HYDERABAD													
Secunderabad, 44 years	Mean	0·3	0·2	0·7	0·7	1·4	3·7	6·0	5·7	5·2	3·3	0·8	0·3
	Max.	4·3	2·2	5·4	2·7	11·3	9·9	18·8	17·9	13·9	9·0	5·4	3·4
	Min.	...	...	...	...	...	...	1·8	1·0	...	...	...	...
MYSORE													
Shimoga, 50 years	Mean	0·1	0·1	0·3	1·8	3·3	4·7	6·6	4·2	3·1	5·0	1·2	0·4
	Max.	2·5	3·8	1·7	6·0	8·0	22·0	25·9	10·1	9·5	15·5	8·5	2·9
	Min.	...	...	...	...	...	...	1·1	0·6	0·3	0·3	...	...

		Jan.	Feb.	March.	April.	May.	June.	July.	August.	Sept.	Oct.	Nov.	Dec.
Bangalore, 51 years	Mean	0·2	0·1	0·6	1·3	5·0	3·2	4·0	5·9	6·3	6·4	1·9	0·7
	Max.	2·9	1·9	6·4	4·2	15·5	8·3	11·4	14·2	16·0	19·5	9·9	5·8
	Min.	...	...	...	...	0·1	0·6	...	0·8	1·1	0·7	...	...
Mysore, 50 years	Mean	0·1	0·1	0·7	2·2	5·6	1·9	2·3	3·2	3·9	6·4	1·6	0·5
	Max.	1·3	2·5	5·2	7·2	12·7	5·5	15·3	8·5	11·9	13·5	9·5	5·8
	Min.	...	...	...	...	...	...	0·2	...	0·3	0·5	...	...
Coorg													
Mercara, 23 years	Mean	0·3	...	1·1	2·2	6·1	25·8	42·1	25·7	12·3	8·0	2·6	0·6
	Max.	3·7	0·6	3·7	5·3	24·6	61·5	101·6	41·3	24·5	14·7	11·0	2·6
	Min.	...	...	...	0·1	1·5	4·8	14·3	14·1	4·6	1·6	...	...
Madras													
Vizianagram, 24 years	Mean	0·2	0·5	0·8	0·4	2·6	4·3	5·0	6·6	7·7	8·4	2·2	1·6
	Max.	2·6	5·5	7·5	2·0	12·8	9·8	9·1	14·4	17·4	25·9	11·3	30·0
	Min.	...	...	...	..	...	1·2	1·0	...	1·8	...	...	...
Rajahmundry, 25 years	Mean	0·2	0·3	0·3	0·9	3·3	4·5	7·2	6·6	7·1	6·5	1·7	0·2
	Max.	3·0	3·1	2·0	5·5	8·6	12·1	15·1	18·0	15·1	22·6	7·6	3·6
	Min.	...	...	...	...	0·2	0·5	2·1	0·8	1·2	...	...	...
Masulipatam, 24 years	Mean	0·3	0·1	0·3	0·1	1·7	4·4	5·6	6·0	6·5	8·8	4·0	0·7
	Max.	1·6	1·3	4·4	0·5	9·5	8·6	16·2	19·9	12·6	22·3	14·9	4·1
	Min.	...	...	...	...	...	0·8	1·0	1·1	1·7	...	...	...
Bellary, 33 years	Mean	0·1	...	0·6	0·8	1·8	1·8	1·3	2·3	3·7	3·9	1·0	0·3
	Max.	2·0	0·9	7·6	4·2	5·6	5·6	7·3	6·2	10·9	12·8	6·3	5·0
	Min.	...	...	...	...	...	...	...	...	0·6	...	...	...

		Jan.	Feb.	March.	April.	May.	June.	July.	August.	Sept.	Oct.	Nov.	Dec.
Cuddapah, 34 years .	Mean	0·1	...	0·2	0·3	1·6	2·6	3·5	5·1	5·8	5·3	3·1	0·7
	Max.	1·5	0·2	4·2	2·6	7·4	7·6	14·4	13·3	30·6	13·4	11·5	4·8
	Min.	...	...	...	...	...	...	0·6	...	...	...	...	...
Madras, 74 years .	Mean	1·0	0·3	0·4	0·6	2·2	2·1	3·8	4·4	4·7	10·8	13·7	5·1
	Max.	8·6	6·3	6·8	6·4	23·3	8·6	11·8	9·7	14·9	37·7	33·5	22·8
	Min.	...	...	...	...	...	...	0·3	0·8	0·5	0·8	0·4	...
Trichinopoly, 33 to 35 years . . .	Mean	1·0	0·5	0·7	1·8	3·8	1·3	2·2	4·4	5·3	7·8	5·2	3·1
	Max.	10·8	4·8	7·4	10·2	12·3	4·6	9·9	19·1	11·3	26·0	13·9	12·1
	Min.	...	...	...	...	...	...	...	0·3	...	0·2	0·4	...
Coimbatore, 24 to 26 years . . .	Mean	0·3	0·1	0·6	1·8	2·6	1·8	1·3	1·2	1·2	5·7	3·4	1·1
	Max.	2·[illegible]	1·3	2·4	5·4	7·8	5·1	3·6	5·1	3·3	13·0	13·7	4·2
	Min.	...	...	...	...	...	...	...	...	...	1·4	0·6	...
Madura, 26 years .	Mean	0·7	0·4	0·6	2·0	2·8	1·6	1·7	4·7	4·5	8·7	5·1	2·2
	Max.	8·0	2·0	4·6	6·8	7·2	6·9	6·4	13·8	19·4	20·4	11·2	13·5
	Min.	...	...	...	...	0·3	...	...	1·0	0·7	2·5	0·5	...
Tuticorin, 21 years .	Mean	1·[illegible]	0·6	0·8	1·3	0·9	0·1	0·2	0·4	0·7	3·7	8·1	2·2
	Max.	7·9	4·4	3·9	7·8	6·0	1·9	1·6	3·0	6·0	10·9	18·0	13·2
	Min.	...	...	...	...	...	...	...	...	...	...	0·9	...
Mangalore, 28 to 34 yrs.	Mean	0·2	0·1	0·1	2·0	8·1	37·8	37·9	23·1	11·3	8·0	1·9	0·5
	Max.	4·2	1·8	0·5	4·9	31·5	69·9	70·4	43·9	29·5	17·0	6·9	5·9
	Min.	...	...	...	...	0·5	22·6	11·3	8·8	4·9	0·9	...	...
Cochin, 40 to 41 years	Mean	0·9	0·7	2·1	4·4	12·7	30·7	22·7	12·4	9·4	12·1	5·1	1·9
	Max.	5·2	3·8	17·9	11·2	27·2	49·5	44·3	29·4	22·5	24·0	10·8	6·7
	Min.	...	...	...	0·9	0·6	14·0	8·1	3·0	1·7	3·1	0·1	...

CEYLON

		Jan.	Feb.	March.	April.	May.	June.	July.	August.	Sept.	Oct.	Nov.	Dec.
Jaffina, 16 years	Mean	2·3	1·5	1·2	2·1	2·3	1·0	0·7	1·2	2·8	9·3	14·8	9·9
	Max.	9·9	6·0	6·1	5·5	5·4	8·5	3·1	3·9	6·1	23·6	30·4	25·7
	Min.	0·1	...	...	...	...	...	...	...	...	1·5	5·5	4·0
Batticaloa, 16 to 17 yrs.	Mean	8·1	3·3	3·2	1·5	2·0	1·2	0·8	2·8	2·3	5·5	12·9	15·7
	Max.	26·9	10·1	8·9	4·0	4·8	3·6	4·9	6·7	5·4	10·1	25·6	28·2
	Min.	1·2	...	0·1	...	...	0·1	...	...	0·4	2·3	3·0	6·3
Colombo, 17 years	Mean	3·0	1·7	5·5	8·8	13·2	8·2	5·5	4·5	4·9	12·9	12·7	6·4
	Max.	12·6	5·2	13·6	19·1	25·9	20·0	28·0	19·5	21·8	19·2	28·8	17·7
	Min.	...	...	0·8	2·6	4·3	0·6	1·3	0·2	1·0	2·8	2·0	2·4
Kandy, 17 years	Mean	5·0	2·6	3·1	6·9	6·7	9·1	7·8	6·2	5·8	11·3	11·6	8·4
	Max.	16·8	5·5	9·0	15·5	14·3	16·5	15·5	13·9	18·4	17·4	22·5	21·9
	Min.	0·4	...	0·5	3·5	1·0	1·8	2·4	2·5	1·4	7·2	5·1	1·0
BURMAH													
Akyab, 29 to 31 years	Mean	0·1	0·2	0·5	1·6	12·2	51·6	51·0	38·6	23·0	12·4	3·9	0·6
	Max.	1·5	3·6	2·5	10·9	45·1	97·6	87·5	65·7	48·2	27·6	22·3	7·4
	Min.	...	...	...	...	2·0	22·1	22·9	24·1	9·9	1·3	...	...
Thyet Myo, 17 years	Mean	...	...	0·1	0·7	5·3	7·9	8·0	8·5	7·8	4·9	2·3	...
	Max.	0·1	0·3	1·0	3·5	12·3	12·5	16·2	20·1	19·2	10·1	8·7	0·2
	Min.	...	...	...	...	0·9	5·1	2·4	4·5	4·3	1·7	...	...
Tounghoo, 17 years	Mean	...	0·2	...	1·5	6·6	13·4	17·5	18·1	11·8	7·4	1·4	0·2
	Max.	0·2	2·4	0·3	5·6	15·0	21·1	23·7	27·0	18·0	13·0	4·0	1·7
	Min.	...	...	...	...	0·9	7·6	7·4	11·4	6·6	0·8	...	...

		Jan.	Feb.	March.	April.	May.	June.	July.	August.	Sept.	Oct.	Nov.	Dec.
Shwegyin, 14 years	Mean	...	0·5	0·3	2·3	10·9	30·0	33·8	32·5	19·4	10·1	2·0	0·1
	Max.	0·2	3·1	2·1	10·5	31·0	54·3	55·7	45·8	30·5	17·7	6·8	1·4
	Min.	...	...	...	...	1·4	16·8	11·7	22·0	9·6	2·2	...	...
Rangoon, 17 years	Mean	0·2	0·1	0·1	1·8	10·9	18·4	21·3	18·6	16·0	8·1	3·4	0·1
	Max.	1·9	1·6	0·5	4·9	27·2	32·4	35·7	29·7	25·9	16·3	15·3	0·9
	Min.	...	...	...	...	1·1	11·5	7·4	9·5	7·9	3·1	...	...
Moulmein, 37 years	Mean	...	0·1	0·1	3·0	19·7	38·4	43·9	43·0	30·3	8·4	1·5	0·1
	Max.	0·5	1·0	1·7	9·0	37·8	65·0	68·4	72·2	63·4	19·7	10·7	0·3
	Min.	...	...	...	...	4·9	21·4	20·6	13·9	15·0	1·2	...	...
Tavoy, 29 to 30 years	Mean	0·2	0·6	0·7	4·0	16·6	39·9	46·0	41·9	33·7	10·5	2·3	0·1
	Max.	2·7	7·0	4·8	19·2	35·5	66·5	67·0	68·1	59·3	24·5	9·8	1·0
	Min.	...	...	...	...	1·6	22·5	18·0	20·0	12·5	1·0	...	...
ANDAMANS AND NICOBARS													
Port Blair, 19 years	Mean	0·9	1·3	0·4	2·4	15·9	17·9	16·5	15·2	19·6	11·8	9·5	5·3
	Max.	2·9	8·6	3·3	8·2	26·5	29·3	24·5	34·3	29·9	20·3	20·0	15·7
	Min.	...	...	...	...	4·8	11·5	6·1	7·5	12·9	7·3	2·4	...
Nancowry, 13 to 14 yrs.	Mean	2·0	1·3	2·1	5·3	12·1	12·7	12·6	12·6	10·6	13·5	11·8	12·2
	Max.	8·4	5·3	8·4	11·3	20·6	20·5	26·8	17·0	17·0	25·8	27·0	34·4
	Min.	0·3	...	0·3	0·4	2·6	4·6	4·1	6·2	4·5	7·8	3·1	2·2

APPENDIX IV

Directions for the Use of the Storm Signals at Ports around the Bay of Bengal, issued by the Government of India

1. The storm signals for use at ports around the Bay of Bengal (excepting at the stations on the Hooghly River) will be of two kinds for the daytime :—

No. 1. A ball as bad weather signal.
„ 2. A drum as storm signal
with corresponding lamp signals for night, viz.—
„ 3. Three lamps suspended on the same line one above the other (in a vertical series) as bad weather signal.
„ 4. Two lamps similarly suspended as storm signal.

2. The ball or three lamps will be hoisted to indicate that a cyclone is in existence on the bay without reference to its position and probable course.

3. The drum or two lamps will be hoisted to indicate that the storm is probably approaching the port.

4. The ball or three lamps will be hoisted on the receipt of telegraphic information from the Calcutta Meteorological Office.

5. When the ball or three lamps is hoisted, port officers must rely on their own judgment whether the drum or two lamps should be hoisted or not, and if so, when. If possible, later information will be telegraphed from the Calcutta Office, but they must not reckon on receiving such. The indications of the barometer and of the direction and force of the wind, together with the general aspect of the weather, must be their chief guides. Some general memoranda are hereto appended to assist them in forming a judgment.

Memoranda on Storm Prognostics

1. Cyclones have been known to occur in the Bay of Bengal

in all months of the year except February, but not between the middle of January and the last week in March; and from the middle of December to the middle of April such cyclones as are on record have been restricted to the southern half of the bay.

2. Cyclones that reach the west (Indian) coast of the bay, follow a course between west by north and north-north-west. On the approach of a cyclone, the wind is generally at first from north-east. If it strengthens from this direction, becoming more squally while the barometer falls rapidly, it is probably approaching the port. But if the wind backs beyond north, the barometer falling moderately, it is passing to the east, and the centre bears about east when the wind is about north-west or north-north-west. If, on the other hand, the wind veers towards east, with a moderate fall of the barometer, the cyclone has crossed the coast line to the south of the station by the time the wind is from east.

3. Cyclones that reach the Arakan coast follow a course between north and north-east. On the approach of a cyclone, the wind is generally from north-east or east-north-east. If it strengthens from between these directions, becoming more easterly with a rapidly falling barometer, it is probably approaching the port. But if the wind veers to south-east with a moderate fall of the barometer, the storm is passing up the bay, and the centre bears about due west when the wind is at south-east or south-south-east. If the wind backs to north, with a similar change in the barometer, the storm centre has reached the coast to the south of the station.

4. Cyclones that reach the coasts of Pegu and Moulmein are formed in the Andaman sea, and follow a course between north and north-east. The indications of the wind and barometer may be interpreted as in the previous paragraph.

5. In judging of the direction of the storm centre, it must be borne in mind that the winds do not blow round the centre in a circle, but spirally inwards. When the centre is still at some distance, it bears about 12 points to the right, if the observer faces the wind. As it approaches nearer, it bears about 10 points to the right.

6. In judging of the fall of the barometer, it must be borne in mind that the barometer falls regularly about a tenth of an inch from 9.30 A.M. to 4.30 P.M., and rather less from 10 P.M. to 3.30 A.M., and rises by similar amounts from 4 to 9.30 A.M., and from 5 to 10 P.M. These changes, which are regular, must be allowed for.

APPENDIX V

Notifications of the Government of Bengal relative to Storm Signals for the Ports of Bengal

The 4th August 1887.—The following arrangements for weather signals for the approaches to the Port of Calcutta, and for the port itself, have been sanctioned by the Lieutenant-Governor, and are published for general information, in supersession of those published in the notification, dated the 5th August 1882, at pages 678-80 of the *Calcutta Gazette* of the 9th August 1882 :—

Weather Signals for the approaches to the Port of Calcutta, and for the Port of Calcutta

1. When a severe cyclonic storm (probably accompanied by a storm-wave) is believed to be advancing to the Bengal coast and across Bengal, one or other of the following signals will be hoisted at the flagstaff near the Lighthouse in Saugor Island, at the flagstaff near the telegraph office, Mud Point, and at the flagstaff near the telegraph office, Diamond Harbour, and also at the following positions in the port, viz. the flagstaff upon the Port Commissioners' Office, on a flagstaff at Shalimar Point, Seebpore, opposite the Government Dockyard, Kidderpore, and on a flagstaff at the telegraph office, Budge Budge, on receipt of instructions telegraphed by the Meteorological Reporter to the Government of Bengal :—

A ball indicates the existence of a cyclonic storm of undetermined intensity and magnitude in the Bay of Bengal, which will either certainly cross the coast to the south of a line between Chittagong and False Point, or which may approach the Bengal coast, but is as yet too far distant to enable its line of advance to be determined.

A cone indicates the early probable passage northward, to the eastward of Saugor Island and west of Chittagong, of the vortex of a cyclonic storm of great intensity and magnitude. It is advisable that sailing vessels, with or without steam, and deep-laden or slow steaming steam-vessels, should not proceed to sea, but remain in the river till the storm has reached the coast and passed inland. The wind at the mouth of the Hooghly will probably haul from north-east through north to north-west, etc.

An inverted cone indicates the early probable passage northward, to the west of Saugor Island and north of False Point, of the vortex of a cyclonic storm of great intensity and magnitude. No vessels should go to sea, and masters and pilots of vessels outward-bound should be guided by the appearance of the weather and height of the barometer in deciding whether it is advisable to proceed below Diamond Harbour or Mud Point. The wind at the mouth of the Hooghly will probably veer from north-east through east to south-east, etc.

A drum indicates the approach of a cyclonic storm of great intensity and magnitude to the mouth of the Hooghly, and which will probably advance to Calcutta. Masters and pilots in charge of vessels are cautioned not to put to sea from Saugor Island, or to proceed down from Diamond Harbour, and they should make their vessels as snug and secure as possible. The masters of vessels in the port should take the special precautions for safety laid down in the Port Rules.

There will probably be a storm-wave, but it should be carefully remembered that its height and destructive effect will depend quite as much upon the state and character of the tide when the cyclonic centre reaches the coast, as upon the depression at the centre, or the intensity and extent of the storm.

2. When a small cyclonic storm (usually not accompanied by a storm-wave) is believed to have formed in the bay, and to be advancing towards the coast, one or other of the following signals will be hoisted, for the information of pilots and mariners, at the staff near the Lighthouse at Saugor Island, at the flagstaff at Mud Point, and at the flagstaff near the telegraph office, Diamond Harbour, and also at the following stations in the port, viz. on a flagstaff upon the Port Commissioners' Office, on a flagstaff at Shalimar Point, Seebpore, and on a flagstaff at the telegraph office, Budge Budge, on the receipt of instructions telegraphed from the Meteorological Office, Calcutta, by the Meteorological Reporter to the Government of Bengal :—

Two cones, the upper one inverted, indicate the existence of a cyclonic storm of small extent in the Bay of Bengal, which will probably reach and cross the coast of the bay south of a line joining Chittagong and False Point.

Two cones, the lower one inverted, indicate the existence of a cyclonic storm of small extent, which will probably or certainly reach and cross the coast of the bay north of a line joining Chittagong and False Point, but the probable path of the vortex of which cannot be determined more exactly from the land observations.

A ball below a cone indicates the probable passage northwards, to the east of Saugor Island and west of Chittagong, of a cyclonic storm of small extent and intensity, of the kind which usually forms in the rainy season. Vessels may proceed to sea if the height of the barometer, state of the sea and weather are such as to lead masters and pilots to infer that there is no danger. The wind at the mouth of the Hooghly will probably haul from north-east through north to north-west, etc.

A ball below an inverted cone indicates the probable passage northwards, to the west of Saugor Island and north of False Point, of a cyclonic storm of small extent and intensity of the kind which usually forms in the rainy season. The wind at the mouth of the Hooghly will probably veer from north-east through east to south-east, etc. As these easterly winds will raise a heavy swell and produce a strong westerly set in the channel at the Sandheads, it is advisable that none but fast steamers in light trim should put to sea, and those only if the weather appearances and state of the sea, etc., are not too unfavourable.

A ball below a drum indicates the approach towards Saugor roads of a cyclonic vortex of small extent, of the kind which forms during the rainy season. It is advisable that no vessels except fast steamers in light trim should put to sea until the wind direction and force, the state of weather and sea, and the rise of the barometer, indicate that the storm has either broken up or passed inland. Cyclonic storms in the Bay of Bengal of small extent, it should be remembered, sometimes blow with hurricane force and raise a high sea near their centres.

3. The following storm warning signals will be hoisted during the prevalence of cyclonic storms in the Bay of Bengal at the signalling stations on the Hooghly previously mentioned, viz. Saugor Island, Mud Point, and Diamond Harbour, and at the following signalling stations in the port, viz. on a flagstaff upon the Port Commissioners' Office, on a flagstaff at Shalimar Point, Seebpore, and on a flagstaff at the telegraph office, Budge Budge, between sunset and sunrise :—

Three lights exhibited on the flagstaff at equal distances indicate the existence of a cyclonic storm in the north of the bay.

Two lights, in a vertical line, indicate the early approach of a cyclonic storm to the Bengal coast, or to the Port of Calcutta if the signal is hoisted in the port.

The following arrangements for weather signals have been sanctioned for the Ports of Orissa, viz. Pooree, False Point, Balasore, and Chandbally by the Lieutenant-Governor, and are published for general information :—

Signals for the Orissa Ports.—Storm signals will be hoisted at the ports named above by the port officers on the receipt of instructions from the Bengal Meteorological Office, Calcutta, to give notice to the shipping and the general public at these ports of the approach of a cyclone (which may or may not, according to the state of the tides and other conditions, be accompanied by a destructive storm-wave). The signals for the Port of Pooree will be hoisted at the flagstaff on the beach; for the Port of False Point at the flagstaff at Hookeytollah; for the Port of Balasore at the flagstaff near the Custom House; and for the Port of Chandbally at the flagstaff near the Custom House.

The following are the signals which will be used at the Orissa Ports, and are identical with those in use at the Burmah and Madras Ports :—

A ball indicates that a cyclone has formed in the centre or north of the bay, and that it is advancing in some direction between north and west, and will probably strike and cross the coast of Orissa.

A drum indicates that a cyclone is in the immediate neighbourhood of the coast of Orissa, and will shortly cross that coast. (It may or may not be accompanied by a destructive storm-wave. The height of the storm-wave depends partly upon the extent of the cyclone, the barometric depression at the centre, and the state of the tides at the time when the cyclone reaches the coast.)

Three lights in a vertical line indicate that a cyclone has formed in the centre and north of the bay, and that it is advancing in some direction between north and west, and will probably strike and cross the coast of Orissa.

Two lights in a vertical line indicate that a cyclone is in the immediate neighbourhood of the coast of Orissa, and will shortly cross that coast. (It may or may not be accompanied by a destructive storm-wave.)

The following arrangements for weather signals for the Port

of Chittagong have been sanctioned by the Lieutenant-Governor, and are published for general information :—

Signals for the Port of Chittagong.—The following storm signals will be used within the limits of the Port of Chittagong to give notice to the shipping and general public of the early approach of a cyclone with its attendant storm-wave. They will be hoisted on the flagstaff at the Sudder Ghât and on the yardarm of the flagstaff at Jooldia on the receipt of instructions from the Meteorological Office :—

A ball indicates that a cyclone, of which the centre is in the north-east of the bay, is advancing towards the mouth of the River Karnafuli, and will probably advance towards Chittagong.

A drum indicates that a cyclone, which may or may not be accompanied by a destructive storm-wave, is approaching Chittagong. (It should be remembered that the force and destructive effect of a storm-wave will depend very considerably upon the state of the tides at the time when the storm centre strikes the Chittagong coast.)

Three lights in a vertical line indicate that a cyclone, of which the centre is in the north-east angle of the bay, is advancing towards the mouth of the river, and will probably advance towards Chittagong.

Two lights in a vertical line indicate the early approach of a cyclonic storm to the Port of Chittagong.

INDEX

THE END